Microbial Processes in Environmental Engineering

emerald PUBLISHING
ice Publishing

Microbial Processes in Environmental Engineering:

From Theory to Practice

Edited by Rajneesh Kumar,
Ram Sharan Singh and Maulin P Shah

Published by Emerald Publishing Limited,
Floor 5, Northspring, 21-23 Wellington Street, Leeds LS1 4DL.

ICE Publishing is an imprint of Emerald Publishing Limited

Other ICE Publishing titles:
Basic Water Treatment, Sixth Edition
Chris Binnie, Martin Kimber and Hugh Thomas
ISBN 978-0-7277-6334-1

A catalogue record for this book is available from the British Library.

ISBN 978-1-83708-327-5

Cover photo: shutterstock.com/g/CatalinIliescu

Commissioning Editor: Michael Fenton
Content Development Editor: Ryan Molyneux
Books Production Lead: Emma Sudderick
Project Manager: Rajachitra S
Typeset by KnowledgeWorks Global Limited
Index created by David Gaskell

Printed and bound by CPI Group (UK) Ltd, Croydon, CR0 4YY

Contents

Preface

Microbial processes lie at the heart of modern environmental engineering and provide innovative, efficient and sustainable pathways to overcome the most pressing ecological challenges. *Microbial Processes in Environmental Engineering: From Theory to Practice* combines foundational concepts and advanced developments that illustrate how microbial systems can be used to transform environmental management across diverse ecosystems and industrial sectors.

This book provides an integrated perspective on microbial roles in wastewater treatment, bioremediation, pollution control and sustainable bioenergy production. It explores the mechanisms, interactions and ecological dynamics of microorganisms that enable the degradation of pollutants, transformation of contaminants and enhancement of environmental quality. It bridges theoretical understanding with practical implementation, highlighting microbial solutions that contribute directly to environmental sustainability and engineering innovation.

A core component of this book is the detailed discussion of wastewater treatment and microbial processes, including advanced strategies, such as bioaugmentation, biostimulation and biofiltration, for improved treatment performance. The chapters also cover microbial degradation of complex pollutants, removal of heavy metals and optimisation of biological treatment systems to meet the growing demands of industrial and municipal wastewater management.

The book also examines emerging microbial technologies aimed at tackling next-generation environmental challenges. Topics include sustainable biohydrogen production through microbial fuel cells, nanobiomaterials for enhancing bioelectrochemical systems and microbial bioremediation of emerging contaminants. Furthermore, the role of microbial biosensors in environmental monitoring, the ecological significance of endophytes in pollution mitigation and the microbial contributions to climate change mitigation are comprehensively addressed.

In addition to aquatic systems, the book covers microbial processes in soil remediation, offering insights into soil health restoration and contaminant removal. Specialised topics such as microbial plastic and microplastic biodegradation, and the complexities of marine microbial corrosion and biofouling, provide an expanded view of microbial applications across environmental domains.

Students in environmental engineering, civil engineering, microbiology, biotechnology and related disciplines will be benefitted. By combining scientific principles, technological innovation and real-world applications, *Microbial Processes in Environmental Engineering: From Theory to Practice* aims to inspire interdisciplinary collaboration and foster the development of transformative microbial solutions for global environmental challenges.

Editors

Rajneesh Kumar

Dr Rajneesh Kumar is a dedicated researcher who focuses on wastewater treatment, where he has made significant contributions, including the utilisation of plant waste as biosorbents to remove chromium from water. He is currently serving as a National Post-Doctoral Fellow in the Department of Chemical Engineering and Technology at the Indian Institute of Technology (BHU), Varanasi. He obtained his PhD from IIT Guwahati, identifying tools and techniques for assessing recovery potential of aerobic sludge biomass stressed with copper, amoxicillin, chlorpyrifos and piggery wastewater. He has laid a strong interdisciplinary foundation for his research in environmental biotechnology and pollution control. With over four years of research experience, Dr Kumar has made significant contributions to sustainable wastewater treatment, emerging pollutant removal and microbial process optimisation. He has authored and co-authored more than 20 peer-reviewed journal articles in reputed international journals and has also contributed to several book chapters. Dr Kumar is an Early Career Editorial Board Member of Emerging Contaminants and a guest editor for a special issue in *Explora: Environment and Resource*. He actively reviews manuscripts for leading journals in the field. Dr Kumar's work integrates biotechnological innovation, environmental microbiology and process engineering toward advancing eco-efficient wastewater treatment technologies.

Ram Sharan Singh

Dr Ram Sharan Singh is a distinguished professor in the Department of Chemical Engineering and Technology at IIT (BHU), Varanasi, with a robust academic and research background. Holding a PhD from IIT (BHU) with a focus on the degradation of volatile organic compounds using supported film bioreactors, he has built an illustrious career in biodegradation, bioremediation and environmental biotechnology. His professional journey includes roles such as GET at Tata Chemicals, lecturer at BITS Pilani and BIET Jhansi, and various academic positions at IIT (BHU) Varanasi, culminating in his current role as a professor at IIT (BHU) since 2015. With an impressive H-index of 47, over 7800 citations and an i10 index of 123, Dr Singh's research contributions are highly regarded. He has 19 years of total research experience, including 13 years post-doctoral, and has mentored numerous post-doctoral fellows, PhD candidates and MTech students. His research primarily focuses on bioaugmentation, biofiltration, synthesis of biodiesel and the application of biochar and petroleum-degrading consortia for wastewater treatment. Dr Singh's work has significantly impacted the field of environmental engineering, providing sustainable solutions for pollution control and fostering future professionals and researchers.

Maulin P Shah

Dr Maulin P Shah has emerged as a globally recognised and highly influential researcher in environmental microbiology, with an unwavering dedication to scientific inquiry spanning more than two decades. His research excellence is firmly rooted in exploring the intricate and dynamic relationships between environmental pollution and microbial diversity in wastewater ecosystems, employing both cultivation-dependent and advanced cultivation-independent (molecular) approaches. Through this integrated framework, Dr Shah has significantly enhanced the understanding of microbial structure, function and resilience in polluted environments. His pioneering research has made substantial and lasting contributions to key domains of environmental microbiology, including biological wastewater treatment, biodegradation, bioremediation and phytoremediation of environmental pollutants. Dr Shah's scholarly impact is reflected in an extraordinary publication record, comprising over 350 peer-reviewed research papers published in highly reputed national and international scientific journals. These publications have significantly advanced scientific knowledge and continue to serve as critical references for researchers, academicians and professionals worldwide. Beyond research articles, Dr Shah holds a distinguished editorial legacy, having served as editor for more than 250 internationally acclaimed scientific books. Further extending his influence, Dr Shah actively serves on the editorial boards of several top-ranked international journals, where he contributes to maintaining high standards of scientific quality, ethics and innovation. Through his research, publications and editorial leadership, Dr Shah stands as a commanding figure in environmental microbiology, driving scientific progress, fostering global collaboration and advancing sustainable solutions for environmental protection.

Contributors

Aditi Singh
Department of Chemical Engineering and Technology, IIT BHU Varanasi, India
Email: aditisingh.rs.che24@itbhu.ac.in

Anand Kumar
Guangdong Ocean University, Zhanjiang, China
Email: ak9661126@gmail.com

Anita Rani
Department of Chemistry, IEC University, Himachal Pradesh, India
Email: caoudhary.anita@gmail.com

Anjali Mishra
Department of Chemical Engineering and Technology, IIT BHU Varanasi, India
Email: anjalimishra.rs.che23@itbhu.ac.in

Aparna Pandey
Central Salt and Marine Chemicals Research Institute, Bhavnagar, India
Email: aparnapandey3376@gmail.com

Ashok K Dubey
Netaji Subhas University of Technology (NSUT), New Delhi, India
Email: akdubey@nsut.ac.in

Ashwini S Thakre
Department of Chemical Engineering, VNIT Nagpur, India
Email: ashthakre@gmail.com

Aubrey Dickon Chigwada
Department of Environmental Sciences, University of South Africa, Gauteng, South Africa
Email: aubreydchi@gmail.com

Chimdi Mang Kalu
Department of Environmental Sciences, University of South Africa, Gauteng, South Africa
Email: ekalucm@unisa.ac.za

Diwakar Z Shende
Department of Chemical Engineering, VNIT Nagpur, India
Email: dzshende@che.vnit.ac.in

Djaber Tazdaït
Department of Nature and Life Sciences, University of Algiers, Benyoucef Benkhedda, Algeria & BIOGEP Laboratory, Algiers, Algeria
Email: d.tazdait@univ-alger.dz

Kailas L Wasewar
Department of Chemical Engineering, VNIT Nagpur, India
Email: k_wasewar@rediffmail.com

Komal Pandey
Amity University Uttar Pradesh, Lucknow, India
Email: kpandey@lko.amity.edu

Kunal Biswas
Centre for Nanoscience and Nanotechnology, Sathyabama Institute of Science and Technology, Chennai, India
Email: kunal.sapiens@gmail.com

Lilian Maureen Oluchi Oguoma
Department of Biotechnology, Federal University of Technology, Owerri, Nigeria
Email: mauoguoma@gmail.com

Memory Tekere
Department of Environmental Science, University of South Africa, Gauteng, South Africa
Email: tekerm@unisa.ac.za

Odangowei Inetiminebi Ogidi
Department of Biochemistry, Bayelsa Medical University, Yenagoa, Nigeria
Email: ogidiodangowei@gmail.com

Padmapriya Subbiah
Madurai Kamaraj University, Madurai, India
Email: priyaartsubbiah@gmail.com

Pooja Dhiman
Maharaja Agrasen University, Himachal Pradesh, India
Email: pooja.dhimanmanshi@gmail.com

Prabhakar Singh
Department of Biotechnology, Sathyabama Institute of Science and Technology, Chennai, India
Email: prabhakar.singh.bte@sathyabama.ac.in

Raja Gowri Swaminathan
Department of Biotechnology, Sathyabama Institute of Science and Technology, Chennai, India
Email: rajagowriswaminathan@gmail.com

Rajneesh Kumar
Department of Chemical Engineering and Technology, IIT BHU Varanasi, India
Email: krajneesh.pf.che24@itbhu.ac.in

Ram Sharan Singh
Department of Chemical Engineering and Technology, IIT BHU Varanasi, India
Email: rssingh.che@itbhu.ac.in

Rym Salah Tazdaït
BIOGEP, École Nationale Polytechnique, Algiers, Algeria
Email: rym.salah@g.enp.edu.dz

Sajitha Menon K
Government Victoria College, Palakkad, Kerala, India
Email: sajithamenonk@gmail.com

Saqib Hassan
Department of Biotechnology, Sathyabama Institute of Science and Technology, Chennai, India
Email: saqib.hassan.bte@sathyabama.ac.in

Sasikala Subbiah
Sri Krishna College of Engineering, Vellore, India
Email: sasisubbiah.sk@gmail.com

Selvarani Subbiah
Thiagarajar College, Madurai, India
Email: selvarani_zool@tcarts.in

Shelly Singh
Netaji Subhas University of Technology (NSUT), New Delhi, India
Email: shelly.singh.bse20@nsut.ac.in

Shalini Singh Visen
Amity University Uttar Pradesh, Lucknow, India
Email: ssingh29@amity.edu

Shilpa Sharma
Netaji Subhas University of Technology (NSUT), New Delhi, India
Email: shilpa.sharma@nsut.ac.in

Sompalli Bhavana
Department of Biotechnology, Surana College (Autonomous), Bengaluru, India
Email: sompallibhavana.biotech@suranacollege.edu.in

Sourish Bhattacharya
Central Salt and Marine Chemicals Research Institute, Bhavnagar, India
Email: sourishb.csmcri@csir.res.in

Subashini Ravikrishnan
Department of Biotechnology, Sathyabama Institute of Science and Technology, Chennai, India
Email: subashalu03@gmail.com

Sudha Suman
Department of Chemical Engineering and Technology, IIT BHU Varanasi, India
Email: sudhasuman.rs.che23@itbhu.ac.in

Rajneesh Kumar, Ram Sharan Singh and Maulin P Shah
ISBN 978-1-83708-327-5
https://doi.org/10.1108/978-1-83708-326-820261001

Chapter 1
Wastewater treatment and microbial processes

Kunal Biswas[1], Subashini Ravikrishnan[2], Raja Gowri Swaminathan[2], Prabhakar Singh[2] and Saqib Hassan[2]

[1]*Centre for Nanoscience and Nanotechnology, International Research Centre, Sathyabama Institute of Science and Technology, Chennai, India*
[2]*Department of Biotechnology, Sathyabama Institute of Science and Technology, Chennai, India*

Abstract

Wastewater treatment is an important environmental process that aims at improving the quality of water for release into natural ecosystems or reuse in different applications. Growing volumes of industrial, agricultural and urban activities continue to pose serious challenges to the effective treatment methods which will mitigate impacts on the environment and ensure public health is protected. This chapter focuses on the role that microbial processes play in the treatment of wastewater and the efficiency, effectiveness and sustainability of these processes. Biological treatment of wastewater involves microbial processes catalysed by diverse communities of microorganisms. Such processes involve various mechanisms, including hydrolysis, fermentation, nitrification, denitrification and dephosphatation, each of which plays a vital role in removing organic matter and nutrients from wastewater. This chapter discusses the different types of microbial systems used in wastewater treatment, such as activated sludge, biofilms and anaerobic digesters, together with the biochemical pathways used by microorganisms to convert pollutants into innocuous byproducts as mentioned in Figure 1.1 (prepared in biorender). In addition, the chapter reviews the innovative developments in microbial biotechnology, including the application of genetically engineered microorganisms and the use of bioaugmentation strategies to improve treatment efficiency. It addresses the challenges of optimising microbial performance under fluctuating environmental conditions and the necessity of monitoring and controlling operational parameters to ensure effective treatment outcomes. By realising the connection between wastewater treatment and microbial processes, this chapter provides an in-depth view of how the utilisation of microbial activity leads to sustainable water management practices and, by extension, improved water quality and environmental conservation.

Keywords: Agricultural; biotechnology; efficiency; microbial; wastewater

1.1. Introduction

Environmental engineering is increasingly relying on microbial processes to provide innovative solutions to some of the most pressing current environmental issues. According to Rittmann and McCarty (2001), the most well-known application of microorganisms is in wastewater treatment, where they break down organic pollutants to provide cleaner water with less of a detrimental environmental impact (shown in Figure 1.1). According to Dalmau *et al.* (2017), this biological treatment approach not only uses fewer pesticides but also ensures that nutrients like nitrogen and phosphorus are recycled, potentially preventing eutrophication in aquatic environments. Furthermore, developments in metabolic engineering have enabled the production of altered microbial strains with enhanced abilities to digest certain pollutants, offering customised treatment choices in contaminated environments (Choudhary *et al.*, 2020).

The increased emphasis on sustainability has also highlighted the importance of bacteria in carbon cycling: Jansson and Taş (2012) state that microbial communities in soils and sediments are crucial to the decomposition of organic matter because they release nutrients that support plant life and absorb carbon dioxide from the atmosphere, reducing the effects of climate change. The environmental impact of industrial operations can be decreased by using microorganisms in soil bioremediation to remove pesticides, heavy metals and petroleum hydrocarbons from contaminated land (Bashir *et al.*, 2018). Microorganisms facilitate biogeochemical cycles such as sulfur reduction and nitrogen fixation, which directly affect water quality, agriculture and biodiversity. According to Fierer *et al.* (2012), these cycles are essential to preserving ecological stability and health. This chapter investigates the intricacies of microbial activity and the potential large-scale solutions these microscopic organisms could provide for environmental issues. It looks at various biotechnological advancements that use microorganisms to produce energy and monitor the environment, such as biosensors and microbial fuel cells (Logan, 2008). It also looks at the future prospects of microbial biotechnology because the capacity to modify microorganisms for particular environmental tasks opens up new possibilities for environmentally responsible management.

1.2. Fundamentals of wastewater treatment

1.2.1 Types of wastewater

The composition, source and possible environmental effects of wastewater allow for its general classification into several types. To choose the appropriate microbiological techniques and technology for efficient treatment, one must be aware of the various types of wastewater. In environmental engineering, wastewater is typically divided into three categories: industrial, agricultural and domestic. The kind of wastewater determines the requirements for microbiological treatment. The water used in houses for activities like cleaning, cooking, bathing and toileting is known as domestic wastewater, or sewage. According to Metcalf and Eddy (2014), this kind of wastewater usually comprises suspended particles, organic debris, nutrients (mostly nitrogen and phosphorus) and microorganisms like bacteria and viruses.

Primary, secondary and tertiary treatment stages are frequently used to treat wastewater from households. Microbial activities are primarily responsible for the breakdown of organic pollutants in secondary treatment; trickling filters and activated sludge systems are frequently employed to promote microbial development and degradation (Tchobanoglous *et al.*, 2003).

Figure 1.1 Graphical abstract of microbial processes in environmental engineering

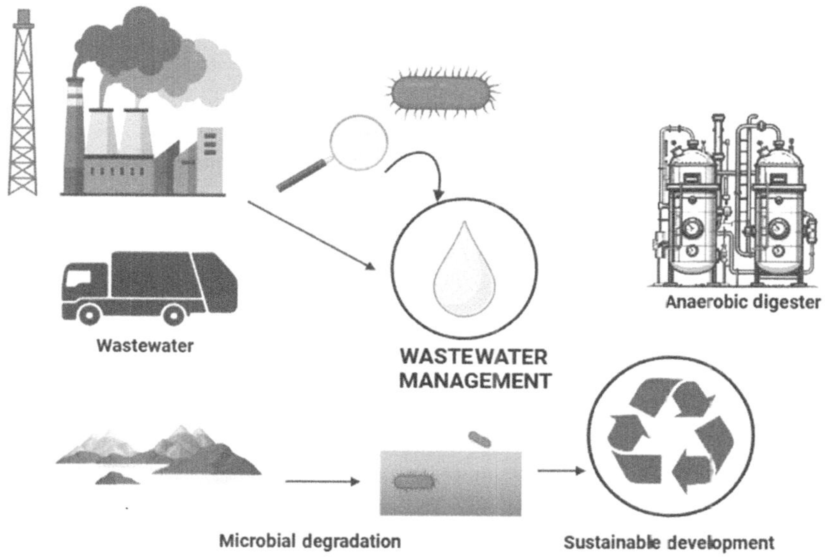

These technologies convert organic contaminants into less hazardous, simpler chemicals by using the natural metabolic processes of bacteria, fungi and protozoa.

Manufacture procedures, food processing, chemical manufacture and other industrial operations all produce industrial effluent. With greater concentrations of hazardous wastes, oils, solvents, heavy metals and toxic chemicals, this kind of wastewater frequently has a more complex makeup than residential wastewater (Joo *et al.*, 2016). Advanced biological, chemical and physical treatment methods may be used in specialised systems for industrial effluents. Industrial wastewater is cleaned by microbial processes including biodegradation and bioaccumulation; some bacteria may break down complex substances like solvents, heavy metals and petroleum hydrocarbons (Lal *et al.*, 2020). Using certain microbial consortia or genetically modified microorganisms with improved metabolic pathways for pollutant breakdown can improve the bioremediation of industrial wastewater (Bhatnagar *et al.*, 2015).

Numerous processes, including animal rearing, fertiliser and pesticide application, and irrigation runoff, result in the production of agricultural wastewater. Along with organic waste, pathogens and suspended particles, this wastewater frequently has significant concentrations of nutrients, especially nitrogen and phosphorus (Basset *et al.*, 2018). Aquatic ecosystems may be disrupted and algal blooms may result from untreated agricultural runoff causing eutrophication in adjacent water bodies. By decomposing organic pollutants and transforming excess nutrients into less harmful forms, microbial mechanisms are essential in reducing the consequences of agricultural runoff (Schroder and Smit, 2018). Specifically, phosphorus-removing bacteria can reduce phosphorus levels, while microbial nitrogen cycling, which includes nitrification and denitrification processes, can reduce nitrogen concentrations in agricultural wastewater (Zhu *et al.*, 2019). In the treatment of agricultural effluents, artificial wetlands and other microbially based natural treatment methods have shown promise (Vymazal, 2013).

Stormwater runoff can be a major source of pollution, particularly in urban areas, even though it is not usually considered wastewater. Roads, building sites and other surfaces release a range of pollutants into stormwater, such as oils, heavy metals, sediments and nutrients (Hunt and DeBusk, 2017). Because microorganisms may break down organic contaminants, extract nutrients and lower the presence of pathogens, microbial treatment of stormwater has gained attention. Filtration systems that increase microbial activity, such as bioretention systems and bioreactors, are commonly used to regulate stormwater. These systems use microbial biofilms to absorb and break down pollutants (Zhou *et al.*, 2020).

1.2.2 Key treatment objectives

The primary objective of wastewater treatment is to lower toxins to levels that are safe for the environment and public health. To eliminate impurities, wastewater is treated using a mix of physical, chemical and biological techniques. Among these, microbial processes are very important for attaining therapy results that are long-lasting, economical and effective. Reducing organic matter, eliminating pathogens, managing sediments and sludge, and removing nutrients (phosphorus and nitrogen) are the main goals of wastewater treatment. These goals can be accomplished with the aid of microbial activity.

One of the main objectives of wastewater treatment is the elimination of organic pollutants. Organic chemicals have the potential to disturb aquatic life by reducing the amount of oxygen in receiving water bodies. These compounds are mostly found in human waste, food scraps and industrial waste. Through activities including fermentation, anaerobic digestion and aerobic respiration, microorganisms are essential to the biological cleanup of organic contaminants. For instance, in activated sludge systems, microorganisms – primarily bacteria – either store organic molecules in their biomass or break them down into simpler components like carbon dioxide and water (Metcalf and Eddy, 2014). Parameters such as chemical oxygen demand (COD) and biological oxygen demand (BOD), which show how much oxygen is needed to break down organic waste, are frequently used to gauge the extent of organic matter clearance.

To keep aquatic habitats from becoming eutrophic, wastewater treatment must remove nutrients, especially phosphorus and nitrogen. Water quality and biodiversity are negatively impacted by eutrophication, which is the over-enrichment of water bodies with nutrients that causes oxygen deprivation and toxic algal blooms (Schroder and Smit, 2018). The elimination of phosphorus and nitrogen is mostly the result of microbial action. Nitrifying bacteria oxidise ammonia to make nitrites and nitrates, while denitrifying bacteria transform nitrates into nitrogen gas, which is then discharged into the environment (Jansson and Taş, 2012). In biological phosphorus removal systems, phosphorus is removed by precipitation with metal ions or microbial phosphorus uptake, both of which are aided by phosphate-accumulating organisms (PAOs) (Mino *et al.*, 1998). These microbial pathways help meet regulatory standards for wastewater discharge by reducing nutrient concentrations to allowable levels.

Numerous pathogens, such as bacteria, viruses and protozoa, are frequently found in wastewater and, if left unchecked, can seriously endanger public health. Viral inactivation, competition and predation are some of the ways that microbial activity significantly lowers pathogen concentrations. In secondary treatment systems, pathogens can be immediately consumed by beneficial microorganisms like nematodes and protozoa, while bacteria can degrade toxic deoxyribonucleic acid (DNA) and ribonucleic acid (RNA) (Gürle *et al.*, 2021). To attain a high level of pathogen elimination in the final effluent, further disinfection methods such as ozonation, ultraviolet (UV) irradiation or chlorination are usually needed (Snyder and Kümmerer, 2020). Frequently, indications that imitate the presence of dangerous bacteria, like enterococci or faecal coliforms, are used to track how well these therapies are working.

The control of sludge and sediments is another important goal in wastewater treatment. The primary and secondary treatment phases of the process generate solids that must be extracted, stabilised and subsequently recycled or disposed of. Primary sludge is the result of the sedimentation process used in primary treatment to remove big particles. Excess microbial biomass is what makes up secondary sludge, which is produced by biological activity during secondary treatment. To prevent health hazards and environmental damage, these substances must be properly controlled. Anaerobic digestion, in which microorganisms break down organic waste without oxygen and produce biogas (methane) as a byproduct, is the main way that microbial activities can stabilise sludge (Khan *et al.*, 2016). Furthermore, aerobic processes like composting further stabilise sludge, lowering its volume and enhancing its safety for land application or disposal (Metcalf and Eddy, 2014).

Resource recovery from wastewater is becoming an increasingly important goal in contemporary wastewater treatment. This includes employing treated wastewater for industrial or agricultural use, recovering nutrients like nitrogen and phosphate, and producing biogas through sludge digestion. Microbial fuel cells (MFCs), in which bacteria convert organic molecules in wastewater into electrical power, are another method that microbial activity can aid energy recovery (Logan, 2008). Furthermore, by eliminating contaminants and producing beneficial byproducts at the same time, microbial fuel cells and other biotechnological advancements provide a sustainable wastewater treatment choice that promotes a circular economy model (Zhang *et al.*, 2019). To reduce the environmental effect and increase the sustainability of wastewater management, resource recovery must be used in wastewater treatment.

1.3. Microorganisms in wastewater treatment

1.3.1 Types of microorganisms

Reducing infections, changing nutrients and breaking down organic contaminants are the main functions of microorganisms in biological wastewater treatment. Members of several taxonomic groupings, including bacteria, fungi, protozoa and algae, are among the diverse range of microorganisms used in wastewater treatment. These bacteria are essential to accomplishing the main treatment goals of wastewater management, which include the elimination of organic debris, the removal of nutrients and the control of pathogens. They achieve this by using a range of metabolic processes to break down contaminants. An extensive summary of the many kinds of microorganisms utilised in wastewater treatment procedures is given in this section. The most common and varied class of microorganisms utilised in wastewater treatment are bacteria. Numerous biological functions are made possible by them, including the breakdown of organic matter, the elimination of phosphorus and nitrogen, and the prevention of infections. Bacteria can be divided into categories according to their oxygen needs and metabolic activities, which affect how they respond during particular treatment techniques.

1.3.1.1 Aerobic bacteria

These bacteria thrive in oxygen-rich settings and are primarily responsible for the decomposition of organic waste in activated sludge systems. They break down complex organic chemicals into simpler ones, such as carbon dioxide and water, using oxygen (Tchobanoglous *et al.*, 2003). *Pseudomonas*, *Aeromonas* and *Bacillus* are popular aerobic bacterial genera used in wastewater treatment. These bacteria significantly reduce the BOD, which is essential for secondary treatment processes, by breaking down the organic components in wastewater (Metcalf and Eddy, 2014).

1.3.1.2 Anaerobic bacteria

Anaerobic bacteria are essential to the stabilisation of sludge by anaerobic digestion because they carry out biological functions without oxygen (Khan *et al.*, 2016). By converting organic resources into carbon dioxide and methane, these bacteria aid in the creation of biogas. This technique uses genera like *Clostridium* and *Methanobacterium* to reduce the amount of waste sludge and turn it into sustainable energy. Because they convert nitrates into nitrogen gas, which makes nitrogen removal easier, anaerobic bacteria are crucial to the denitrification process (Jansson and Taş, 2012).

1.3.1.3 Facultative bacteria

Microorganisms known as facultative bacteria are able to survive in both anaerobic and aerobic settings. They may transition between aerobic and anaerobic conditions based on the oxygen content and are frequently discovered in the mixed liquor of activated sludge systems. Common facultative bacteria in wastewater treatment systems that aid in the breakdown of organic contaminants and the extraction of nutrients are *Escherichia coli* and *Enterobacter* (Dalmau *et al.*, 2017).

Fungi play a minor but important part in wastewater treatment, particularly in the breakdown of complex organic materials like cellulose, lignin and other tough materials. In specialised treatment systems like biofilters and artificial wetlands, fungi can break down organic pollutants with the aid of extracellular enzymes (Ramsay and Enari, 2000). Filamentous fungi such as *Phanerochaete* and *Trametes* are highly effective at decomposing complex pollutants, such as aromatic compounds that are commonly found in industrial effluent (Ramsay and Enari, 2000). Fungi are also used in bioremediation because of their capacity to decompose both organic and inorganic pollutants in disturbed environments.

In some situations, fungi have been shown to help cleanse wastewater that contains heavy metals by immobilising or altering the metals (Chakraborty and Patel, 2015) as shown in Figure 1.2 (prepared in biorender). They are appropriate for managing more difficult effluents from sectors including paper and pulp, textiles and pharmaceuticals because of their capacity to break down complex organic molecules. Single-celled eukaryotic creatures called protozoa are crucial to the treatment of wastewater, especially in secondary treatment procedures. By inhibiting the growth of particular bacterial species and preserving a balanced microbial

Figure 1.2 Microorganism in water

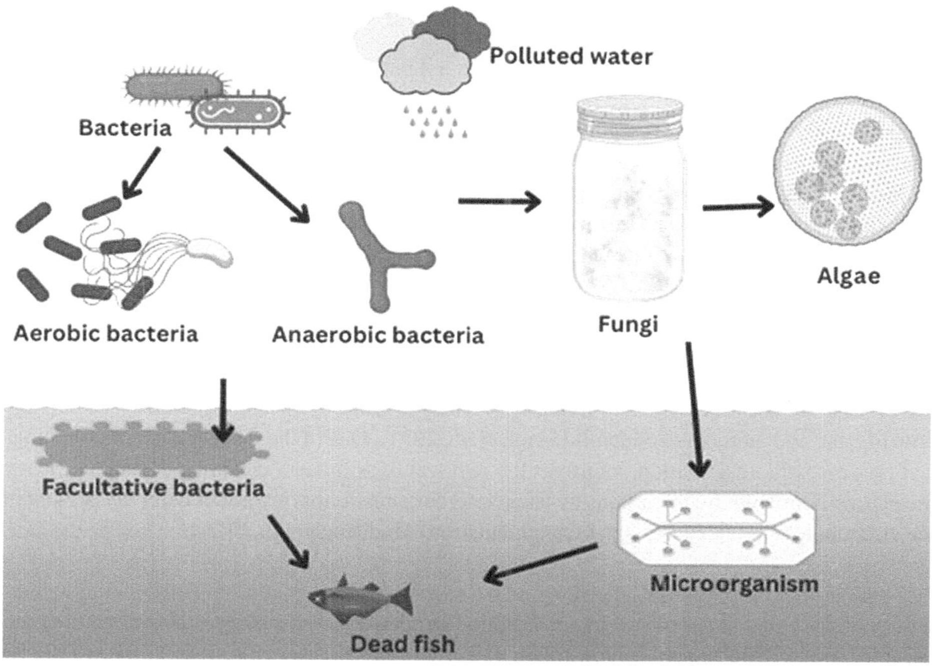

community, eating bacteria aids in the regulation of microbial populations in the human gut (Rao and Sharma, 2008).

Protozoa like *Paramecium*, *Amoeba* and *Vorticella*, commonly found in activated sludge systems, enhance the overall effectiveness of biological treatment processes by feeding on bacteria and organic particles. Protozoa help reduce the amount of suspended particles in wastewater and regulate bacterial populations. Because they are predators, they break down aggregates of organic debris, resulting in cleaner effluent (Metcalf and Eddy, 2014). Because they help release minerals like phosphate and nitrogen that other bacteria in the system may eat, protozoa are also crucial to the nutrition cycle.

Algae – especially microalgae – have the capacity to absorb nutrients like phosphorus and nitrogen, which are frequent pollutants in wastewater, and are playing an increasingly significant role in wastewater treatment (Vymazal, 2013). In integrated systems, such as artificial wetlands or algae-based biofilms, microalgae like *Chlorella* and *Scenedesmus* are employed to both produce oxygen and remove nutrients (Chadwick and Caperchione, 2015).

Algae contribute to wastewater treatment by absorbing nutrients during their growth. They can then be harvested for resource recovery – for example, fertilisers or biofuels – making wastewater treatment more sustainable (Rittmann and McCarty, 2001). In addition to eliminating nutrients, algae can help reduce infections by generating antimicrobial compounds during particular stages of development (Zhou *et al.*, 2020). Algae are a promising way to remove pollutants from wastewater, especially in systems designed for nutrient-rich effluents.

Although viruses are typically not directly engaged in the breakdown of pollutants, they can be used as indicators of pathogen prevalence and affect microbial populations in treatment systems, which makes them significant in wastewater treatment. Wastewater viral load is often associated with human diseases, such as enteric viruses, which pose a threat to public health. Therefore, monitoring viruses such as *Rotavirus* and *Enterovirus* is essential for assessing the effectiveness of treatment protocols, especially in tertiary care where pathogen elimination is a primary goal (Snyder and Kümmerer, 2020).

1.3.2 Role in pollutant degradation

Microbial degradation of pollutants is crucial to environmental engineering because it facilitates the breakdown of toxic substances into less harmful or benign forms. The primary mediators of this process are bacteria, fungi and archaea, which use a range of metabolic pathways to degrade both organic and inorganic pollutants (Singh *et al.*, 2021). One of the most studied microbiological processes, bioremediation, improves the removal of pollutants including pesticides, heavy metals and petroleum hydrocarbons by using microorganisms that are either naturally present in the environment or added through bioaugmentation (Megharaj *et al.*, 2018).

To break down contaminants, microorganisms employ a range of mechanisms, including aerobic and anaerobic metabolism. In aerobic biodegradation, where oxygen acts as an electron acceptor, organic pollutants such as phenols and polycyclic aromatic hydrocarbons (PAHs) are broken down by enzymes such as monooxygenases and dioxygenases (Jin *et al.*, 2019). On the other hand, anaerobic biodegradation occurs in environments with low oxygen, when microorganisms use alternative electron acceptors like nitrate, sulfate or iron to break down pollutants such as nitroaromatic compounds and chlorinated solvents (Löffler *et al.*, 2017).

In addition to metabolic activities, microbial consortia have a key role in the efficiency of degradation. Co-metabolism, in which microorganisms oxidise poisons unintentionally while metabolising another primary substrate, has been successfully used to break down complex pollutants such as trichloroethylene (TCE) and persistent organic pollutants (POPs) (Semprini, 2020). Additionally, one possible tactic for dealing with persistent pollutants is the use of genetically engineered microorganisms (GEMs), which are made to boost biodegradation ability (Sayler and Ripp, 2022).

Microbial degradation is used in environmental engineering for industrial effluent management, soil bioremediation and wastewater treatment. Microbial communities are used in bioreactors and artificial wetlands to break down organic contaminants and nitrogenous chemicals in wastewater (Vymazal, 2018). Similarly, biofilters that remove volatile organic compounds (VOCs) from industrial pollutants use microbial biofilms (Delhoménie and Heitz, 2021).

Environmental changes, the bioavailability of pollutants and possible toxicity effects on microbial populations are some of the obstacles that microbial pollutant degradation must overcome despite its efficacy (El Fantroussi and Agathos, 2019). To increase the effectiveness of pollutant degradation in environmental engineering, research is still being conducted on fortifying microbial consortia, creating resilient microbial strains and combining microbial processes with physicochemical treatment techniques.

1.4. Biological treatment processes

1.4.1 Activated sludge process

Ardern and Lockett first described the activated sludge process, a popular biological wastewater treatment method, for biological sewage remediation in 1914 (Ardern and Lockett, 1914). In order to assimilate or oxidise the dissolved organic matter found in influent wastewater, the process relies on heterotrophic microbial biomass (Ardern and Lockett, 1914). Over the past 70 years, it has attracted a lot of engineering attention for wastewater treatment and is considered the traditional standard (Alleman and Prakasam, 1985a, 1985b).

The procedure includes several steps that are necessary for treatment. Because they convert liquid and particulate organic materials into cell mass, microorganisms perform a crucial function. Because the process depends on aerobic microorganisms using waste organic matter as their substrate – which entails oxygen – aerobic conditions are crucial. Additionally, an aeration tank permits adequate detention time, guaranteeing a thorough mixing of the influent wastewater and activated sludge. A consistent and timely supply of oxygen is required to maintain these aerobic conditions and stop the activated sludge from settling. Wastewater is mixed with a colony of microorganisms in an aeration tank to create activated sludge.

The addition of oxygen or air promotes the growth of aerobic bacteria, which eat organic matter. These bacteria form flocs, which are collections of microorganisms that can be removed from the water and facilitate the effective decomposition of organic matter. Following aeration, the wastewater and activated sludge mixture is pumped to a sedimentation tank or clarifier, where the active sludge settles and separates from the treated water. Some of the settled sludge, also referred to as return-activated sludge, is recycled back into the aeration tank to feed the microbial population, while waste-activated sludge, also known as surplus sludge, is periodically collected for treatment or disposal.

Technologies for cyclic activated sludge can be improved over time to support effective and long-lasting water treatment. Globally, activated sludge-based biological wastewater treatment plants have revolutionised water sanitation despite growing urbanisation and industrial development. The process of activated sludge can also be described as a system in which biological flocs are continuously recycled to come into contact with and oxidise organic materials. The effectiveness of wastewater treatment facilities generally remains relatively stable even when influent flow increases and COD and BOD levels fluctuate.

1.4.2 Trickling filters

Trickling filters, which use microbial action to break down organic contaminants, are one of the most popular biological treatment methods for wastewater treatment. In this method, wastewater is dispersed across a fixed bed of coarse media, such as rocks, plastic or other human-made materials as shown in Figure 1.3 (prepared in biorender).

The microbial biofilm that forms on the media surface is crucial to the nitrification and decomposition of organic waste (Tchobanoglous *et al.*, 2014). Both aerobic and anaerobic microbial activity are used in the trickling filter's treatment procedure. Natural or artificial aeration adds oxygen to wastewater as it flows through a filter, promoting the growth of aerobic bacteria that decompose organic pollutants (Metcalf and Eddy, 2017). The microbial biofilm's bacteria, fungi and protozoa cooperate to decompose organic materials and change ammonia into nitrate (Rittmann and McCarty, 2001). The excess biomass and the sloughed-off biofilm are subsequently washed away with the effluent, which is often treated with secondary clarity (Bitton, 2011). The rates of hydraulic and organic loading can be used to distinguish between different kinds of trickling filters.

While low-rate trickling filters have lower loading rates, longer contact durations and higher treatment efficiency, high-rate trickling filters may process bigger organic loads but may need more treatment steps (Henze *et al.*, 2008). Due to their effects on microbial activity and biofilm formation, temperature, hydraulic loading rate and medium type all affect the performance of trickling filters (Gerardi, 2006). Although trickling filters offer their own benefits, they can have drawbacks, including clogging from too much biofilm growth and problems controlling odours (Grady *et al.*, 2011). Novel inventions have emerged to maximise performance, such as hybrid trickling filters and rotating biological contactors (RBCs), which combine attached and suspended growth processes to maximise treatment efficiency (Mara and Horan, 2003).

Generally speaking, trickling filters remain a cost-effective and sustainable option for biological wastewater treatment, particularly in small- to medium-sized treatment facilities. They are an economical method for removing organic matter and nitrifying wastewater treatment facilities due to their ability to support complex microbial communities (Crittenden *et al.*, 2012).

1.4.3 Anaerobic treatment

A popular biological method for treating wastewater, particularly for highly organic waste streams, is anaerobic treatment. This process uses various microbial communities and takes place without oxygen to break down organic materials through a sequence of biochemical events that eventually result in biogas, which is primarily made up of carbon dioxide and methane (Metcalf and Eddy, 2017). Anaerobic treatment has several advantages over aerobic systems, including lower energy demand, reduced sludge creation and the possibility of energy recovery through methane synthesis (Lettinga, 1995). The four primary phases of anaerobic digestion are hydrolysis, acidogenesis, acetogenesis and methanogenesis (Angelidaki *et al.*, 2011). Hydrolytic bacteria use a process called hydrolysis to break down complex organic compounds including proteins, lipids and carbohydrates into simpler monomers.

Acidogenesis is the process by which fermentative bacteria transform these monomers into volatile fatty acids, alcohols, hydrogen and carbon dioxide (Batstone *et al.*, 2002). During acetogenesis, acetogenic bacteria further break down these intermediates to create carbon dioxide, hydrogen and acetate, which are then used as substrates by methanogens during the last stage, methanogenesis. Methanogenic archaea produce biogas, which can be used for energy recovery, by converting hydrogen and acetate into carbon dioxide and methane (Gerardi, 2003). Wastewater can be treated using a variety of anaerobic treatment technologies, such as anaerobic lagoons, anaerobic filters and upflow anaerobic sludge blanket (UASB) reactors (Speece, 2008).

Figure 1.3 Trickling filter

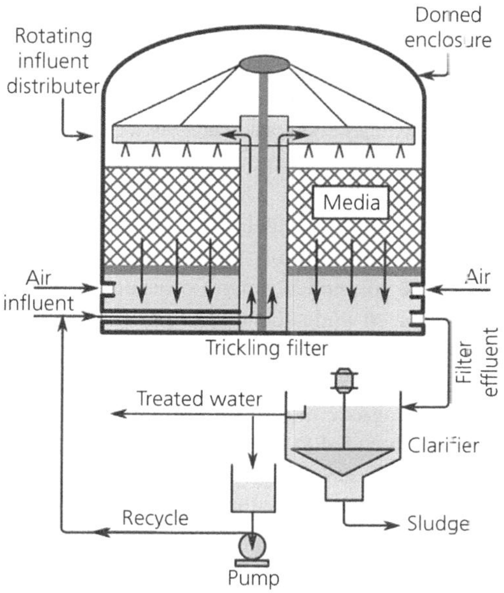

UASB reactors are among the most popular and efficient systems due to their short hydraulic retention durations and high organic load removal efficiency. They maintain a deep microbial sludge bed that allows organic materials to break down in order to remediate wastewater that travels uphill (Lettinga and Hulshoff Pol, 1991). Anaerobic treatment is especially well-suited for treating wastewater with a high organic content, such as industrial effluent from breweries, food processing plants, and pulp and paper mills (Chernicharo, 2007). Long starting times, toxicity to dangerous compounds and process instability brought on by changes in pH and temperature, however, can all affect performance (Ahring, 2003). Despite these drawbacks, reactor design and process improvement continue to increase the efficacy and suitability of anaerobic treatment technologies in wastewater treatment (McCarty *et al.*, 2011).

1.4.4 Biofilms and their applications

Because biofilms provide a permanent and effective home for the microbial populations that break down pollutants, they are an essential part of biological wastewater treatment systems. Biofilms are microbial colonies structured in a self-generated extracellular polymeric substance (EPS) that sticks to surfaces like filter media or reactor walls (Flemming and Wingender, 2010). Because of their remarkable resilience to contaminants, environmental changes and antimicrobial agents, these microbial consortia are highly successful at treating wastewater (Costerton *et al.*, 1995). In biological wastewater treatment procedures, biofilms help in nitrification, denitrification and the decomposition of organic materials. The multilayered structure of biofilms supports the simultaneous occurrence of numerous metabolic activities and the existence of facultative, anaerobic and aerobic microorganisms (Stoodley *et al.*, 2002).

While aerobic bacteria oxidise ammonia and organic pollutants in the outer layers of biofilms, which are usually rich in oxygen, anaerobic or facultative bacteria are employed for denitrification and other reductive reactions in the core layers, which have low oxygen levels (Xavier *et al.*, 2005). Many biofilm-based technologies, such as trickling filters, RBCs and moving bed biofilm reactors (MBBRs), are used to treat wastewater (Rittmann and McCarty, 2001). In trickling filters, wastewater flows over a layer of media where contaminants are broken down by biofilm. Through frequent contact with air and wastewater, as well as spinning discs, which are the site of biofilm formation, RBCs efficiently remove organic material (Morgenroth *et al.*, 2000).

For more efficient treatment of both municipal and industrial wastewater facilities, MBBRs use suspended plastic carrier parts that allow the biofilms to acquire a broad surface area (Ødegaard, 2006). Compared to suspended growth procedures, biofilm-based systems have several advantages, such as reduced sludge generation, shock load resilience and high microbial retention (Wuertz *et al.*, 2003). System performance may be impacted by issues such as clogging, biofilm detachment and uneven biomass distribution (Van Loosdrecht *et al.*, 2002). New developments in biofilm engineering, including the application of nanomaterials and quorum-sensing inhibitors, are intended to enhance biofilm stability and increase wastewater treatment efficiency (Singh *et al.*, 2006). In general, biofilms are a critical part of biological treatment systems, providing an environmentally friendly and effective method of wastewater treatment based on increased microbial activity and the degradation of pollutants (Flemming *et al.*, 2016).

1.5. Chemical and physical treatment methods

1.5.1 Coagulation and flocculation

In order to remove suspended solids, colloidal particles and organic materials from wastewater, two essential physicochemical treatment processes are flocculation and coagulation. They aid in the removal of tiny particles through filtration or sedimentation by promoting their aggregation (Gregory and Duan, 2001).

While flocculation uses gradual agitation to help the destabilised particles aggregate into larger flocs, coagulation involves using coagulants to destabilise suspended particles (Bratby, 2016). Dosing wastewater with chemical coagulants such as ferric chloride, polyaluminium chloride or aluminium sulfate (alum) produces coagulation. These coagulants reduce repulsion and enable the colloidal particles to form aggregates by neutralising their surface charges (Duan and Gregory, 2003). The coagulant dose, pH and mixing conditions all affect coagulation efficiency. Because different coagulants react best at different pH levels, pH adjustment must be done correctly (Zouboulis and Tzoupanos, 2010).

The next step after coagulation is flocculation, which involves slowly mixing wastewater to promote the formation of bigger, more soluble flocs. To improve the process, flocculants are added, such as synthetic polymers like polyacrylamides or natural polymers like starch and chitosan, which agglomerate and bridge particles together (Bolto and Gregory, 2007). Stable flocs that are easily removed by sedimentation or filtration are produced as a result of the effective particle collision made possible by the gentle agitation that flocculation facilitates (Renault *et al.*, 2009).

Coagulation and flocculation find extensive application in wastewater treatment facilities for the elimination of turbidity, heavy metals, organic contaminants and pathogens (Matilainen *et al.*, 2010). These processes are most effective in the treatment of industrial effluents – for example, textile, paper and food processing industries, where high suspended solid and organic matter concentrations are found (Teh *et al.*, 2016). However, operational challenges like excessive sludge production and optimal coagulant dosing can influence process efficiency (Jarvis *et al.*, 2005).

New developments in environmentally friendly coagulants, including plant-based and biodegradable polymers, are being investigated to enhance sustainability and minimise chemical reliance in wastewater treatment (Okuda *et al.*, 2001). In general, coagulation and flocculation are basic treatment processes, greatly enhancing water quality by increasing the removal of fine particles and impurities prior to further treatment processes (Duan and Gregory, 2003).

1.5.2 Disinfection techniques

Disinfection is a crucial step in wastewater treatment that aims to eliminate or render pathogenic bacteria inactive in order to guarantee that the treated effluent is safe for the environment and public health. Wastewater treatment facilities employ a variety of disinfection techniques, from chemical to physical approaches, each with its own mechanisms and effectiveness (Metcalf and Eddy, 2017).

Due to its affordability and ability to provide a residual disinfection, chlorination is the most widely used chemical disinfection method. Chlorine, whether in the form of sodium hypochlorite, calcium hypochlorite or chlorine gas, affects the components of microbial cells and disrupts vital enzyme functions, leading to cell lysis (White, 2010). However, chlorination produces harmful disinfection by products (DBPs) such as haloacetic acids (HAAs) and trihalomethanes (THMs), whose effects on health have raised concerns (Richardson *et al.*, 2007). Before discharge, excess chlorine must usually be chemically dechlorinated using substances like sodium bisulfite (Hua and Reckhow, 2007).

One of the most popular physical disinfection methods is UV irradiation, which stops the reproduction of germs by destroying their DNA (Zimmer and Slawson, 2002). UV disinfection produces no harmful byproducts and is highly effective against bacteria, viruses and protozoa. However, a number of factors, including contact time, UV irradiation and water turbidity, affect how effective it is (Mamane *et al.*, 2007). The lack of residual protection – that is, any recontamination after treatment could lead to microbial regrowth – is one of the drawbacks of UV disinfection (Linden *et al.*, 2012).

Another high-level disinfection technique is ozonation, which uses ozone (O_3), a very potent oxidising agent, to break down bacteria cell walls and disrupt metabolic activities (Von Gunten, 2003a). According to Li *et al.* (2008), ozone is highly effective against a wide range of infections, including *Giardia* and *Cryptosporidium*, which are resistant to chlorine. Compared to chlorination, ozonation also reduces the formation of harmful DBPs. However, ozone manufacturing is more expensive because it requires an on site system of production due to its short half-life and high energy consumption (Gottschalk *et al.*, 2010).

The effectiveness of the treatment, cost, operational adaptability and environmental impact all influence the disinfection technique. The most often used technique is still chlorination, but UV and ozone technologies are becoming more and more popular due to their effectiveness and lower likelihood of producing hazardous byproducts (Shannon *et al.*, 2008). To enhance microbial inactivation and maximise wastewater treatment performance, advancements in combined disinfection techniques, such as UV–chlorine and ozone–hydrogen peroxide treatment, are also being researched (Wang *et al.*, 2016).

1.5.3 Advanced oxidation processes

By generating very reactive hydroxyl radicals (•OH), advanced oxidation processes (AOPs) are a group of chemical treatment systems that oxidise organic pollutants and microorganisms in wastewater (Glaze *et al.*, 1987). Because of the radicals' strong oxidising power, stable organic contaminants, including industrial chemicals, insecticides and medications, can be broken down into harmless or biodegradable byproducts (Oturan and Aaron, 2014). Fenton's process, photocatalysis, ozonation and electrochemical oxidation are a few AOPs used to clean wastewater.

Fenton's process is a reaction that produces hydroxyl radicals, which effectively break down organic contaminants when hydrogen peroxide (H_2O_2) and ferrous ions (Fe^{2+}) come into contact (Neyens and Baeyens, 2003). Despite requiring pH adjustment and producing iron sludge as a byproduct, the technique is well-liked due to its simplicity and effectiveness (Kavitha and Palanivelu, 2004).

Another potential AOP is photocatalysis, particularly with titanium dioxide (TiO_2), which employs visible or UV light to activate the catalyst and produce hydroxyl radicals that eliminate pollutants (Fujishima *et al.*, 2000). Although the method is highly effective in eliminating medications and substances that disrupt hormones from wastewater, it has drawbacks, such as catalyst deactivation and the need for excellent light penetration (Herrmann, 1999).

One AOP that uses ozone (O_3) to speed up the oxidation of organic pollutants is ozonation, which generates hydroxyl radicals (Von Gunten, 2003b). Because its creation is energy-intensive and short-lived, ozone requires on site generation and specific advanced reactor designs, yet it may effectively oxidise a large number of distinct contaminants and greatly improve biodegradability (Andreozzi *et al.*, 1999).

According to Martínez-Huitle and Ferro (2006), electrochemical oxidation is a novel AOP that uses anodic oxidation to create hydroxyl radicals and other reactive species for the destruction of pollutants. Although electrode stability and energy consumption are issues, electrochemical oxidation offers great efficiency, minimal chemical consumption and less sludge creation (Moreira *et al.*, 2017).

As tertiary treatment methods, AOPs are increasingly being utilised to remove non-biodegradable pollutants and improve water quality before it is released or used again. Hybrid AOPs, such as UV/H_2O_2, ozone/Fenton and electro-Fenton, are being researched to increase treatment efficiency and overcome the limitations of single processes, even if each process has advantages and disadvantages (Ikehata and Gamal El-Din, 2004).

1.6. Innovative approaches to microbial biotechnology

1.6.1 Genetic engineering of microorganisms

The genetic engineering of microorganisms has revolutionised microbial biotechnology by enabling the optimisation of metabolic pathways for improved pollutant degradation, environmental remediation and resource recovery. Using metabolic engineering, synthetic biology and recombinant DNA technologies, researchers have successfully modified microbial genomes to boost their efficiency in industrial bioprocessing, bioremediation and biofuel generation (Singh *et al.*, 2022). Environmental engineers are increasingly using GEMs to break down hard-to-degrade pollutants such as heavy metals, plastic debris and POPs (Ramos *et al.*, 2019).

One of the most important advances in genetic engineering is the use of CRISPR-Cas9 genome editing to modify microbial strains for better bioremediation. For instance, hydrocarbons, pesticides and dangerous chemical compounds have been broken down by *Pseudomonas* putida and *Bacillus subtilis* using CRISPR technology (Johns *et al.*, 2021). Synthetic biology approaches have also enabled the development of microbial biosensors capable of detecting and degrading certain environmental contaminants. By fusing genetic circuits with pollutant-responsive promoters, scientists have engineered bacteria like *Escherichia coli* to behave as bio-indicators for heavy metal contamination (Wang *et al.*, 2020).

Another innovative approach is the genetic modification of microorganisms to enhance heavy metal bioremediation. *Cupriavidus metallidurans* and *Ralstonia eutropha* are two bacteria that have been altered to express metal-binding proteins in order to immobilise and detoxify dangerous metals such as lead, mercury and cadmium (Nies, 2021). Additionally, the insertion of genes encoding for phytochelatins and metallothioneins has improved bioaccumulation efficiency and microbial resistance in wastewater treatment systems (Jiang *et al.*, 2019).

According to Yoshida *et al.* (2016), microbial genetic engineering has also enhanced the degradation of microplastics and plastics. *Ideonella sakaiensis* and other naturally occurring bacteria that break down polyethylene terephthalate (PET) have undergone genetic modification to break down plastic more quickly, perhaps providing a solution to the world's plastic pollution problem. Bioelectrochemical systems (BES), which use modified electroactive bacteria to treat wastewater and produce bioelectricity, are also utilising these developments in microbial genetic engineering (Logan and Rossi, 2020).

Future research will focus on biocontainment strategies, such as confinement plasmids and synthetic kill-switches, to prevent the inadvertent spread of altered microorganisms in natural settings (Lee *et al.*, 2018). The continued advancement of genetic engineering in microbial biotechnology could have significant implications for applications of sustainable environmental engineering. Despite these positive advancements, concerns remain regarding the environmental discharge of GEMs, regulatory limitations and potential ecological risks.

1.6.2 Use of microbial fuel cells

A cutting-edge technique in microbial biotechnology, MFCs use the metabolic activity of microorganisms to generate bioelectricity and purify wastewater. MFCs function by using electroactive bacteria, such as *Shewanella oneidensis* and *Geobacter sulfurreducens*, to transfer electrons to an anode during the breakdown of organic pollutants (Logan *et al.*, 2019). Through an external circuit, these electrons make their way to the cathode, where they reduce oxygen or other electron acceptors to create an electrical current (Santoro *et al.*, 2017).

One major advantage of MFCs is their dual application in wastewater treatment and bioenergy production. According to research, MFCs effectively remove organic matter, heavy metals and emerging contaminants from wastewater while producing sustainable energy (Pant *et al.*, 2020). Additionally, changes like biocathodes and mediator-based electron transfer have improved pollutant degradation and electricity output by increasing the efficiency of microbial electron transfer (Rabaey and Rozendal, 2019).

Innovative uses of MFCs include their incorporation into biosensors for water quality monitoring, which provide real-time contaminant detection by varying power generation according to pollutant levels (Zhang *et al.*, 2021). Additionally, the creation of microbial electrolysis cells (MECs), a variation of MFCs, enables the synthesis of hydrogen gas, hence increasing their use in the production of renewable energy (Escapa *et al.*, 2019).

Despite their potential, barriers like low power output, high electrode material costs and scale-up limitations keep them from being widely used (You *et al.*, 2020). Research on enhancing microbial electron transfer, enhancing electrode designs and integrating MFCs with other treatment techniques is still ongoing for better environmental applications. The future of MFCs in environmental engineering depends on their further development as energy-efficient, ecologically acceptable wastewater treatment techniques.

1.6.3 Bioaugmentation techniques

Bioaugmentation is a sophisticated microbial biotechnology approach used in environmental engineering to enhance the degradation of pollutants through the introduction of specific microbial strains into areas that are polluted. This approach has garnered a lot of interest since it can increase the effectiveness of bioremediation in wastewater treatment, industrial effluent management and soil decontamination (Singh *et al.*, 2021). Bioaugmentation speeds up the breakdown of organic and inorganic pollutants, such as hydrocarbons, pesticides and heavy metals, by adding GEMs or pre-adapted GEMs into natural microbial populations (Tyagi *et al.*, 2020).

Introducing particular microbial consortia to improve the removal of phosphorus, nitrogen and resistant contaminants is one of the main uses of bioaugmentation in wastewater treatment. For instance, nitrogen removal has risen when denitrifying and ammonia-oxidising bacteria (AOB) are added to municipal wastewater treatment plants (Chen *et al.*, 2018). In a similar manner, petroleum hydrocarbons and polycyclic aromatic hydrocarbons (PAHs) in contaminated soil and groundwater have been effectively broken down by *Pseudomonas* and *Bacillus* species in bioaugmentation (Mishra *et al.*, 2019). Bioaugmentation is crucial for the industrial treatment of xenobiotics, such as dyes, chlorinated solvents and pharmaceutical residues. Modified strains of *Dehalococcoides mccartyi* can be introduced into contaminated aquifers to totally dechlorinate toxic solvents like TCE and perchloroethylene (PCE)

(Holliger *et al.*, 2019). Furthermore, microbial consortia for heavy metal bioremediation have been formed using bacteria that produce biosorbents, such as *Cupriavidus metallidurans*. This enhances lead, mercury and cadmium immobilisation and detoxification in mining-affected areas (Nies, 2021).

The use of microencapsulation and bio-carriers, which improve microbial survival and activity in challenging environmental circumstances, is a recent development in bioaugmentation. It has been demonstrated that immobilising bacteria on polymeric carriers, charcoal and alginate beads increases the effectiveness of biodegradation by creating a safe habitat for microbial development (Jiang *et al.*, 2020). Furthermore, to create synthetic microbial consortia with improved degradation and stress tolerance, genetic engineering techniques are being used into bioaugmentation procedures (Wang *et al.*, 2021).

Notwithstanding its benefits, bioaugmentation has drawbacks, including subpar field-scale results, microbial competition and shifting weather patterns. Two strategies being investigated to increase microbial endurance and effectiveness in practical applications are quorum sensing regulation and adaptive laboratory evolution (ALE) (Giri *et al.*, 2020). Bioaugmentation is an emerging technique that has promise for enhancing pollution degradation and ensuring long-term environmental repair.

1.7. Challenges in wastewater treatment

1.7.1 Operational challenges

Wastewater treatment is an essential process in environmental engineering that mostly relies on microbial communities to remove contaminants. Nonetheless, a variety of operational concerns affect the efficiency, sustainability and cost-effectiveness of wastewater treatment plants (WWTPs). Among these challenges include microbial imbalances, sludge management, energy use and the appearance of new contaminants (Gude, 2017).

The instability of the microbial community is a serious issue that may reduce the efficacy of treatment. Microbial consortia are essential to the breakdown of organic contaminants and nutrients in biological treatment methods like activated sludge systems. System failures and decreased efficacy in removing pollutants might result from microbial equilibrium being upset by hydraulic shocks, dangerous compounds and changing wastewater composition (Luo *et al.*, 2019). Furthermore, too many filamentous bacteria can thicken the sludge, which might impact sedimentation and raise operating expenses (Katsou *et al.*, 2017). Since wastewater treatment generates a lot of excess sludge that must be appropriately disposed of, sludge management is another problem. Traditional sludge treatment methods including anaerobic digestion, composting and burning have issues with high operating costs, greenhouse gas emissions and disposal needs (Tyagi and Lo, 2019). Researchers are investigating innovative methods such as sludge reduction technologies and resource recovery strategies like biogas and nutrient recovery to mitigate these challenges (Muga and Mihelcic, 2008).

Aeration processes in WWTPs require between 50 and 60% of the energy used, hence energy consumption is another important operational issue. It has been proposed to optimise aeration systems, create energy-efficient microbial consortia and integrate bioelectrochemical systems like MFCs in order to reduce energy needs and enhance sustainability (Khan *et al.*, 2020).

The emergence of new contaminants, like microplastics, personal care products and pharmaceuticals, is a growing issue. Because traditional treatment techniques are often inefficient at getting rid of these pollutants, they build up in the environment and provide health risks (Tran *et al.*, 2018). Research is being conducted on membrane bioreactors (MBRs), AOPs and bioaugmentation with particular microbial strains to improve the efficacy of pollution removal (Ahmed *et al.*, 2017).

1.7.2 Environmental impact

Although wastewater treatment is essential for protecting human health and the environment, issues with how these systems operate can have a significant detrimental impact on the ecosystem. Inefficiencies in microbiological processes, sludge disposal, energy use and pollutant removal lead to water contamination, greenhouse gas emissions and ecosystem deterioration (Gude, 2017).

One of the most dangerous environmental problems is the release of untreated or partially treated wastewater into natural water bodies due to process failures, system overload or microbial imbalances. Insufficient microbial breakdown of organic pollutants and nutrients can lead to eutrophication, which causes algal blooms, oxygen depletion and a decrease in aquatic biodiversity (Luo *et al.*, 2019). Excessive levels of nitrogen and phosphorus in wastewater effluents have been linked to significant ecological changes, including dead zones in coastal regions (Conley *et al.*, 2018).

Another problem is the accumulation of new contaminants, such as heavy metals, herbicides, microplastics and pharmaceuticals, that are often too difficult to remove using conventional treatment techniques. These pollutants persist in aquatic environments, bioaccumulate in food chains and pose a threat to human and wildlife health (Tran *et al.*, 2018) – for example, antibiotic residues in wastewater contribute to the global antimicrobial resistance epidemic by promoting the proliferation of antibiotic-resistant bacteria (ARB) (Michael *et al.*, 2021).

Sludge disposal and greenhouse gas emissions are further environmental concerns. Wastewater treatment generates large amounts of sludge, which can contaminate soil and water if not managed appropriately. Conventional sludge disposal methods like landfilling and incineration release methane (CH_4) and carbon dioxide (CO_2), two of the primary contributors to climate change (Gålfalk and Bastviken, 2025). Furthermore, with a potential for global warming 300 times greater than that of CO_2, nitrous oxide (N_2O) emissions from biological nitrogen removal operations are a significant contributor to climate change (Kampschreur *et al.*, 2018).

The excessive energy consumption of WWTPs also contributes to carbon emissions. Aeration systems alone can account for up to 60% of a WWTP's energy usage, which worsens environmental effects and increases reliance on fossil fuels (Khan *et al.*, 2020). Sustainable alternatives, such as MFCs, anaerobic digestion for the production of biogas and integrated energy recovery systems, are being researched to mitigate these effects (Zhou *et al.*, 2021).

Wastewater treatment systems are using advancements in bioaugmentation, MBRs, artificial intelligence (AI)-driven process optimisation and designed wetlands to get around these challenges and improve efficiency while minimising environmental harm. Implementing circular economy techniques, such as resource reuse and nutrient recovery, can further reduce the ecological impact of wastewater treatment (Batstone *et al.*, 2015).

1.7.3 Regulatory considerations

Wastewater treatment is essential to environmental protection, and regulatory frameworks are required to ensure compliance with water quality requirements and public health safety. However, a number of challenges exist for the implementation and enforcement of wastewater regulations, including shifting pollutant profiles, technological limitations and regional variations in infrastructure and administration (Daughton, 2019).

One of the biggest regulatory challenges is establishing and implementing stringent discharge limits for contaminants including nutrients, heavy metals, medications and emerging pollutants. Traditional rules primarily address levels of BOD, total nitrogen and phosphate; however, more recent contaminants such as microplastics and antibiotic-resistant genes (ARGs) provide challenging issues for monitoring and treatment. Existing WWTPs are often not designed to effectively remove these pollutants, resulting in a mismatch between technological capabilities and regulatory requirements (Michael *et al.*, 2021).

Effective pollution control is often hampered by antiquated or inconsistent wastewater regulations. Diverse legal systems across different nations and jurisdictions result in different requirements for wastewater treatment. Untreated or inadequately treated wastewater is frequently released into natural water bodies in underdeveloped countries due to a lack of financing for wastewater infrastructure upgrades and lax regulations (UNEP, 2020). To provide standard environmental protection measures and encourage best practices in wastewater treatment, more legislation and international collaboration are needed.

Sludge management rules represent an extra problem because wastewater treatment produces a significant number of biosolids that need to be properly disposed of or repurposed. Due to worries about heavy metal accumulation, disease persistence and greenhouse gas emissions, sludge use in agriculture, landfill disposal and incineration is strictly regulated. To lessen the environmental impact of sludge disposal, recent legislation promotes resource recovery techniques such as phosphorus extraction and biogas production (Tyagi and Lo, 2019).

Regulations pertaining to energy efficiency and carbon footprint are becoming important factors in wastewater treatment. Aeration and advanced oxidation are two energy-intensive processes used in many WWTPs that raise carbon emissions and operating costs. To lessen the environmental impact of the sector, regulatory agencies are promoting sustainable wastewater treatment methods such as anaerobic digestion, MFCs and energy-efficient aeration systems (Gude, 2017).

With the introduction of digital monitoring and compliance reporting, regulatory enforcement is evolving. Modern technologies including blockchain-based compliance tracking, AI-driven process control and real-time sensors are being used to increase regulatory efficiency and transparency (Zhou *et al.*, 2021). These advancements help WWTPs achieve discharge criteria while improving operational resilience and overall process optimisation.

1.8. Sustainability and future perspectives

1.8.1 Sustainable practices in wastewater management

Sustainable wastewater management is essential for resource recovery, long-term economic viability and environmental preservation. Given the growing global water scarcity and pollution, sustainable practices place a high priority on reducing energy consumption, recovering precious resources and minimising environmental consequences (Corcoran *et al.*, 2010). Holistic approaches – like decentralised treatment facilities, nature-based solutions and circular economy approaches – are increasingly being considered since they can make wastewater more sustainable (Guest *et al.*, 2009). Energy efficiency is one of the most important aspects of sustainable wastewater treatment. Energy-hungry traditional WWTPs have become more sustainable thanks to anaerobic digestion, MFCs and energy-efficient aeration techniques.

By recovering biogas through anaerobic digestion, fossil fuels are no longer necessary to produce power and heat (Appels *et al.*, 2008). MFCs are seen as a promising future technology for energy-positive or energy-neutral wastewater treatment, in which electroactive microorganisms transform organic matter into electrical power (Logan and Rabaey, 2012). Resource recovery plays a critical role in sustainable wastewater management. Ammonia stripping and struvite precipitation are two methods that can be used to recover nitrogen and phosphorus from wastewater and turn them into fertilisers (Cordell *et al.*, 2011). According to Mohanty *et al.* (2020), wastewater sludge can also be composted or converted into biochar, which lowers waste production and improves soil quality.

Water reuse is also growing in popularity; in some locations, treated wastewater is even utilised for irrigation, industrial cooling and drinking water supply (Shannon *et al.*, 2008). Examples of nature-based solutions that provide competitive alternatives to traditional wastewater treatment include artificial wetlands and green infrastructure. These systems replicate natural processes to eliminate pollutants while fostering biodiversity and carbon storage (Vymazal, 2011) – for example, artificial wetlands can efficiently eliminate contaminants while using less energy, which makes them suitable for decentralised wastewater treatment in rural and underdeveloped regions (Wu *et al.*, 2014). Emerging trends in sustainable wastewater management include digitalisation, sophisticated monitoring systems and AI optimisation.

While using less energy and chemicals, WWTP efficiency can be raised with real-time data analytics and AI-assisted process control (Olsson *et al.*, 2005). Integrating wastewater treatment with renewable energy sources, such as solar or wind power, can also boost sustainability (Liu *et al.*, 2018). In sustainable wastewater management, resource recovery, energy efficiency and nature-based solutions are generally gaining traction. The circular economy concept and the implementation of cutting-edge technologies are essential for the sustainability of wastewater treatment over the long run (Verstraete and Vlaeminck, 2011).

1.8.2 Emerging technologies and research directions

The goal of new wastewater treatment technologies is to lessen their negative effects on the environment while increasing sustainability, efficiency and resource recovery. Wastewater management has been transformed by novel microbiological processes, nanotechnology, bioelectrochemical systems and AI-based treatment techniques, opening up new creative opportunities (Shannon *et al.*, 2008). Anaerobic membrane bioreactors (AnMBRs) and MFCs are two examples of microbial technology-based wastewater treatment techniques that are becoming increasingly energy-efficient. MFCs are a sustainable alternative to conventional energy-dependent treatment methods because they use electroactive microorganisms to capture organic waste and produce power (Logan and Rabaey, 2012).

By combining membrane filtering and anaerobic digestion, AnMBRs generate high-quality wastewater and produce biogas for energy recovery (Lin *et al.*, 2013). These methods could be used to create wastewater treatment facilities that are energy-neutral or even energy-positive (Chen *et al.*, 2020). Nanotechnology is another fascinating field of wastewater treatment, especially for contamination removal and water purification. Examples of nanomaterials that can remove organic pollutants, heavy metals and new contaminants including microplastics and drugs include metal-organic frameworks (MOFs), graphene-based membranes and photocatalytic nanoparticles (Qu *et al.*, 2013). Nanobiosensors are also being developed to assess water quality in real time to efficiently regulate treatment operations (Wen *et al.*, 2017).

Two bioelectrochemical systems that are being developed as novel wastewater treatment methods are MECs and electro-fermentation. By utilising microbial metabolism to transform wastewater into hydrogen gas, MECs provide a sustainable way to produce biohydrogen (Cusick *et al.*, 2010). Electro-fermentation improves the bioconversion of organic substrates into advantageous bio-based products, such as biofuels and bioplastics, by fusing electric fields with microbial metabolism (Rodriguez *et al.*, 2015). AI and machine learning (ML), which optimise operational efficiency and predictive maintenance, are revolutionising wastewater treatment. AI-based models provide real-time data analysis, process automation and early equipment failure or treatment inefficiency identification (Olsson *et al.*, 2005).

AI and the Internet of Things (IoT) together should improve resource efficiency and reduce operational costs in intelligent wastewater treatment facilities (Li *et al.*, 2021). In order to achieve sustainable water management, future research will prioritise decentralised wastewater systems, hybrid treatment technologies and the concepts of the circular economy. Innovations in carbon capture technologies, bioengineered microbial consortia and wastewater energy recovery will accelerate the transition to a more sustainable and environmentally friendly wastewater treatment paradigm (Verstraete and Vlaeminck, 2011).

1.9. Conclusions

Improved microbiological processes, physical and chemical treatment methods and innovative sustainable solutions have drastically changed wastewater treatment. Innovative, reasonably priced and ecologically friendly wastewater treatment solutions are now essential as the world's pollution and water shortage problems worsen.

Microbial activities, such as biofilms, trickling filters and activated sludge processes, are crucial to biological wastewater treatment because they are very effective at breaking down organic materials and extracting nutrients. The capacity of anaerobic treatment methods to effectively treat high-strength wastewater and generate biogas has also drawn a lot of attention. By getting rid of suspended particles, pathogens and stubborn contaminants, physical and chemical treatment methods like coagulation, flocculation, disinfection and advanced oxidation processes support biological treatment.

A multi-barrier system that can produce high-quality effluent is created by combining the various treatment methods. In particular, advanced oxidation procedures have shown promise in degrading new contaminants and micropollutants that are challenging to remove using traditional methods. Furthermore, the use of disinfection techniques including UV treatment, ozonation and chlorination has increased the safety of treated water and decreased the risk of waterborne infections.

The primary factor propelling wastewater management toward resource recovery, energy saving and the implementation of circular economy principles is sustainability. Technologies including anaerobic digestion, microbial fuel cells and bioelectrochemical systems offer opportunities for producing energy from wastewater and offer ways to break free from reliance on fossil fuels. In order to extract valuable nutrients for use in agriculture, such as nitrogen and phosphorus, nutrient recovery techniques are being researched concurrently. Low-energy solutions that mimic natural cleaning processes are provided by nature-based solutions, such as artificial wetlands and green infrastructure, which also have additional environmental benefits.

Wastewater treatment will be revolutionised in the future by emerging technologies like nanotechnology, AI and smart sensors. AI-powered process optimisation combined with real-time monitoring will improve treatment outcomes, reduce costs and raise operational effectiveness. More flexible and reliable water management techniques will be made possible by the integration of smart sensor networks with decentralised wastewater systems. Furthermore, developments in synthetic biology and bioengineered microbial consortia hold out the possibility of developing extremely effective, tailored bioremediation techniques.

In conclusion, wastewater treatment is changing to become more technologically advanced, efficient and sustainable. Even if there are still issues with energy consumption, treatment costs and new contaminants, continuous research and development will increase the efficiency of treatment systems. The future of wastewater treatment can be in line with global environmental and public health goals to provide a cleaner and more resilient water future by integrating microbiological processes with cutting-edge technologies and sustainable practices.

Acknowledgments

The authors are thankful to Department of Biotechnology, Sathyabama Institute of Science and Technology, Chennai, India and Centre for Nanoscience and Nanotechnology, International Research Centre, Sathyabama Institute of Science and Technology, Chennai, India, for carrying out the work.

REFERENCES

Ahmed MB, Zhou JL, Ngo HH and Guo W (2017) Adsorptive removal of antibiotics from water and wastewater: progress and challenges. *Science of the Total Environment* **532**: 112–126.

Ahring BK (2003) Perspectives for anaerobic digestion. *Biomass and Bioenergy* **26**(5): 331–340.

Alleman JE and Prakasam TBS (1985a) Refinement of the activated sludge process. *Environmental Science and Technology* **19**(2): 127–135.

Alleman JE and Prakasam TBS (1985b) Reflections on the activated sludge process. *Water Pollution Control Federation* **57**(5): 394–399.

Andreozzi R, Caprio V, Insola A and Marotta R (1999) Advanced oxidation processes (AOP) for water purification and recovery. *Catalysis Today* **53**(1): 51–59.

Angelidaki I, Ahring BK and Ellegaard L (2011) Anaerobic digestion and methane production process. *Biomass Bioenergy* **35**(7): 1415–1421.

Appels L, Baeyens J, Degrève J and Dewil R (2008) Principles and potential of the anaerobic digestion of waste-activated sludge. *Progress in Energy and Combustion Science* **34**(6), 755–781.

Ardern E and Lockett WT (1914) Experiments on the oxidation of sewage without the aid of filters. *Journal of the Society of Chemical Industry* **33**(10): 523–539.

Bashir S, Raza W and Shahzad R (2018) *Bioremediation of Hazardous Waste in Soil: Microbial Processes and Technologies*. Springer, New York, NY, USA.

Basset L, Spataro A and Colantonio D (2018) Characterization and treatment of agricultural wastewater for reuse in irrigation. *Water* **10**(10), 1412.

Batstone DJ, Hülsen T, Mehta CM and Keller J (2015) Platforms for energy and nutrient recovery from domestic wastewater: a review. *Chemosphere* **140**: 2–11.

Batstone DJ, Keller J, Angelidaki I *et al.* (2002) The IWA Anaerobic Digestion Model No 1 (ADM1). *Water Science and Technology* **45**(10): 65–73.

Bhatnagar A, Gupta P and Kothari R (2015) Bioremediation of industrial wastewater using microbial processes. *Applied Microbiology and Biotechnology* **99**(16): 6609–6619.

Bitton G (2011) *Wastewater Microbiology*. Wiley, Hoboken, NJ, USA.

Bolto B and Gregory J (2007) Organic polyelectrolytes in water treatment. *Water Research* **41**(11), 2301–2324.

Bratby J (2016) *Coagulation and Flocculation in Water and Wastewater Treatment*. IWA Publishing, London, UK.

Chadwick D and M (2015) Microalgae in wastewater treatment: a review of current technologies and future perspectives. *Environmental Science and Pollution Research* **22**(11): 8419–8430.

Chakraborty S and Patel R (2015) *Fungal Bioremediation of Wastewater Contaminants*. Springer, New York, NY, USA.

Chen J, Li J, Liu Y and Cheng J (2018) Bioaugmentation for nitrogen removal in wastewater treatment: a review. *Bioresource Technology* **270**: 720–731.

Chen X, Feng Y, Liu K and Zhang X (2020) Anaerobic membrane bioreactors: applications and developments in energy-efficient wastewater treatment. *Renewable and Sustainable Energy Reviews* **134**: 110–174.

Chernicharo CAL (2007) *Anaerobic Reactors*. IWA Publishing, London, UK.

Choudhary A, Singh R and Singh S (2020) Bioremediation of industrial pollutants using engineered microbial strains: a comprehensive review. *Science of the Total Environment* **703**: 134848.

Conley DJ, Paerl HW, Howarth RW *et al.* (2018) Controlling eutrophication: nitrogen and phosphorus. *Science* **323**(5917): 1014–1015.

Corcoran E, Nellemann C, Baker E *et al.* (2010) *Sick Water? The Central Role of Wastewater Management in Sustainable Development*. United Nations Environment Programme.

Cordell D, Drangert JO and White S (2011) The story of phosphorus: global food security and food for thought. *Global Environmental Change* **19**(2): 292–305.

Costerton JW, Lewandowski Z, Caldwell DE, Korber DR and Lappin-Scott HM (1995) Microbial biofilms. *Annual Review of Microbiology* **49**: 711–745.

Crittenden JC, Trussell RR, Hand DW, Howe KJ and Tchobanoglous G (2012) *MWH's Water Treatment: Principles and Design*. Wiley, Hoboken, NJ, USA.

Cusick RD, Logan BE and Liu G (2010) Electricity generation from wastewater using microbial fuel cells. *Environmental Science and Technology* **44**(21): 7971–7976.

Dalmau M, Henson M and De Lange S (2017) Nutrient removal and recovery from wastewater using biological systems. *Water Research* **116**: 191–206.

Daughton CG (2019) Wastewater surveillance for population-wide public health: the potential of wastewater-based epidemiology. *Science of the Total Environment* 619–620: 409–425.

Delhoménie MC and Heitz M (2021) Biofiltration of air: a review. *Critical Reviews in Biotechnology* **41**(3): 567–589.

Duan J and Gregory J (2003) Coagulation by hydrolyzing metal salts. *Advances in Colloid and Interface Science* 100–102: 475–502.

El Fantroussi S and Agathos SN (2019) Is bioaugmentation a feasible strategy for pollutant removal and site remediation? *Current Opinion in Microbiology* **47**: 98–104.

Escapa A, Mateos R, Martínez EJ and Blanes J (2019) Microbial electrolysis cells: an overview of research and potential applications. *Bioelectrochemistry* **131**: 107377.

Fierer N, Bradford MA and Jackson RB (2012) *Toward an ecological classification of soil biota. Ecology* **93**(7): 1951–1966.

Flemming HC and Wingender J (2010) The biofilm matrix. *Nature Reviews Microbiology* **8**(9): 623–633.

Flemming HC, Neu TR and Wingender J (2016) *The Perfect Slime: Microbial Extracellular Polymeric Substances (EPS)*. IWA Publishing, London, UK.

Fujishima A, Rao TN and Tryk DA (2000) Titanium dioxide photocatalysis. *Journal of Photochemistry and Photobiology C: Photochemistry Reviews* **1**(1): 1–21.

Gålfalk M and Bastviken D (2025) In situ observations reveal underestimated greenhouse gas emissions from wastewater treatment with anaerobic digestion – sludge was a major source for both CH_4 and N_2O. *Environmental Science and Technology* **59**: 18146–18155.

Gerardi MH (2003) *The Microbiology of Anaerobic Digesters*. Wiley, Hoboken, NJ, USA.

Gerardi MH (2006) *Wastewater Bacteria*. Wiley, Hoboken, NJ, USA.

Giri S, Pandey B and Tripathi M (2020) Recent developments in adaptive laboratory evolution for bioaugmentation applications. *Applied Microbiology and Biotechnology* **104**(9): 3879–3892.

Glaze WH, Kang JW and Chapin DH (1987) The chemistry of water treatment processes involving ozone, hydrogen peroxide, and ultraviolet radiation. *Ozone: Science and Engineering* **9**(4): 335–352.

Gottschalk C, Libra JA and Saupe A (2010) *Ozonation of Water and Waste Water: A Practical Guide to Understanding Ozone and its Applications*. Wiley, Hoboken, NJ, USA.

Grady CPL, Daigger GT, Love NG and Filipe CDM (2011) *Biological Wastewater Treatment*. CRC Press, Boca Raton, FL, USA.

Gregory J and Duan J (2001) Hydrolyzing metal salts as coagulants. *Pure and Applied Chemistry* **73**(12): 2017–2026.

Gude VG (2017) Energy and water independence for sustainable wastewater treatment systems. *Renewable and Sustainable Energy Reviews* **67**: 459–472.

Guest JS, Skerlos SJ, Barnard JL *et al.* (2009) A new planning and design paradigm to achieve sustainable resource recovery from wastewater. *Environmental Science and Technology* **43**(16): 6126–6130.

Gürle G, Akın N and Arslan E (2021) Pathogen reduction and microbial processes in wastewater treatment. *Science of the Total Environment* **751**: 141766.

Henze M, van Loosdrecht MCM, Ekama GA and Brdjanovic D (2008) *Biological Wastewater Treatment: Principles, Modelling, and Design*. IWA Publishing, London, UK.

Herrmann I (1999) Advances in the degradation of organic pollutants in water. *Water Research* **33**(8): 2136–2146.

Holliger C, Wohlfarth G and Diekert G (2019) Reductive dechlorination in the energy metabolism of anaerobic bacteria and its application in bioremediation. *FEMS Microbiology Reviews* **43**(4): 364–393.

Hua S and Reckhow DA (2007) Comparison of disinfection by-product formation from chlorine, chloramine, and ozone. *Environmental Science and Technology* **41**(5): 1451–1456.

Hunt WF and DeBusk KM (2017) Stormwater management and the role of microorganisms in mitigating urban pollution. *Environmental Management* **59**(4): 551–559.

Ikehata K and Gamal El-Din M (2004) Ozone treatment of organic pollutants in drinking water. *Environmental Science and Technology* **38**(10): 2195–2201.

Jansson JK and Taş N (2012) Soil microbial ecology and the nitrogen cycle. *Soil Biology and Biochemistry* **52**: 85–91.

Jarvis P, Yates M, Nowak B and Jenkins J (2005) Removal of micro-pollutants from wastewater using adsorption. *Water Research* **39**(4): 728–737.

Jiang C, Wang X, Fan W and Li Y (2019) Genetic engineering strategies for enhancing heavy metal bioremediation by microorganisms. *Environmental Pollution* **252**(A): 1079–1089.

Jiang Y, Li S and Wu Z (2020) Advances in microbial immobilization techniques for bioaugmentation in wastewater treatment. *Environmental Science and Technology* **54**(16): 10460–10474.

Jin D, Kong X, Cui F *et al.* (2019) Biodegradation of polycyclic aromatic hydrocarbons by marine-derived fungi: a review. *Environmental Science and Pollution Research* **26**(28): 29032–29047.

Johns NI, Blazejewski T, Gomes AL and Wang HH (2021) Engineering microorganisms using CRISPR-Cas systems for bioremediation applications. *Nature Communications* **12**(1): 3438.

Joo JH, Cho JW and Kwon H (2016) Industrial wastewater treatment and its microbiological aspects. *Environmental Engineering Science* **33**(3): 188–199.

Kampschreur MJ, Temmink H, Kleerebezem R, Jetten MS and van Loosdrecht MC (2018) Nitrous oxide emission during wastewater treatment. *Water Research* **43**(17): 4093–4103.

Katsou E, Malamis S and Haralambous KJ (2017) Sludge reduction technologies in wastewater treatment plants. *Journal of Environmental Management* **203**: 821–829.

Kavitha V and Palanivelu K (2004) Electrochemical treatment of industrial wastewater. *Chemical Engineering Journal* **105**(1): 45–52.

Khan AA, Rehman SU and Khan MZ (2016) Anaerobic digestion of sewage sludge: a review. *Renewable and Sustainable Energy Reviews* **55**: 27–40.

Khan MU, Lee J and Kim J (2020) Advances in microbial fuel cells for wastewater treatment. *Bioelectrochemistry* **134**: 107500.

Lal R, Babu A and Sharma A (2020) Bioremediation of industrial wastewater with microbial processes: recent advancements and prospects. *Journal of Environmental Management* **262**: 110376.

Lee JW, Chan CT, Slomovic S and Collins JJ (2018) Next-generation biocontainment systems for engineered organisms. *Nature Chemical Biology* **14**(6): 530–537.

Lettinga G (1995) Anaerobic digestion and wastewater treatment systems. *Antonie van Leeuwenhoek* **67**: 3–28.

Lettinga G and Hulshoff Pol LW (1991) UASB-process design for various types of wastewaters. *Water Science and Technology* **24**(8): 87–107.

Li J, Wei D, Song C, Shi H and Zhang X (2021) Application of nanomaterials in wastewater treatment: a review. *Environmental Science: Nano* **8**(1): 128–148.

Li X, Zuo R and Qian X (2008) Advanced oxidation processes for the degradation of organic contaminants. *Journal of Hazardous Materials* **153**(1-2): 49–57.

Lin Y, Guo M, Shah N and Stuckey DC (2013) Anaerobic digestion and membrane bioreactors: are they a suitable option for wastewater treatment and resource recovery? *Bioresource Technology* **143**: 142–150.

Linden KG, Blanchard SL and Liao CH (2012) Advancements in disinfection and wastewater treatment: the potential for UV. *Water Research* **46**(2): 460–468.

Liu Y, Tay JH, Show KY and Tang PY (2018) Microbial granulation in wastewater treatment: current status and future perspectives. *Biotechnology Advances* **36**(4): 1098–1110.

Löffler FE, Ritalahti KM and Zinder SH (2017) Microbial reductive dechlorination of halogenated ethenes and halogenated ethanes. *Biodegradation* **28**(2): 147–168.

Logan BE (2008) *Microbial Fuel Cells*. Wiley-Blackwell, Hoboken, NJ, USA.

Logan BE and Rossi R (2020) Bioelectrochemical systems for sustainable wastewater treatment and energy recovery. *Nature Reviews Microbiology* **18**(5): 307–319.

Logan BE, Rossi R, Ragab A and Saikaly PE (2019) Electroactive microorganisms in bioelectrochemical systems. *Nature Reviews Microbiology* **17**(5): 307–319.

Logan BE and Rabaey K (2012) Conversion of wastes into bioelectricity and chemicals by using microbial electrochemical technologies. *Science* **337**(6095): 686–690.

Luo J, Xia S and Wang Z (2019) Microbial community dynamics in wastewater treatment systems: impact of environmental and operational factors. *Water Research* **157**: 106–119.

Mamane H, Sillanpää M and Lassi U (2007) Removal of organic contaminants from wastewater using advanced treatment technologies. *Environmental Technology* **28**(12): 1553–1563.

Mara D and Horan N (2003) *The Handbook of Water and Wastewater Microbiology*. Elsevier Science, Oxford, UK.

Martínez-Huitle CA and Ferro S (2006) Electrochemical oxidation of organic pollutants for water treatment. *Environmental Science and Technology* **40**(7): 2224–2234.

Matilainen A, Sjöblom J, Lassi U and Sillanpää M (2010) Removal of natural organic matter in drinking water treatment by coagulation: a review. *Environmental Technology* **31**(3): 231–241.

McCarty PL, Bae J and Kim J (2011) Domestic wastewater treatment as a net energy producer–can this be achieved? *Environmental Science and Technology* **45**(17): 7100–7106.

Megharaj M, Ramakrishnan B, Venkateswarlu K, Sethunathan N and Naidu R (2018) Bioremediation approaches for organic pollutants: a critical perspective. *Environment International* **134**: 105–118.

Metcalf and Eddy (2014) *Wastewater Engineering: Treatment and Resource Recovery* (5th edn.). McGraw-Hill Education, New York, NY, USA.

Metcalf and Eddy (2017) *Wastewater Engineering: Treatment and Resource Recovery*. McGraw-Hill Education, New York, NY, USA.

Michael I, Rizzo L, McArdell CS *et al.* (2021) Urban wastewater treatment plants as hotspots for antibiotic-resistant bacteria and genes spread into the environment: a review. *Science of the Total Environment* **447**: 345–360.

Mino T, Van Loosdrecht MCM and Heijnen JJ (1998) Microbial phosphorus removal in the activated sludge process. *Water Research* **32**(5): 1395–1407.

Mishra S, Lin Z and Pang S (2019) Bioremediation of petroleum hydrocarbons in contaminated soil using microbial consortia: a review. *Environmental Technology and Innovation* **14**: 100311.

Mohanty SK, Swain C, Jha SK and Patel AK (2020) Wastewater treatment using sustainable electrochemical techniques: opportunities and challenges. *Journal of Environmental Chemical Engineering* **8**(5): 104176.

Moreira F, Ferreira J and Almeida A (2017) Recent advancements in electrochemical treatment of wastewater. *Environmental Chemistry Letters* **15**(3): 353–365.

Morgenroth E, Wilderer PA and Reinhard W (2000) Controlled biomass detachment in a pilot-scale biofilm airlift suspension reactor. *Water Research* **34**(3): 773–781.

Muga HE and Mihelcic JR (2008) Sustainability of wastewater treatment technologies. *Journal of Environmental Management* **88**(3): 437–447.

Neyens E and Baeyens J (2003) A review of classic and novel technologies for the treatment of wastewater. *Waste Management* **23**(1): 99–106.

Nies DH (2021) Microbial heavy-metal resistance: biochemistry and biotechnological applications. *Applied Microbiology and Biotechnology* **105**(7): 2825–2837.

Ødegaard H (2006) Innovations in wastewater treatment: the moving bed biofilm process. *Water Science and Technology* **53**(9): 17–33.

Okuda T, Miyahara M and Kawai M (2001) Evaluation of ozone for removal of organic micropollutants from municipal wastewater. *Water Science and Technology* **43**(1): 39–46.

Olsson G, Nielsen M, Yuan Z and Lynggaard-Jensen A (2005) *Wastewater Treatment Systems: Modelling, Diagnosis and Control*. IWA Publishing, London, UK.

Oturan MA and Aaron JJ (2014) Advanced oxidation processes in water and wastewater treatment. *Science of the Total Environment* **468**: 172–187.

Pant D, Van Bogaert G, Diels L and Vanbroekhoven K (2020) A review of the substrates used in microbial fuel cells for sustainable energy production. *Bioresource Technology* **255**: 300–310.

Qu X, Alvarez PJJ and Li Q (2013) Applications of nanotechnology in water and wastewater treatment. *Water Research* **47**(12): 3931–3946.

Rabaey K and Rozendal RA (2019) Microbial electrosynthesis—revisiting the electrical route for microbial production. *Nature Reviews Microbiology* **18**(3): 190–203.

Ramos JL, Molina-Henares AJ and Daddaoua A (2019) Engineering microorganisms for environmental applications: bioremediation and biomonitoring. *Current Opinion in Biotechnology* **57**: 45–52.

Ramsay A and Enari T (2000) *Fungal bioremediation of industrial pollutants*. *Bioresource Technology* **74**(2): 145–150.

Rao AS and Sharma A (2008) *Protozoa in Wastewater Treatment: Role in Bioremediation*. Springer, New York, NY, USA.

Renault F, Sancey B, Badot PM and Crini G (2009) Chitosan for coagulation/flocculation processes – an eco-friendly approach. *European Polymer Journal* **45**(5): 1337–1348.

Richardson SD, McDonald T, Kus J, Ulrich J and Thruston AD (2007) The analysis of drinking water contaminants by mass spectrometry. *Environmental Science and Technology* **41**(18): 6544–6551.

Rittmann BE and McCarty PL (2001) *Environmental Biotechnology: Principles and Applications*. McGraw-Hill Education, New York, NY, USA.

Rodriguez C, Alsina MA, Pocurull E and Barceló D (2015) Advances in liquid chromatography–tandem mass spectrometry for determining emerging contaminants in the environment. *Trends in Analytical Chemistry* **72**: 15–28.

Santoro C, Arbizzani C, Erable B and Ieropoulos I (2017) Microbial fuel cells: from fundamentals to applications. *Journal of Power Sources* **356**: 225–244.

Sayler GS and Ripp S (2022) Field applications of genetically engineered microorganisms for bioremediation processes. *Current Opinion in Biotechnology* **38**(4): 1–6.

Schroder JL and Smit AL (2018) Microbial processes in the treatment of agricultural wastewater. *Agricultural Water Management* **201**: 122–132.

Semprini L (2020) The role of co-metabolism in the degradation of chlorinated solvents. *Water Research* **174**: 115635.

Shannon MA, Bohn PW, Elimelech M, Georgiadis JG and Mariñas BJ (2008) Science and technology for water purification in the coming decades. *Nature* **452**(7185): 301–310.

Singh A, Jain A and Sharma VK (2021) Microbial metabolism in biodegradation of environmental contaminants. *Frontiers in Microbiology* **12**: 678903.

Singh A, Sharma VK and Jain A (2022) Advances in synthetic biology for environmental engineering. *Frontiers in Microbiology* **13**: 876954.

Singh PK, Parsek MR, Greenberg EP and Welsh MJ (2006) A component of innate immunity prevents bacterial biofilm development. *Nature* **417**(6888): 552–555.

Snyder SA and Kümmerer K (2020) Water disinfection and its effects on public health. *Environmental Toxicology and Chemistry* **39**(8): 2137–2147.

Speece RE (2008) *Anaerobic Biotechnology and Bioengineering*. Archae Press, Nashville, TN, USA.

Stoodley P, Sauer K, Davies DG and Costerton JW (2002) Biofilms as complex differentiated communities. *Annual Review of Microbiology* **56**: 187–209.

Tchobanoglous G, Burton FL and Stensel HD (2003) *Wastewater Engineering: Treatment and Reuse* (4th edn.). McGraw-Hill Education, New York, NY, USA.

Tchobanoglous G, Burton FL and Stensel HD (2014) *Wastewater Engineering: Treatment and Reuse*. McGraw-Hill Education, New York, NY, USA.

Teh JS, Yusoff MM, Mohamed M and Hussain MA (2016) Electrocoagulation as a wastewater treatment: a review. *Environmental Science and Pollution Research* **23**(7): 6342–6358.

Tran NH, Urase T and Kusakabe O (2018) A comprehensive study on the removal of pharmaceuticals from municipal wastewater by biological treatment processes. *Water Research* **63**: 245–257.

Tyagi M, da Fonseca MMR and de Carvalho CCCR (2020) Bioaugmentation and biostimulation strategies for bicremediation of oil-contaminated environments: a review. *Biodegradation* **31**(4): 465–487.

Tyagi VK and Lo SL (2019) Sludge management in wastewater treatment plants: a review. *Environmental Science and Technology* **53**(4): 1696–1712.

UNEP (United Nations Environment Programme) (2020) Progress on wastewater treatment – Global update and acceleration needs. *United Nations Environment Programme Report*.

Van Loosdrecht MCM, Hooijmans CM, Brdjanovic D and Heijnen JJ (2002) Biological wastewater treatment: principles, modelling and design. *Water Science and Technology* **45**(6): 1–11.

Verstraete W and Vlaeminck SE (2011) ZeroWasteWater: short-cycling of wastewater resources for sustainable cities of the future. *International Journal of Sustainable Development and World Ecology* **18**(3): 253–264.

Von Gunten U (2003a) Ozone treatment of drinking water: part I. *Science of the Total Environment* **303**(3): 69–80.

Von Gunten U (2003b) Ozone treatment of drinking water: part II. *Science of the Total Environment* **303**(3): 81–89.

Vymazal J (2013) The use of constructed wetlands for wastewater treatment in the Czech Republic: a review. *Ecological Engineering* **61**: 3–14.

Vymazal J (2018) Constructed wetlands for wastewater treatment: five decades of experience. *Environmental Science and Technology* **52**(9): 4203–4222.

Vymazal J (2011) Constructed wetlands for wastewater treatment: five decades of experience. *Environmental Science and Technology* **45**(1): 61–69.

Wang X, Wang Y and Li J (2021) Engineering microbial consortia for bioremediation and environmental sustainability. *Current Opinion in Biotechnology* **67**: 166–172.

Wang Y, Yang Q and Li X (2020) Genetically engineered biosensors for heavy metal detection and bioremediation. *Environmental Science and Technology* **54**(21): 13358–13371.

Wang Z, Zhang S, Zhang Y and Wang L (2016) Removal of pharmaceutical pollutants from water using carbon nanotubes. *Environmental Pollution* **214**: 216–224.

Wen Q, Li X, Feng Y and Li W (2017) Bioremediation of heavy metal-contaminated wastewater using microbial electrochemical technologies. *Bioresource Technology* **241**: 1152–1165.

White S (2010) An overview of the treatment of organic contaminants in water. *Journal of Environmental Engineering* **136**(6): 607–613.

Wu S, Wallace S, Brix H *et al.* (2014) Treatment of industrial effluents in constructed wetlands: challenges, operational strategies and overall performance. *Environmental Pollution* **192**: 245–262.

Wuertz S, Bishop PL and Wilderer PA (2003) *Biofilms in Wastewater Treatment: An Interdisciplinary Approach*. IWA Publishing, London, UK.

Xavier JB, de Kreuk MK, Picioreanu C, van Loosdrecht MCM and Kleerebezem R (2005) Multi-scale individual-based model of microbial and bioconversion dynamics in aerobic granular sludge. *Environmental Science and Technology* **39**(8): 2894–2900.

Yoshida S, Hiraga K, Takehana T *et al.* (2016) A bacterium that degrades and assimilates poly(ethylene terephthalate). *Science* **351**(6278): 1196–1199.

You J, Greenman J, Melhuish C and Ieropoulos I (2020) Future applications of microbial fuel cells in microbial biotechnology. *Biotechnology Advances* **42**: 107578.

Zhang F, Ge Z, Grimaud J, Hurst J and He Z (2021) Integrating microbial fuel cells into the wastewater treatment process: energy, treatment efficiency, and challenges. *Water Research* **185**: 116299.

Zhang H, Zhang H and Li Y (2019) Microbial fuel cells for wastewater treatment and energy recovery: a review. *Bioresource Technology* **271**: 269–278.

Zhou J, Wang Y and Zhang Y (2021) AI-driven wastewater treatment process optimization: challenges and opportunities. *Water Research* **203**: 117533.

Zhou X, Tang Y and Zhang Y (2020) Algae-based technologies for wastewater treatment and pathogen removal. *Environmental Pollution* **260**: 113997.

Zhu L, Zhang X and Wu M (2019) Nitrogen and phosphorus removal from agricultural wastewater by microbial bioremediation. *Environmental Science and Pollution Research* **26**(3): 2321–2334.

Zimmer B and Slawson R (2002) The removal of organic matter from water by adsorption techniques. *Water Research* **36**(8): 2191–2201.

Zouboulis AI and Tzoupanos ND (2010) Polymer coagulation: a review on the recent developments in the use of synthetic polymers and biopolymers in water treatment. *Journal of Chemical Technology and Biotechnology* **85**(10): 1325–1336.

Microbial Processes in Environmental Engineering: From Theory to Practice

Rajneesh Kumar, Ram Sharan Singh and Maulin P Shah
ISBN 978-1-83708-327-5
https://doi.org/10.1108/978-1-83708-326-820261002

Chapter 2
Bioaugmentation and biostimulation: strategies in wastewater treatment

Chimdi Mang Kalu and Memory Tekere

Department of Environmental Science, College of Agriculture and Environmental Sciences University of South Africa, Johannesburg, South Africa

Abstract

The environmental conditions in the stages of wastewater treatment plants (WWTPs) play a crucial role in the effectiveness of contaminant removal as well as influence the microbial communities that are pivotal in biological contaminant removal. Bioaugmentation, which involves the addition of a specific microorganism or microbial consortia capable of degrading contaminants in a media, and biostimulation, which involves the modification of the physicochemical treatment environment to create a conducive environment for the microorganisms to proliferate and decontaminate the waste effluents, are promising strategies in wastewater treatment and other biotreatment processes. Individual application of bioaugmentation or biostimulation is often not sufficient to produce the required contaminant removal, leading to the application of the two strategies concurrently. This chapter provides information on the beneficial inherent microbial communities and the role of the application of the bioaugmentation or biostimulation strategies in WWTPs. Information on the relevant biostimulation interventions that can enhance the proliferation and activities of the contaminant-removing microbial communities is discussed. Information on the application of bioaugmentation and biostimulation and the understanding of the gaps existing in the individual application or co-application of the strategies for a sustainable solution to wastewater treatment is provided.

Keywords: Wastewater; wastewater treatment; microorganisms; bioaugmentation; biostimulation

2.1. Introduction

Industrialisation and urbanisation, as well as the increase in anthropogenic activities, have resulted in the production and discharge of high volumes of wastewater with varying contaminants that could pose serious health and environmental problems globally (Long *et al.*, 2023). WWTPs are charged with the responsibility of ensuring effective treatment of wastewaters from industrial, domestic and agricultural activities before the final treated water is discharged to the appropriate receptacle.

To actualise the goal of effective wastewater treatment, diverse stages of treatment are put in place in many WWTPs. According to Menghini *et al.* (2023) and Ehalt Macedo *et al.* (2021), 58 502 WWTPs globally, as documented in the HydroWASTE global database, employ the following treatment stages: pre-treatment, equalisation, primary sedimentation, biological treatment, secondary settling and disinfection. The primary aim of adopting these stages of treatment is to ensure that proper decontamination of the organic and inorganic contaminants present in the wastewater occurs.

Although these stages of treatment have been effective, their sustainability in light of the quantity and diverse pollutant burden has been a challenging situation, especially in the biological treatment stage (which involves the application of microorganisms in the breaking down of contaminants, including organic matter). The biological component of wastewater treatment relies heavily on microbial competence and adaptation to growth prevailing conditions for the microbial effectiveness to realise the treatment objectives.

The introduction of bioaugmentation and biostimulation in the biological treatment stages has shown to be a promising approach in the provision of sustainable biological wastewater treatment processes (Muter *et al.*, 2017). Bioaugmentation by nature involves the addition of microbial communities either from historically contaminated sites or genetically modified microbial communities to support the degradation of contaminants present in the contaminated environment, such as wastewater (Adams *et al.*, 2015; Sarkar and Bhattacharjee, 2025). Biostimulation involves the modulation of environmental conditions that include limiting or increasing nutrients and electron acceptors, such as phosphorus, nitrogen, oxygen or carbon, to enhance the growth of microbial communities existing in the contaminated environment and promote their bioremediation potentials (Adams *et al.*, 2015; Perfumo *et al.*, 2007; Saidi *et al.*, 2025).

Biostimulation could be done by the addition of both organic and inorganic nutrients to the contaminated environment, including wastewater. Details of the roles of bioaugmentation and biostimulation in the wastewater treatment, examples of microbial communities that have been employed as bioagents and the applied biostimulants that enhanced the growth of inherent microbial communities of the wastewater, and possible examples of studies done combining these two approaches in wastewater treatment, are summarised.

2.2. Overview of conventional wastewater treatment process

Table 2.1 provides a summary of the wastewater treatment stages, operational objectives and conditions, and limitations. Pre-treatment of wastewater involves the physical screening of the waste with the aim of removing coarse materials, grits and oil in the waste. This process prepares the wastewater for the next stage of treatment. According to Li *et al.* (2019) and Saravanan *et al.* (2021), the pre-treatment stage could also cut across chemical and physical methods depending on the nature of wastewater or the sources of wastewater – for example, petrochemical wastewater could be passed through two pre-treatment stages, such as oil/water separation and photocatalytic organics degradation (Li *et al.*, 2019).

Table 2.1 Treatment stages, operational objectives and conditions, advantages and limitations (continued on next page)

Treatment stages	Purpose	Operational conditions	Benefits/advantages	Limitations
Pre-treatment	Removal of solid and some dissolved particles. Modification of the particle size to create good surface area that will facilitate the next process. Promote pollutant adsorption.	Adequate treatment time. Appropriate variable concentrations. Appropriate temperature, pH and other environmental conditions depending on the wastewater.	Enhance speedy remediation process by ensuring the produced effluent in the stage is in suitable condition for the next stage. Promote the efficiency of the next stage of treatment.	There are no general operational conditions for all types of wastewaters. This implies that different wastewater demands are present for different operational conditions, and this could be challenging in a situation where there is a mixture of different wastewaters. The cost of purchasing some equipment could be high.
Equalisation	Improvement of the secondary treatment process of wastewater. To promote the movement of appropriate quantities of wastewater to the next stage of treatment.	Appropriate temperature. A well-defined tank size. Properly designed bar screens, grit chambers and comminutors to suit different wastewater type.	It ensures a constant flow of wastewater into the next stage of treatment at an appropriate rate.	The cost of developing an adequate equalisation system. Lack of expert personnel to carry out and maintain the whole process.
Primary sedimentation	Removal of organic matter and partly microbial load.	Appropriate flow conditions. A well-designed sedimentation tank. A regulated flow rate, retention time and surface load.	Production of treating effluent in a good condition to ensure success in the preceding stage.	Success of the previous stage could influence the sedimentation process. Cost of developing the primary sedimentation system. Maintenance of the system.

Table 2.1 Continued

Biological treatment	Decomposition of organic matter as well as other contaminants.	An environment with good recycle flows, an aeration system, good retention times, nutritional and other environmental conditions to promote the growth of biological agents.	Enhanced the removal of microbial contaminants as well as organic pollutants.	The cost of developing a good bioreactor. The challenge of biological agents' selection.
Secondary settling	Removal of bio-flocs.	Appropriate flow conditions. A well-designed sedimentation tank. A regulated flow rate, retention time and surface load.	Enhanced removal of bio-flocs. Production of treating effluent in a good condition to ensure success in the preceding stage.	Cost of developing the primary sedimentation system.
Disinfection	Decontamination of wastewater from microbial contaminants.	Conditions varied with the type of disinfectant such as chlorine, ultraviolet light radiation, chloramines and ozone.	Enhanced the removal of biological indicators/ sanitisation of effluent.	Cost of developing disinfection system. Efficiency in the removal of unculturable pathogens in the wastewater.

For the oil/water separation technique, a modified steel mesh/glass microfibre filter (constituting a coupling membrane distillation approach) was employed for the removal of oil emulsions from the petrochemical wastewater (Li *et al.*, 2019). This process was immediately followed by the addition of TiO_2 P25 as a photocatalyst during the photocatalysis process for the decomposition of organic compounds remaining after the first pre-treatment stage (Li *et al.*, 2019). This photocatalysis stage was also carried out under ultraviolet light radiation to promote the inactivation of any microbial communities present in the petrochemical wastewater.

Wastewater, such as pulp and paper mill wastewater, distillery wastewater, slaughterhouse wastewater, dairy wastewater and cheese whey wastewater, requires a specific pre-treatment for effective treatment (Liew *et al.*, 2020). Liew *et al.* (2020) regarded the above-mentioned wastes as high-strength wastewaters because of their compositions, such as oil/grease, carbohydrate, protein and lignin, which require specific pre-treatment for the effective treatment of the wastewaters.

The pre-treatment processes of these high-strength wastewaters involve physical, chemical and biological approaches. The physical approach includes the following: milling, chipping and grinding, which enhance the reduction of particle size, providing increased surface area for other processes to take place and a reduction in the degree of polymerisation (Hendriks and Zeeman, 2009; Jiang *et al.*, 2025). The chemical approach takes into consideration acid and alkaline pre-treatment with the aim of removing lignin and promoting the recovery of glucose from cellulose and hence promoting better hydrolysis (Mosier *et al.*, 2005). The biological pre-treatment approach involves the addition of enzymes or microbial consortia that have the ability to degrade the organic matter (Liew *et al.*, 2020; Parawira, 2012).

Equalisation is a technique that improves the effectiveness of wastewater treatment processes. It takes into consideration the flow rate of wastewater through the stages of treatment. According to Gantz *et al.* (2020), it ensures that the flow rate, pollutant levels and the temperature of the wastewater are maintained optimally to reduce the downstream effects of these parameters. Equalisation also involves the storage of wastewater in a constructed large basin located downstream of the bar screens, grit chambers and comminutors to ensure that the wastewater is pumped into the treatment plants at a constant rate (Gantz *et al.*, 2020; Iloms *et al.*, 2025).

The settlement of wastewater in settling tanks for hours to promote the formation of primary sludge and clarified liquid effluent needed for the next stage of treatment constitutes primary sedimentation (Iloms *et al.*, 2025). Jover-Smet *et al.* (2017) pointed out that the primary settlement has been widely used in the removal of suspended solids through gravitational settling techniques. Primary settlement has also formed a crucial part of the biological wastewater treatment process and sludge treatment process (Patziger and Kiss, 2015; Patziger *et al.*, 2016). Through the primary settlement techniques, the removal of suspended solids is enhanced, leading to a corresponding decrease in the organic load measured or expressed as biochemical oxygen demand (BOD) or chemical oxygen demand (COD).

The treatment stage that involves the addition of microbial communities, improving the survival of the beneficial inherent microorganisms or the application of any other living organisms that promote the removal of contaminants in different wastewaters is termed the biological treatment stage (Iloms *et al.*, 2025). The biologically activated sludge process is widely employed in the treatment of industrial and municipal wastewaters. According to Gentile *et al.* (2007), Iloms *et al.* (2025) and Wang *et al.* (2011. 2014), biologically activated sludge constitutes activated sludge tanks, sedimentation tanks, various recycle flows, and an aeration system designed in such a way that it enhances the addition of functional microbial communities that can promote the degradation of toxins, organic pollutants and other nutrients from the wastewater. After biological treatment, the flocs formed (bio-flocs) are removed through secondary settlement.

Disinfection involves the addition of disinfectants such as chlorine, chloroamine, ultraviolet light and ozone to inhibit the growth of pathogenic microbial communities or completely eradicate the microorganisms (Bertanza *et al.*, 2021; Papa *et al.*, 2021). All these treatment stage processes are believed to alter the microbial community structures in wastewater, leading to the uncertainty of the survival of the beneficial microbial communities capable of enhancing the degradation of pollutants in the wastewater. Hence, continuous monitoring of the microbial communities of the wastewater strategically becomes pertinent to understand the right time for bioaugmentation and biostimulation for the maximisation of treatment processes. Figure 2.1 summarises the information on the wastewater treatment plant stages and bioaugmentation as well as biostimulation application in the biological treatment stage.

2.3. Bioaugmentation and applications

In wastewater reactors, bioaugmentation proceeds with the addition of microbial consortia (groups of microbial communities that interact closely through symbiotic relationships) or selected strains to enhance the catabolism of the pollutants present in the wastewater (Adams *et al.*, 2015; Herrero and Stuckey, 2015; Saltymakova *et al.*, 2025). Bioaugmentation could be done by using culture collections, commercially available microbial consortia, as well as indigenous or exogenous strains (Herrero and Stuckey, 2015; Saltymakova *et al.*, 2025).

A lot of effort has been put into developing strains that can be used as seed to benefit wastewater treatment processes (Saltymakova *et al.*, 2025). The commercially available microbial consortia could be a mixture of microbial populations that can make their own food (autotrophs), depend on available nutrients to survive (heterotrophs), and be freeze-dried, frozen or liquid-prepared (Herrero and Stuckey, 2015). For effective biological remediation, adoption of a bioaugmentation approach and certain selection procedures/criteria should be taken into consideration in the selection of microbial communities to be used in the remediation process. According to Yu and Mohn (2002) and Shan *et al.* (2025), the microbial communities to be used must be able to degrade the contaminants through their microbial catabolic processes even when clouded by potential inhibitors, have the ability to persist and compete favourably

Figure 2.1 Figure summarising the information collated in this study

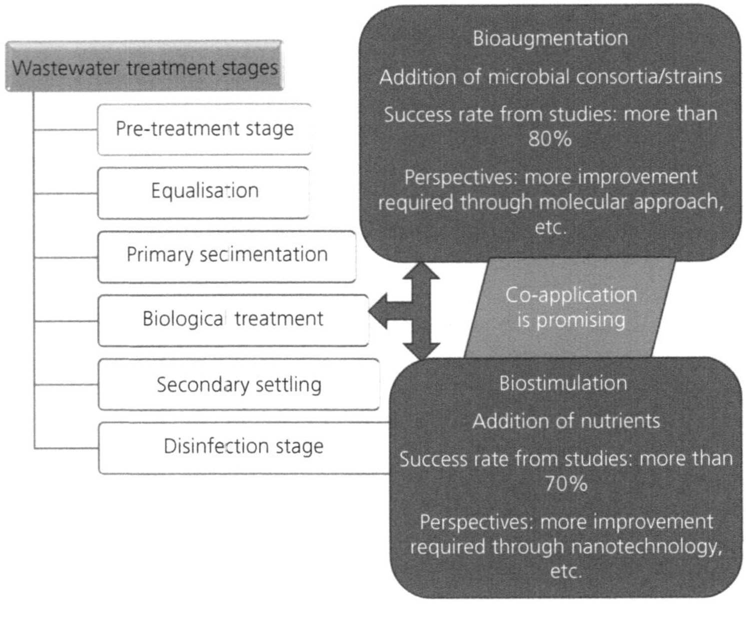

with other microbial communities in the contaminated environment, and have the ability to symbiotically interact with other beneficial microorganisms in the environment to ensure effective remediation.

2.4. Wastewater treated with bioaugmentation

Different wastewaters contain varied compositions based on the contaminants contained. Furthermore, the varied contaminant composition of the wastewater is crucial in the determination of microorganisms that could metabolically proliferate in the wastewater. Hence, understanding the microbial diversities of the wastewater will help to provide direction on the microbial consortia that could be used in the bioaugmentation process in wastewater treatment (Wang *et al.*, 2014). Table 2.2 provides a summary of the wastewater that bioaugmentation has been applied to in their treatment.

Table 2.2 Wastewater sources treated using bioaugmentation approach (continued on next page)

Wastewater type	Type of contaminants	Microbial communities added	References
Industrial wastewater (petrochemical wastewater)	Ammonia and polycyclic aromatic hydrocarbons (PAHs)	Immobilised bacteria such as *Nitrobacter* and *Nitrosomonas* spp.	Raper *et al.*, 2018
Palm oil effluent (agro-industrial wastewater)	Organic nitrogen	Mixed culture of microbial communities containing *Nitrosomonas*, *Nitrobacter*, and *Nitrococcus* spp.	Onyla *et al.*, 2001
Prawn aquaculture wastewater	Total ammoniacal nitrogen (TAN)	Immobilised nitrifying bacteria such as *Nitrobacter* and *Nitrosomonas* spp.	Shan and Obbard, 2001
Petrochemical wastewater	Nitrogenous contaminants	*Pseudomonas*, *Bacillus*, *Acinetobacter*, *Flavobacterium* and *Micrococcus*	Ma *et al.*, 2009
Petrochemical wastewater	Alkaline phenol	*Pseudomonas* sp. JY-2	Qu *et al.*, 2011
Coal gasification wastewater	Phenol	Phenol-degrading bacteria such as *Stenotrophomonas*, *Klebsiella* and *Enterobacter*	Fang *et al.*, 2013
2-fluorophenol wastewaters	2-fluorophenol	2-fluorophenol-degrading strain such as *Rhodococcus* sp. FP1	Duque *et al.*, 2011
p-nitrophenol (PNP) wastewaters	p-nitrophenol	PNP-degrading bacteria such as *Acinetobacter* and *Arthrobacter* spp.	Martín-Hernández *et al.*, 2012

Table 2.2 Continued

Wastewater type	Type of contaminants	Microbial communities added	References
Industrial wastewater	Waste oil	Oil-degrading bacterial consortium such as *Mycobacterium, Proteiniphilum, Nocardioides, Luteimicrobium* and *Azospirillum*	Jemli *et al.*, 2025
Municipal wastewater	Cephalexin	*Glutamicibacter nicotianae* S2 and *Herbaspirillum seropedicae* S8	Shan *et al.*, 2025

2.5. Advantages and limitations of bioaugmentation applications in wastewater treatment

Bioaugmentation is also referred to as seeding a contaminated wastewater undergoing biological treatment with the application of microbial consortia from the similarly contaminated environment or genetically modified microorganisms (Chettri *et al.*, 2024; Ferraro *et al.*, 2021). Kuppusamy *et al.* (2020) explained that the genetically modified microbial cells used in bioaugmentation are developed through the addition of specific genes into the microbial cells by the genetic engineering approach. According to Tyagi *et al.* (2011) and Lyon and Vogel (2013), genetically modified microbial consortia in wastewater enhance the indigenous microbial communities' ability to degrade contaminants either through enzymatic processes or vertical/horizontal gene transfer. Tyagi *et al.* (2011) underscore that the advantage of the bioaugmentation approach in the treatment of wastewater is the immediate biodegradation action by the microbial consortia once applied in the wastewater.

Another advantage of bioaugmentation is the inexpensive nature of the treatment because it can be an in situ remediation approach, with no addition of chemicals harmful to humans and other organisms; sophisticated equipment and labour are not necessarily required immediately after the microbial consortia are added (Chettri *et al.*, 2024).

Kumar *et al.* (2018) pointed out that the instability of the seed or the genetically modified microbial cells and the ethical concerns in their development and application are the major drawbacks to the use of these modified microbial cells despite their beneficial role in the remediation of wastewater through bioaugmentation. Another drawback to the application of bioaugmentation is the uncertainty of the quantity of microbial inoculum, as inadequate inoculum can result in the loss of microbial strains to competition, and the addition of a large quantity of the inoculum can result in the alteration of microbial distribution in the treatment environment (Raper *et al.*, 2018). Increased movement of water, especially in the wastewater treatment, could reduce the bioaugmentation process efficiency due to microbial washout or insufficient contact times (Sayed *et al.*, 2021).

2.6. Application of bioaugmentation in treatment of wastewater and success rate

Bioaugmentation has been applied in the treatment of many wastewaters with great success. Cheng *et al.* (2021) investigated the bioaugmentation of tannery wastewater with novel strains BM-S-1 and KCTC11789BP under a full-scale oxic process and observed 95.2% COD removal efficiency. Table 2.3 provides a summary of studies done using the bioaugmentation approach in the treatment of wastewater and the level of success achieved.

Table 2.3 Bioaugmentation approach in wastewater treatment and successes achieved (continued on next page)

Wastewater	Bioaugmented microbial strains/consortia	Result achieved	References
2,4-Dichlorophenol (2,4-DCP) contaminated wastewater in conventional activated sludge	Mixed special cultures containing *Flavobacterium breve* and *Achromobacter* sp.	The efficiency of 2,4-DCP removal rangec from 46.1% to 99.3% based on the three shocks load	Quan *et al.*, 2004
2,4,6-trichlorophenol (2,4,6-TCP) contaminated wastewater	*Desulfitobacterium* strains	A high chlorophenol removal observed alongside mineralisation up to 85% TOC reduction and 80% ecotoxicity diminished	Puyol *et al.*, 2015
Wastewater containing strong pyridine	*Rhizobium* sp. NJUST18	Volumetric degradation of pyridine ranging from 1164.5 mg L^{-1} h^{-1} to 1867.4 mg L^{-1} h^{-1}	Liu *et al.*, 2015
Coke oven wastewater containing quinoline and pyridine	*Paracoccus* sp. BW001 and *Pseudomonas* sp. BW003	Stable pyridine removal efficiencies of 90% and 99%. Quinoline removal efficiency between 81% and 100%	Bai *et al.*, 2011
Coal wastewater containing toxic cyanides	*Cryptococcus humicolus* and an unidentified cyanide-degrading microorganism	14 mg L^{-1} cyanide present in the wastewater reduced to 0.1 mg L^{-1} cyanide making about 93% reduction	Park *et al.*, 2008

Table 2.3 Continued

Wastewater	Bioaugmented microbial strains/consortia	Result achieved	References
Wastewater containing azo dyes	*Pseudomonas putida* (BN-4), *Bacillus cereus* (BN-7), *Stenotrophomonas acidaminiphila* (BN-3) and *P. fluorescence* (BN-5)	Bacterial consortium degraded all the dyes in the wastewater	Khehra *et al.*, 2005
Wastewater from subsurface wastewater infiltration systems containing nitrogenous contaminants	*Pseudomonas, Yersinia, Flavobacterium* and *Acinetobacter*	The bacterial consortia reduced ammonium ion (NH_4+ –N) concentration from 38.61 ± 8.02 mg L^{-1} to 7.95 ± 2.03 mg L^{-1}, while total nitrogen was reduced from 49.47 ± 4.46 mg L^{-1} to 28.17 ± 8.61 mg L^{-1}	Liu *et al.*, 2018
Oil-contaminated wastewater	*Rhodococcus pyridinivorans* F5	79 ± 0.03% of hydrocarbon degradation	Mazumder *et al.*, 2020
Raw industrial wastewater containing phenolic compound	*Achromobacter* sp. C-1	90% biodegradation of a mixture of phenols was achieved	Gracioso *et al.*, 2019
Malachite green (MG) and hexavalent chromium-contaminated wastewater from textile industry	*Aeromonas hydrophila* LZ-MG14	The bacterial strain removed 96.8% of MG (200 mg L^{-1}) and 93.71% of Cr(VI) (0.5 mmol L^{-1})	Ji *et al.*, 2020
Wastewater containing high-strength pyridine	*Rhizobium* sp. NJUST18	Complete removal of pyridine	Hou *et al.*, 2017
Aquaculture effluent	*Stutzerimonas stutzeri* SW22	Removal efficiencies of 76.83%, 75.32% and 65.71% for NH_4+ –N, NO_2^- –N and NO_3–N	Song *et al.*, 2025
Printing and dyeing wastewater (PDW)	*Ochrobactrum anthropi* S1	Removal efficiencies of reactive black 5, Cr(VI), COD and ammonia nitrogen of 95%, 65%, 90% and 85%, respectively	Sun *et al.*, 2025

Table 2.3 Continued

Wastewater	Bioaugmented microbial strains/consortia	Result achieved	References
Cephalexin-contaminated municipal wastewater	*Glutamicibacter nicotianae* S2 and *Herbaspirillum seropedicae* S8	Cephalexin removal of 84.12% (S2) and 82.35% (S8) was achieved	Shan *et al.*, 2025
Municipal wastewater containing 21 pharmaceutical compounds	Activated sludge containing *Proteobacteria* (24.0%), *Bacteroidetes* (19.8%) and *Firmicutes* (13.2%)	14 pharmaceutical compounds were completely removed (about 67%). 70% of concentration of carbamazepine was removed	Muter *et al.*, 2017
Textile wastewater	Mixture of native bacterial and fungal strains	Textile wastewater decolourisation and COD removal yields by over 98%	Khelifi *et al.*, 2009
Textile wastewater containing azo-dye Direct Black 22 (DB22)	*Penicillium chrysogenum*	High decolourisation rate and COD removal	Lanfranconi *et al.*, 2022
Reactive Blue in synthetic textile wastewater	Mixed fungal biomasses (*Rhizopus arrhizus* and *Aspergillus versicolor*)	Removal efficiency of colour and COD were 90.71 and 90%, respectively	Acikgoz *et al.*, 2016

2.7. Biostimulation and applications in wastewater treatment

The biostimulation approach is useful and effective in the bioremediation of the environment, including in wastewater treatment. It involves the stimulation of existing microbial communities' growth and metabolism to enhance their ability to remediate the environment through the addition of various limiting nutrients or adjustment of growth/metabolic limiting factors. According to Dixit *et al.* (2015), phosphorus and nitrogen are mostly the external nutrients added alongside the proper aeration to promote the growth of microbial communities inherent in the wastewater and promote the rate of biodegradation.

In the wastewater environment, there are limiting factors which include temperature, pH, moisture and nutrients that may need to be adjusted for the stimulation of indigenous microbial communities for the enhancement of contaminant degradation (Chettri *et al.*, 2024). Kuppusamy *et al.* (2020) highlighted that for effective stimulation of the microbial communities in the wastewater, the nutrients added need to be readily available to the microbial communities for the bioremediation process to proceed without the nutrients disappearing or being lost in the environment.

2.8. Wastewater types and nutrients applied in biostimulation

Nutrients applied in a contaminated environment, including wastewater, to stimulate the activities of indigenous microbial communities in the biodegradation of the contaminant could be referred to as a biostimulant or biostimulating agent. The biostimulating agent could be organic or inorganic, and their application could depend on the nature of the wastewater, contaminants, as well as the indigenous microbial communities in the contaminated environment (Adams *et al.*, 2015).

Jayasekara *et al.* (2025) highlighted that the addition of a biostimulating agent promotes the growth of microbial communities in wastewater by way of increased optical density, protein content and increased enzymatic activities such as dehydrogenase. In most wastewater treatment plants, it is noted that nitrogen removal can be a rate-limiting step and anoxic conditions are required to promote denitrifying microorganisms (Chu *et al.*, 2021).

In the study done by Chu *et al.* (2021), passing wastewater from the coke industry through consecutive anoxic (without free oxygen) and oxic (oxygen) conditions enhanced the degradation of contaminants. In their study, the addition of oxygen served as a biostimulating agent and operational condition. The authors observed that phenols, quinolines, polycyclic aromatic hydrocarbons and other types of hydrocarbons were completely degraded. The authors further observed that members of bacterial communities belonging to *Proteobacteria*, *Nitrospirae* and *Bacteroidetes* (denitrifying microorganisms) proliferated well in the two conditions and enhanced the degradation of the contaminants.

Studies have also shown that the amount and type of carbon source (glucose, about 19.6 g/L) will impact the performance of denitrifying bacteria that could promote the removal of contaminants in wastewater (Chu and Wang, 2011; Cui *et al.*, 2019). Hence, there is a need for the addition of a biostimulating agent in appropriate quantities that will enhance the proliferation of the microbial communities as well as their ability to metabolise the contaminants. A summary of the results obtained in wastewater treatment enhanced through biostimulation is provided in Table 2.4.

2.9. Co-application of bioaugmentation and biostimulation

A combination strategy where both bioaugmentation and biostimulation approaches are applied concurrently has shown better outcomes when compared to individual applications. With the addition of microbial strains, a burden is placed on the operational system to provide optimal conditions for the microbial community to operate and thrive. Most of these successes have been attributed to the studies done on contaminated soil (Abdulsalam and Omale, 2009; Abdulsalam *et al.*, 2011; Bento et al., 2005).

In wastewater treatment, few studies have reported the impact of the combination strategy on the remediation of contaminants present in wastewater – for example, Muter *et al.* (2017) reported that the combination of the two approaches [activated sludge (containing *Proteobacteria*, *Bacteroidetes* and *Firmicutes*) and sugar beet molasses as a biostimulant] promoted complete degradation of all the 21 pharmaceutical compounds in the municipal wastewater.

Table 2.4 Wastewater treatment through biostimulation and the successes (continued on next page)

Wastewater type	Nutrient added and purpose	Organic or inorganic	Result obtained	References
Industrial wastewater with Thiabendazole (TBZ) contamination	Tunisian hypersaline sediments (THSs). To promote the growth of the microorganisms and enhance their metabolism of the contaminants.	Mixed	>99% TBZ degradation in all reactors was oobserved. Microbial community shift after biostimulation leading to a reduction in *Bacillota* (–64%) and an increase in *Proteobacteria* (+62%).	Saidi *et al.*, 2025
Municipal wastewater containing 21 pharmaceutical compounds	Sugar beet molasses. It serves as a glucose source for the microbial communities.	Mixed	The nutrient enhanced higher removal of erythromycin among other compounds.	Muter *et al.*, 2017
Textile wastewater containing azo-dye Direct Black 22 (DB22)	Glucose, potato starch and potato industry wastewater. It serves as a source of glucose for the growth of the microorganisms and enhances COD removal as well as the azo-dye.	Organic	Potato starch showed the highest decolourisation rate and COD removal.	Lanfranconi *et al.*, 2022
Styrene-containing petrochemical wastewater	Magnetic granular activated carbon (MGAC), cuttlefish bone (CFB) and pumice. To create a good surface area for the interaction between the microorganisms and the contaminants. It also serves as a source of carbon.	Mixed	MGAC removed 85.4%–91.4% styrene which was highest among the biostimulants added.	Qaretapeh *et al.*, 2024

Table 2.4 Continued

Wastewater type	Nutrient added and purpose	Organic or inorganic	Result obtained	References
Purified terephthalic acid (PTA) and dimethyl terephthalate (DMT) wastewater	Ethylene glycol. It serves as a source of carbon for the microorganisms.	Organic	Organic removal rate of 11.0 ± 0.6 kg COD/m^3/d was achieved. Increase in the abundance of aromatic degraders.	Kuroda *et al.*, 2024
Terephthalic acid (TA) wastewater	TiO_2/Fe_3O_4 nanoparticles (NPs, up to 81 days) or ethanol (up to 50 days). It serves as a source of iron and oxygen as well as carbon to promote the growth of the microbial communities and create a good environmental condition for the interaction between the microorganisms and the contaminants.	Inorganic	Complete degradation period of terephthalic acid (TA, 1000 mg L^{-1}) wastewater was reduced	Li *et al.*, 2022
PTA wastewater	Co-substrate (glucose or molasses). It serves as a source of glucose.	Mixed	Glucose destabilised both terephthalic acid and para-toluic acid removal. Molasses improved para-toluic acid removal. Both substrates decreased the abundances of syntrophs and methanogens and increased carbohydrate-fermenting bacteria	Liu *et al.*, 2017

The degradation of indigo dye present in textile wastewater using native microbial communities and NPK fertiliser in combination showed about an 88% reduction of dye (Rendón-Castrillón *et al.*, 2024). Lanfranconi *et al.* (2022) also observed a high decolourisation rate and reduction in COD of textile wastewater containing azo-dye Direct Black 22 (DB22) by *Penicillium chrysogenum* with potato starch added as a biostimulant.

The addition of molasses to PTA wastewater containing methanogenic bacteria recorded an improved para-toluic acid removal (Liu *et al.*, 2017). Bose and Tiwari (2019) explored mine tailing sludge (MTS) and acid mine drainage sludge (AMDS) as biostimulants in combination with septic tank sludge (containing microbial consortia) in the treatment of domestic sewage. The authors observed high COD removal along with increased acidogenic and methanogenic activity as well as improved biomass retention in the bioreactor system.

Tetteh and Rathilal (2021) co-applied four magnetic nanoparticles (CuO, Fe_2O_3, TiO_2, Cu/Fe-TiO_2) and microbial consortia in activated sludge for the degradation of organic contaminants in the wastewater from sugar refineries into water and biogas at a temperature of 35°C and hydraulic retention time of 30 days. The nanoparticles increased biogas (methane) production from 63% to 100% with 75% removal of COD, volatile solids and total suspended solids. The addition of the nanoparticles increased the abundance of the methanogenic microbial communities either through the creation of oxic conditions or the provision of essential elements needed to promote their growth and enhance the adsorptive removal of contaminants and methane production. The combination of these two strategies is promising in the bioremediation of contaminated environments, including wastewater treatment.

However, more studies on the application of the strategies in wastewater treatment taking into consideration the different contaminants in wastewater and inherent conditions within the wastewater that can impede the success of the two strategies are required.

2.10. Bioaugmentation and biostimulation research considerations on the operational strategies

Operational considerations, which include dosing techniques, dosing location, dosing size and regime, play a crucial role in the determination of the sustainability of the process, especially in wastewater treatment. Dosing technique plays a crucial role in the efficiency of the bioaugmentation process (Chettri *et al.*, 2024).

The widely applied dosing technique involves direct application of the microbial inoculum into the treatment bioreactor or vessel. One of the advantages of the direct application, according to Raper *et al.* (2018), is that the microbial inoculum can be applied as and when required for the process without any required modification or operational cost. Studies have shown success in the rate of contaminant degradation by the microbial inoculum when directly applied to the bioreactor (Liu *et al.*, 2018; Mazumder *et al.*, 2020; Parker and Wanner, 2007; Zheng *et al.*, 2019). However, the survival rate of the inocula is very low, as they may not survive the environmental conditions as well as the antagonistic approach of indigenous microorganisms. Hence, it becomes pertinent for the development of techniques that could allow slow release or intermittent introduction of the microbial inoculum (either lyophilised, capsules/pellets and/or immobilised seed strains) at time intervals to ensure sustainability of the remediation process through inoculum survival.

Dosage size is another aspect of bioaugmentation that influences the success of the application of bioaugmentation in wastewater treatment. Sátiro *et al.* (2025) evaluated the impact of the addition of 240 l of activated sludge (source of microbial consortia) in the high-rate pond in the treatment of domestic wastewater. The authors observed that the dosage enhanced the removal of 84.2% total nitrogen, 63.7% phosphorus and 74.7% organic matter. This implied that the survival of the microbial consortia/inocula is dosage-dependent, as the concentration of the inoculum could affect the density of the microbial communities as well as create specific pressure within the environment that has the tendency to create competition (intraspecific and interspecific competition) among the microbial communities, leading to delay or failure in the bioaugmentation process (Bouchez *et al.*, 2000; Ramadan *et al.*, 1990).

In addition, Martín-Hernández *et al.* (2012) highlighted that the dosage size and regime, as well as the cost of application, could vary according to the contaminant/wastewater. This becomes a great concern in the actualisation of sustainable bioaugmentation in diverse wastewater and contaminants. More interdisciplinary studies, particularly using machine learning approaches, are required to identify proper inoculum dosage while minimising the cost of application in various wastewater scenarios, to ensure efficiency and sustainability in various wastewater bioremediation.

Identification of the conditions that favour the growth of the microbial communities is important as it indicates the location (anaerobic, anoxic and aerobic reactor) of application of the inoculum during the wastewater treatment. Jianlong *et al.* (2002)'s work indicated that *Burkholderia picketii*, under aerobic conditions (aerobic reactor), showed greater COD removal in the bioremediation of quinoline present in coking plant wastewater. Quan *et al.* (2004) observed improved bioremediation of 2,4-DCP in wastewater when bioaugmented by sludge obtained from two wastewater treatment plants containing a mixture of microbial communities. The location of the bioreactor containing the microbial consortia played a crucial role, as they observed a 90.3% removal rate when the bioreactor was located after the aeration cell when compared to the bioreactor located before the aeration cell with an 86.2% removal rate. In view of this, pre-knowledge of the survival conditions of the microbial communities needed for bioaugmentation processes for any wastewater treatment becomes pertinent.

Furthermore, nutrient addition is also essential in the bioremediation of wastewater, as this could enhance the proliferation of the microbial communities, ensure their survival and improve bioremediation. The addition of a food source as well as the aerobic condition in the bioreactor containing *Burkholderia picketti* promoted the remediation of quinoline in the wastewater (Jianlong *et al.*, 2002).

Li *et al.* (2022) further applied nanomaterials such as TiO_2/Fe_3O_4 for the treatment of terephthalic acid (TA) wastewater and observed that the contaminant removal was high, indicating the ability of these nanoparticles in wastewater remediation. The nanoparticles provide essential nutrients such as Fe as well as oxic conditions needed for the growth of the microbial consortia to a threshold that they can degrade the contaminants. However, there is a paucity of studies on nanoparticle application in wastewater treatment, taking into consideration different wastewater contaminants.

2.11. Future perspectives

Bioaugmentation and biostimulation have shown great success in the remediation of soil and wastewater contaminated with diverse pollutants, including emerging pollutants (Kuroda *et al.*, 2024; Lanfranconi *et al.*, 2022). Despite the success made in the individual or combined application of bioaugmentation and biostimulation, more strategic integrated work is needed to ensure the sustainability of these two approaches in the efficient remediation of wastewater.

Among the strategic work needed to be done is the strain selection for bioaugmentation. Bitton (2011) highlighted that a strain selected for bioaugmentation must have the ability to adapt to the new environment and survive environmental pressures that include pH, dissolved oxygen, availability of nutrients, temperature, environmental toxicity and other microbial community pressures. In view of this, Raper *et al.* (2018) pointed out that there is the need to understand the conditions within the treatment process and ascertain whether the microbial inoculum should be able to survive, proliferate and assist in the degradation of the contaminants under normal/optimal or varied conditions. Without this initial assessment of the environmental conditions and microbial ability, bioaugmentation may not succeed because of poor survival of the inoculum or devastating competition from the indigenous microbial communities (More *et al.*, 2010).

Ueno *et al.* (2007) underscore the importance of the isolation and selection of indigenous microorganisms in the contaminated environment as bioaugmentation agents, because such microbial agents are well adapted to the environment, making remediation more successful. However, the major drawback could be the ability of indigenous microorganisms to grow sufficiently to a threshold that can enhance their remediation potentials. More research should be geared towards the development of an approach or technology that could holistically consider the variety of wastewater contaminants and develop, through genetic engineering, microbial communities that can withstand varieties of contaminants that compete favourably with the indigenous microbial communities, as well as form mutual relationships with some of the indigenous microorganisms and enhance sustainable remediation.

In summary, an in-depth analysis of the efficiency of varied conditions necessary for bioaugmentation and biostimulation, in the laboratory and in situ studies, that take into consideration the various contaminants in wastewater needs to be considered when designing bioreactors used in the bioaugmentation and biostimulation processes.

2.12. Conclusions and recommendations

An increase in urbanisation, industrialisation and anthropogenic activities leads to increased generation of wastewater containing diverse contaminants in increasingly high concentrations. Also, capacity constraints occur for most treatment plants, as they now have to handle wastewater volumes beyond their design capacity. This creates serious pressure on wastewater treatment plants globally, as they are expected to release final effluents that meet national and international standards.

The biological treatment stages play crucial roles, as they involve the application of microorganisms in the removal of most of the contaminants in the wastewater. Bioaugmentation and biostimulation approaches have been successfully adopted in the biological treatment stage of most treatment plants, directly or indirectly, to ensure proper treatment of the wastewater. From the collated information, both approaches have shown to be promising techniques, either applied individually or in combination.

However, gaps still exist in their application due to some inherent system dynamics within the treatment plants, as well as within the stages of treatment, that could interfere with microbial consortia activities and the applied nutrients. Hence, there is a need to comprehend the conditions in the treatment stages, especially the biological stage where these two approaches are dominantly applied. This will play a crucial role in the prevention of the inoculum being impacted negatively by the inherent factors within the wastewater. It will also guide on the dosing strategy, including the type and location of inoculum, minimising or completely removing the negative environmental conditions that affect the success of the approach. In view of the biostimulants needed to ensure the proliferation and bioremediation of contaminants in wastewater, the application of different nanomaterials, individually or in combination, needs to be considered in the development of nutrients required to boost the growth of microbial consortia in the wastewater.

Acknowledgements

The authors wish to thank the University of South Africa for their support.

REFERENCES

Abdulsalam S and Omale AB (2009) Comparison of biostimulation and bioaugmentation techniques for the remediation of used motor oil contaminated soil. *Brazilian Archives of Biology and Technology* **52**: 747–754.

Abdulsalam S, Bugaje IM, Adefila SS and Ibrahim S (2011) Comparison of biostimulation and bioaugmentation for remediation of soil contaminated with spent motor oil. *International Journal of Environmental Science and Technology* **8**: 187–194.

Acikgoz C, Gül ÜD, Özan K and Borazan AA (2016) Degradation of Reactive Blue by the mixed culture of Aspergillus versicolor and Rhizopus arrhizus in membrane bioreactor (MBR) system. *Desalination and water treatment* **57**(8): 3750–3756.

Adams GO, Fufeyin PT, Okoro SE and Ehinomen I (2015) Bioremediation, biostimulation and bioaugmention: a review. *International Journal of Environmental Bioremediation and Biodegradation* **3**(1): 28–39.

Bai Y, Sun Q, Sun R, Wen D and Tang X (2011) Bioaugmentation and adsorption treatment of coking wastewater containing pyridine and quinoline using zeolite-biological aerated filters. *Environmental Science and Technology* **45**(5): 1940–1948.

Bento FM, de Oliveira Camargo FA, Okeke BC and Frankenberger Jr WT (2005) Diversity of biosurfactant producing microorganisms isolated from soils contaminated with diesel oil. *Microbiological Research* **160**(3): 249–255.

Bertanza G, Boniotti J, Ceretti E *et al.* (2021) Environmental footprint of wastewater treatment: a step forward in the use of toxicological tools. *International Journal of Environmental Research and Public Health* **18**(13): 6827.

Bitton G (2011) *Wastewater Microbiology*. Wiley, Hoboken, NJ, USA, p. 9.

Bose RS and Tiwari MK (2019) Mine sludge waste recycling as bio-stimulant for applications in anaerobic wastewater treatment. *Water Science and Technology* **79**(3): 425–434.

Bouchez T, Patureau D, Juretschko D and Delgenès M (2000) Ecological study of a bioaugmentation failure. *Environmental Microbiology* **2**(2): 179–190.

Cheng Y, Chon K, Ren X *et al.* (2021) Bioaugmentation treatment of a novel microbial consortium for degradation of organic pollutants in tannery wastewater under a full-scale oxic process. *Biochemical Engineering Journal* **175**: 108131.

Chettri D, Verma AK and Verma AK (2024) Bioaugmentation: an approach to biological treatment of pollutants. *Biodegradation* **35**(2): 117–135.

Chu H, Liu X, Ma J *et al.* (2021) Two-stage anoxic-oxic (A/O) system for the treatment of coking wastewater: Full-scale performance and microbial community analysis. *Chemical Engineering Journal* **417**: 129204.

Chu L and Wang J (2011) Comparison of polyurethane foam and biodegradable polymer as carriers in moving bed biofilm reactor for treating wastewater with a low C/N ratio. *Chemosphere* **83**(1): 63–68.

Cui YX, Guo G, Biswal BK, Chen GH and Wu D (2019) Investigation on sulfide-oxidizing autotrophic denitrification in moving-bed biofilm reactors: An innovative approach and mechanism for the process start-up. *International Biodeterioration and Biodegradation* **140**: 90–98.

Dixit R, Wasiullah X, Malaviya D *et al.* (2015) Bioremediation of heavy metals from soil and aquatic environment: an overview of principles and criteria of fundamental processes. *Sustainability* **7**(2): 2189–2212.

Duque AF, Bessa VS, Carvalho MF and Castro PM (2011) Bioaugmentation of a rotating biological contactor for degradation of 2-fluorophenol. *Bioresource Technology* **102**(19): 9300–9303.

Ehalt Macedo H, Lehner B, Nicell J *et al.* (2021) Global distribution of wastewater treatment plants and their released effluents into rivers and streams. *Earth System Science Data Discussions* **2021**: 1–33.

Fang F, Han H, Zhao Q, Xu C and Zhang L (2013) Bioaugmentation of biological contact oxidation reactor (BCOR) with phenol-degrading bacteria for coal gasification wastewater (CGW) treatment. *Bioresource Technology* **150**: 314–320.

Ferraro A, Massini G, Miritana VM *et al.* (2021) Bioaugmentation strategy to enhance polycyclic aromatic hydrocarbons anaerobic biodegradation in contaminated soils. *Chemosphere* **275**: 130091

Gantz RG, Lipták J and Liu DH (2020) Equalization and primary treatment. In *Wastewater Treatment* (pp 123–165). CRC Press, Boca Raton, FL, USA.

Gentile ME, Jessup CM, Nyman JL and Criddle CS (2007) Correlation of functional instability and community dynamics in denitrifying dispersed-growth reactors. *Applied and Environmental Microbiology* **73**(3): 680–690.

Gracioso LH, Vieira PB, Baltazar MP *et al.* (2019) Removal of phenolic compounds from raw industrial wastewater by achromobacter sp. isolated from a hydrocarbon-contaminated area. *Water and Environment Journal* **33**(1): 40–50.

Hendriks AT and Zeeman G (2009) Pretreatments to enhance the digestibility of lignocellulosic biomass. *Bioresource Technology* **100**(1): 10–18.

Herrero M and Stuckey DC (2015) Bioaugmentation and its application in wastewater treatment: a review. *Chemosphere* **140**: 119–128.

Hou C, Shen J, Zhang D *et al.* (2017) Bioaugmentation of a continuous-flow self-forming dynamic membrane bioreactor for the treatment of wastewater containing high-strength pyridine. *Environmental Science and Pollution Research* **24**: 3437–3447.

Iloms EC, Kalu CM, Tekere M *et al.* (2025) Microbial Diversity in Wastewater Sources and Biological Activated Sludge System. *International Journal of Agriculture and Biology* **33**: 330311

Jayasekara UG, Hadibarata T, Hindarti D *et al.* (2025) Environmental bioremediation of pharmaceutical residues: microbial processes and technological innovations: a review. *Bioprocess and Biosystems Engineering* **6**: 1–9.

Jemli M, Karray F, Mansour L *et al.* (2025) Wastewater biotreatment and bioaugmentation for remediation of contaminated sites at an oil recycling plant. *Water Science and Technology* **91**(2): 139–159.

Ji J, Kulshreshtha S, Kakade A *et al.* (2020) Bioaugmentation of membrane bioreactor with Aeromonas hydrophila LZ-MG14 for enhanced malachite green and hexavalent chromium removal in textile wastewater. *International Biodeterioration and Biodegradation* **150**: 104939.

Jiang G, Ameer K, Ramachandraiah K *et al.* (2025) Alkaline, Acid and Acid-Wet Ball Milling Extraction of Dietary Fiber from Kiwi berry (*Actinidia arguta*): Structural, Rheological, Physicochemical and Functional Properties. *Food Bioscience* **21**: 106202.

Jianlong W, Xiangchun Q, Libo W, Yi Q and Hegemann W (2002) Bioaugmentation as a tool to enhance the removal of refractory compound in coke plant wastewater. *Process Biochemistry* **38**(5): 777–781.

Jover-Smet M, Martín-Pascual J and Trapote A (2017) Model of suspended solids removal in the primary sedimentation tanks for the treatment of urban wastewater. *Water* **9**(6): 448.

Khehra MS, Saini HS, Sharma DK, Chadha BS and Chimni SS (2005) Comparative studies on potential of consortium and constituent pure bacterial isolates to decolorize azo dyes. *Water Research* **39**(20): 5135–41.

Khelifi E, Bouallagui H, Touhami Y, Godon JJ and Hamdi M (2009) Enhancement of textile wastewater decolourization and biodegradation by isolated bacterial and fungal strains. *Desalination and Water Treatment* **2**(1-3): 319–25.

Kumar NM, Muthukumaran C, Sharmila G and Gurunathan B (2018) Genetically modified organisms and its impact on the enhancement of bioremediation. *Bioremediation: Applications for Environmental Protection and Management. Energy, Environment, and Sustainability*. Springer, Singapore.

Kuppusamy S, Venkateswarlu K, Megharaj M, Vasudevan N and Lee YB (2020) Biostimulation and bioaugmentation: Modern strategies for the successful bioremediation of contaminated. In *The Handbook of Environmental Remediation: Classic and Modern Techniques*. The Royal Society of Chemistry, London, UK, pp. 299–330.

Kuroda K, Takai M, Sekiguchi T *et al.* (2024) Development of an internal two-stage upflow anaerobic reactor integrating biostimulation strategies to enhance the degradation of aromatic compounds in wastewater from purified terephthalic acid and dimethyl terephthalate manufacturing processes. *Water Research* **258**: 121762.

Lanfranconi I, Ceretta MB, Bertola N, Wolski EA and Durruty I (2022) Textile dyeing wastewater treatment by Penicillium chrysogenum: design of a sustainable process. *Water Science and Technology* **86**(2): 292–301.

Li C, Deng W, Gao C *et al.* (2019) Membrane distillation coupled with a novel two-stage pretreatment process for petrochemical wastewater treatment and reuse. *Separation and Purification Technology* **224**(1): 23–32.

Li H, Zhang W, Xia D *et al.* (2022) Improved anaerobic degradation of purified terephthalic acid wastewater by adding nanoparticles or co-substrates to facilitate the electron transfer process. *Environmental Science: Nano* **9**(3): 1011–1024.

Liew YX, Chan YJ, Manickam S *et al.* (2020) Enzymatic pretreatment to enhance anaerobic bioconversion of high strength wastewater to biogas: A review. *Science of the Total Environment* **713**(15): 136373.

Liu C, Xie J, Song M *et al.* (2018) Nitrogen removal performance and microbial community changes in subsurface wastewater infiltration systems (SWISs) at low temperature with different bioaugmentation strategies. *Bioresource Technology* **250**: 603–610.

Liu M, Wang S, Nobu MK *et al.* (2017) Impacts of biostimulation and bioaugmentation on the performance and microbial ecology in methanogenic reactors treating purified terephthalic acid wastewater. *Water Research* **122**: 308–316.

Liu X, Chen Y, Zhang X *et al.* (2015) Aerobic granulation strategy for bioaugmentation of a sequencing batch reactor (SBR) treating high strength pyridine wastewater. *Journal of Hazardous Materials* **295**: 153–160.

Long Y, Zhou Z, Wen X *et al.* (2023) Microplastics removal and characteristics of a typical multi-combination and multi-stage constructed wetlands wastewater treatment plant in Changsha, China. *Chemosphere* **312**: 137199.

Lyon DY and Vogel TM (2013) *Bioaugmentation for Groundwater Remediation: An Overview. Bioaugmentation for Groundwater Remediation*. SERDP ESTCP Environmental Remediation Technology. Springer, New York, NY, USA.

Ma F, Guo JB, Zhao LJ, Chang CC and Cui D (2009) Application of bioaugmentation to improve the activated sludge system into the contact oxidation system treating petrochemical wastewater. *Bioresource Technology* **100**(2): 597–602.

Martín-Hernández M, Suárez-Ojeda ME and Carrera J (2012) Bioaugmentation for treating transient or continuous p-nitrophenol shock loads in an aerobic sequencing batch reactor. *Bioresource Technology* **123**: 150–156.

Mazumder A, Das S, Sen D and Bhattacharjee C (2020) Kinetic analysis and parametric optimization for bioaugmentation of oil from oily wastewater with hydrocarbonoclastic Rhodococcus pyridinivorans F5 strain. *Environmental Technology and Innovation* **17**: 100630.

Menghini M, Pedrazzani R, Feretti D *et al.* (2023) Beyond the black box of life cycle assessment in wastewater treatment plants: which help from bioassays? *Water* **15**(5): 960.

More TT, Yan S, Tyagi RD and Surampalli RY (2010) Potential use of filamentous fungi for wastewater sludge treatment. *Bioresource Technology* **101**(20): 7691–7700.

Mosier N, Wyman C, Dale B *et al.* (2005) Features of promising technologies for pretreatment of lignocellulosic biomass. *Bioresource Technology* **96**(6): 673–686.

Muter O, Ingus P, Selga T *et al.* (2017) Removal of pharmaceuticals from municipal wastewaters at laboratory scale by treatment with activated sludge and biostimulation. *Science of the Total Environment* **584**: 402–13.

Onyla CO, Uyub AM, Akunna JC, Norulaini NA and Omar AK (2001) Increasing the fertilizer value of palm oil mill sludge: bioaugmentation in nitrification. *Water Science and Technology* **44**(10): 157–162.

Papa M, Paredes L, Feretti D *et al.* (2021) How should ecohazard of micropollutants in wastewater be gauged? Using bioassays to profile alternative tertiary treatments. *Environmental Engineering Research* 26(3).

Parawira W (2012) Enzyme research and applications in biotechnological intensification of biogas production. *Critical Reviews in Biotechnology* **32**(2): 172–86.

Park D, Lee DS, Kim YM and Park JM (2008) Bioaugmentation of cyanide-degrading microorganisms in a full-scale cokes wastewater treatment facility. *Bioresource Technology* **99**(6): 2092–2096.

Parker D and Wanner J (2007) Review of methods for improving nitrification through bioaugmentation. In WEFTEC. *Water Environment Federation*, pp. 5304–5326.

Patziger M and Kiss K (2015) Towards a hydrodynamically enhanced design and operation of primary settling tanks – Results of a long term in situ measurement investigation program. *Water and Environment Journal* **29**(3): 338–345.

Patziger M, Günthert FW, Jardin N, Kainz H and Londong J (2016) On the design and operation of primary settling tanks in state-of-the-art wastewater treatment and water resources recovery. *Water Science and Technology* **74**(9): 2060–2067.

Perfumo A, Banat IM, Marchant R and Vezzulli L (2007) Thermally enhanced approaches for bioremediation of hydrocarbon-contaminated soils. *Chemosphere* **66**(1): 179–184.

Puyol D, Monsalvo VM, Sanchís S *et al.* (2015) Comparison of bioaugmented EGSB and GAC–FBB reactors and their combination with aerobic SBR for the abatement of chlorophenols. *Chemical Engineering Journal* **259**: 277–285.

Qaretapeh MZ, Zonoozi MH and Azarmanesh R (2024) Biostimulation of styrene-containing wastewater in carrier-supported anaerobic bioreactor towards enhancing biogas production and styrene degradation. *Journal of Water Process Engineering* **64**: 105708.

Qu Y, Zhang R, Ma F, Zhou J and Yan B (2011) Bioaugmentation with a novel alkali-tolerant Pseudomonas strain for alkaline phenol wastewater treatment in sequencing batch reactor. *World Journal of Microbiology and Biotechnology* **27**(8): 1919–1926.

Quan X, Shi H, Liu H, Lv P and Qian Y (2004) Enhancement of 2, 4-dichlorophenol degradation in conventional activated sludge systems bioaugmented with mixed special culture. *Water Research* **38**(1): 245–253.

Ramadan MA, El-Tayeb OM and Alexander M (1990) Inoculum size as a factor limiting success of inoculation for biodegradation. *Applied and Environmental Microbiology* **56**(5): 1392–1396.

Raper E, Stephenson T, Anderson DR, Fisher R and Soares A (2018) Industrial wastewater treatment through bioaugmentation. *Process Safety and Environmental Protection* **118**: 178–187.

Rendón-Castrillón L, Ramírez-Carmona M, Ocampo-López C, González-López F, Cuartas-Uribe B and Mendoza-Roca JA (2024) Efficient bioremediation of indigo-dye contaminated textile wastewater using native microorganisms and combined bioaugmentation-biostimulation techniques. *Chemosphere* **353**: 141538.

Saidi N, Erable B, Etchevery L, Cherif A and Chouchane H (2025) Enhanced bioelectrochemical degradation of Thiabendazole using biostimulated Tunisian hypersaline sediments: kinetics, efficiency, and microbial community shifts. *Frontiers in Microbiology* **15**: 1529841.

Saltymakova D, Desmond DS, Smith AF *et al.* (2025) Enhanced crude oil degradation observed in sea ice following bioaugmentation with arctic bacteria. *Marine Environmental Research* **204**: 106942.

Saravanan A, Kumar PS, Jeevanantham S *et al.* (2021) Effective water/wastewater treatment methodologies for toxic pollutants removal: Processes and applications towards sustainable development. *Chemosphere* **280**: 130595.

Sarkar A and Bhattacharjee S (2025) Biofilm-mediated bioremediation of xenobiotics and heavy metals: a comprehensive review of microbial ecology, molecular mechanisms, and emerging biotechnological applications. *3 Biotechnology* **15**(4): 1–30.

Sátiro J, dos Santos Neto A, Tavares J *et al.* (2025) Impact of inoculum on domestic wastewater treatment in high-rate ponds in pilot-scale: assessment of organic matter and nutrients removal, biomass growth, and content. *Algal Research* **86**: 103923.

Sayed K, Baloo L and Sharma NK (2021) Bioremediation of total petroleum hydrocarbons (TPH) by bioaugmentation and biostimulation in water with floating oil spill containment booms as bioreactor basin. *International Journal of Environmental Research and Public Health* **18**(5): 2226.

Shan H and Obbard J (2001) Ammonia removal from prawn aquaculture water using immobilized nitrifying bacteria. *Applied Microbiology and Biotechnology* **57**(5-6): 791–798.

Shan X, Shan Z, Guo H and Ma F (2025) Biodegradation of cephalexin in wastewater by *Glutamicibacter* sp. S2 and *Herbaspirillum* sp. S8: Performance, pathway, genomes and synergistic bio-augmentation. *Chemical Engineering Journal* (**507**): 160783.

Song Y, Zhao L, Huang L *et al.* (2025) Enhanced nitrogen removal through bioaugmentation with Stutzerimonas stutzeri SW22: From denitrification mechanism to optimized sequencing batch reactor. *Journal of Water Process Engineering* **71**: 107252.

Sun J, Yuan M, Zhou H *et al.* (2025) Enhanced printing and dyeing wastewater treatment using anaerobic-aerobic systems with bioaugmentation. *Journal of Hazardous Materials* **486**: 136982.

Tetteh EK and Rathilal S (2021) Application of biomagnetic nanoparticles for biostimulation of biogas production from wastewater treatment. *Materials Today: Proceedings* **45**: 5214–20.

Tyagi M, da Fonseca MM and de Carvalho CC (2011) Bioaugmentation and biostimulation strategies to improve the effectiveness of bioremediation processes. *Biodegradation* **22**: 231–241.

Ueno A, Ito Y, Yumoto I and Okuyama H (2007) Isolation and characterization of bacteria from soil contaminated with diesel oil and the possible use of these in autochthonous bioaugmentation. *World Journal of Microbiology and Biotechnology* **23**: 1739–45.

Wang X, Wen X, Yan H, Ding K and Hu M (2011) Community dynamics of ammonia oxidizing bacteria in a full-scale wastewater treatment system with nitrification stability. *Frontiers of Environmental Science and Engineering in China* **5**: 92–98.

Wang X, Xia Y, Wen X, Yang Y and Zhou J (2014) Microbial community functional structures in wastewater treatment plants as characterized by GeoChip. *PLoS One* **9**(3): e93422.

Yu Z and Mohn WW (2002) Bioaugmentation with the resin acid-degrading bacterium Zoogloea resiniphila DhA-35 to counteract pH stress in an aerated lagoon treating pulp and paper mill effluent. *Water Research* **36**(11): 2793–2801.

Zheng Y, Cheng C, Zhou Z *et al.* (2019) Insight into the roles of packing carriers and ultrasonication in anaerobic side-stream reactor coupled membrane bioreactors: sludge reduction performance and mechanism. *Water Research* **155**: 310–319.

Microbial Processes in Environmental Engineering: From Theory to Practice

Rajneesh Kumar, Ram Sharan Singh and Maulin P Shah
ISBN 978-1-83708-327-5
https://doi.org/10.1108/978-1-83708-326-820261003

Chapter 3
Biofiltration of water pollutants

Odangowei Inetiminebi Ogidi[1], Lilian Maureen Oluchi Oguoma[2] and Anand Kumar[3]

[1]*Department of Biochemistry, Bayelsa Medical University, Yenagoa, Bayelsa State, Nigeria*
0000-0002-9961-893X
[2]*Department of Biotechnology, Federal University of Technology, Owerri, Nigeria*
[3]*College of Food Science and Technology, Guangdong Ocean University, Zhanjiang, China.*

Abstract

Biofiltration is an environmentally friendly method for eliminating contaminants from water, utilising the natural abilities of microorganisms and plant-based systems. This process involves the filtration of contaminated water through biologically active media, such as sand, gravel or artificial substrates, which facilitate microbial colonies that effectively degrade organic pollutants and immobilise inorganic contaminants. A wide range of pollutants, such as harmful microorganisms, heavy metals, nutrients (most notably nitrogen and phosphorus) and organic waste, can be effectively eliminated through the use of biofilters. Hydraulic retention time, filter medium composition and structure, in addition to environmental characteristics such as pH, temperature and oxygen availability, are all factors that influence the efficacy of biofiltration systems. Advancements in biofilter design, particularly through the integration of biochar and nanomaterials, have improved pollutant removal efficiency and expanded the range of treatable contaminants. This technology has been effectively utilised in municipal wastewater treatment, stormwater management and industrial effluent remediation. Biofiltration systems are cost-effective, require minimal maintenance and enhance ecological sustainability by reducing reliance on chemical treatment methods. This chapter aims to optimise biofilter designs and assess the potential of genetically modified bacteria in addressing chronic pollution. Biofiltration represents a viable approach to addressing water pollution, supporting initiatives aimed at enhancing clean water accessibility and promoting environmental conservation.

Keywords: Biofiltration; environment; water; pollutants; biomass

3.1. Introduction

Both people and the ecosystem are in grave danger due to the increasing prevalence of water contamination. This growing environmental and public health crisis partly informed the adoption of the Sustainable Development Goals (SDGs) by the United Nations in 2015, which outline 17 global objectives to be achieved by 2030 to protect the environment, eliminate extreme poverty and promote global prosperity and peace (Ogidi and Izah, 2024).

The sixth objective encompasses the provision of safe drinking water and adequate sanitation, emphasising sustainable water management. Water contamination results from human activities, including population growth, urbanisation and industrialisation. Approximately 3% of Earth's water is freshwater, whereas the remaining 97% is saline, found in the world's oceans (Ogidi and Akpan, 2022). According to Devda *et al.*, (2021), approximately 80% of water contamination was attributed to improper waste disposal, which introduced biological agents, chemical substances and radioactive materials into the water supply, affecting about 3% of the total available water.

Approximately 2 billion individuals consume contaminated water, exposing them to the risk of waterborne diseases. Water contamination is responsible for 485 000 fatalities worldwide (Izah and Ogidi, 2024). Heavy metals and volatile organic compounds (VOCs) represent two major categories of contaminants. Some chemicals are widely recognised as harmful to both humans and the environment due to their carcinogenic nature and toxicity. New technologies are necessary to repair or restore water supplies and guarantee that individuals can live a healthy existence.

The United States Environmental Protection Agency (USEPA) observes that VOCs are organic molecules that are photochemically reactive when exposed to air, highly soluble in air at room temperature and inadequately soluble in water (Okereke *et al.*, 2016c). Industrial solvents, lubricants, paints, petroleum fuels, dry cleaning chemicals, inks, varnishes and cosmetics are among the products that contain VOCs, which include benzene, butane, toluene, esters, propane, pentane, methane, hexane, chlorohydrocarbons, trichlorofluoromethane, ketones, chloroform and acetates (Plaisance *et al.*, 2017).

The petroleum and chemical industries are responsible for most volatile organic compound emissions. VOCs are responsible for environmental pollution. The primary mechanism responsible for the generation of ground-level ozone is the interaction between tropospheric nitrogen oxides and vapour organic molecules. The interaction between ground-level ozone and VOCs results in the formation of smog (Tisserand and Young, 2014). The health of species was adversely affected by significant alterations in weather and environment caused by ground-level ozone and smog.

Heavy metals, characterised by a density ranging from 3.5 to 7 g/cm^3, are detrimental to aquatic ecosystems, even in minimal concentrations, and can lead to the mortality of organisms (Gautam *et al.*, 2014). Mining, industrial emissions related to metals and household usage are the main anthropogenic activities contributing to heavy metal pollution (Briffa *et al.*, 2020; Gautam *et al.*, 2014). Heavy metals such as arsenic, cadmium, chromium, cobalt, copper, iron, lead, mercury, manganese, nickel, selenium, silver, thallium and zinc are among the most significant contaminants of concern in water systems (Azimi *et al.*, 2017).

Silver, lead, mercury, cadmium and arsenic can be detrimental, even in minimal quantities (Engwa *et al.*, 2019; Gautam *et al.*, 2014). The delicate equilibrium of the food chain is gravely endangered by the long-lasting environmental persistence of heavy metals, which are non-biodegradable. The primary method by which heavy metals are introduced into the biosystem is through the consumption of contaminated food and water (Rani and Shanker, 2020). Heavy metals affect the body's functionality by contributing to cellular oxidative stress, which is the result of the conversion of reactive oxygen species into free radicals. By disrupting lipids, proteins and nucleic acids, free radicals induce cell apoptosis (Briffa *et al.*, 2020; Engwa *et al.*, 2019; Madkour, 2020). Death and malignancy can be induced in living entities by even minute levels of heavy metal ions (densities greater than 6 g/cc) (Rani and Shanker, 2020).

3.2. Biofiltration

3.2.1 History of biofiltration techniques

In order to purify contaminated air and water, biofiltration is implemented (Adler, 2001; Boehem *et al.*, 2020). Biofilters, which are more commonly known as biological filters, are employed in the biofiltration process. By adhering to the filter media, biofilters enable bacteria to establish colonies.

Air and water pollutants can be degraded by microorganisms through the process of oxidation. They are capable of absorbing and decomposing a wide variety of air and water contaminants, regardless of whether they are organic or inorganic (Raebitso-Senior *et al.*, 2012). The biotrickling filter, an innovative technology, was first employed for sewage wastewater treatment in England in 1893 (Bressani-Ribeiro *et al.*, 2018). As early as 1923, the German scientist HG Bach reported the use of living microorganisms to biologically oxidise hydrogen sulfide, a toxic compound commonly found in wastewater, thereby establishing the foundational principles of biofiltration (Leson and Winer, 1991). However, the first full-scale engineered biofilter system was not installed until 1953 in Long Beach, California, USA, for the control of sewage odours. Following this milestone, biofiltration systems became increasingly utilised during the 1950s.

The management of hazardous volatile chemicals in industrial contexts was a primary concern for both the USA and Europe. Appropriate filter beds and microbial communities were integrated into the biofiltration system for this objective (Leson and Winer, 1991; Swanson and Loehr, 1997). More than 500 biofilters have shown commercial success in Japan and Europe, exhibiting high operational efficiency (Bohn, 1992; Wani *et al.*, 2012). Biofiltration techniques (Figure 3.1) for the remediation of polluted water and air have been extensively researched and documented in the literature. Reports indicate that biofiltration surpasses other existing pollution control technologies in the reduction of volatile organic compounds, achieving an efficacy of approximately 90% (Fulazzakay *et al.*, 2014; Leson and Winer, 1991; Ralebitso-Senior *et al.*, 2012).

Figure 3.1 Types of biofiltration methods

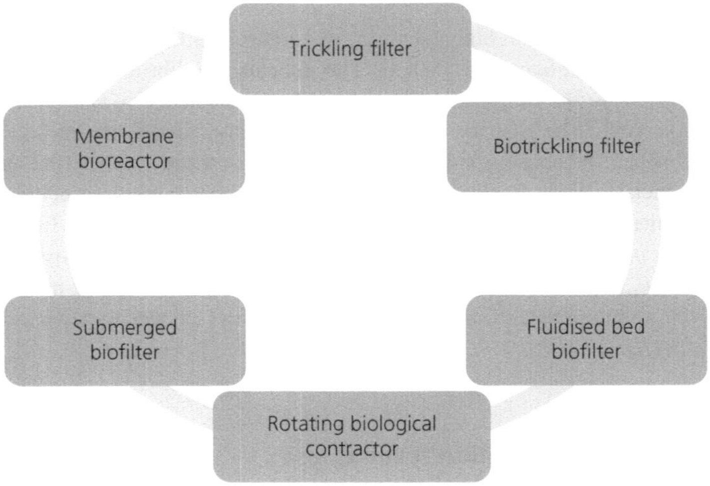

3.2.2 Biofiltration mechanism in the removal of contaminants

Biofiltration is a multi-stage process that is designed to purify air or water by eradicating pollutants. A biofilter system initially receives contaminated air or water as input and subsequently facilitates the absorption of pollutants through the biofilm or cellular membrane of the biofilter bed. Bed media become contaminated during the aqueous phase (Adler, 2001; Schroeder, 2002). The microorganisms obtain their carbon source from the contaminants. The proliferation of microorganism colonies that degrade toxins is a consequence of effective metabolism (Adler, 2001). Lastly, the procedure produces byproducts such as water, biomass and carbon dioxide in addition to the purified air or water. The equation: pollutants + live organisms + oxygen → biomass + water + carbon dioxide illustrates the critical processes of sedimentation and biodegradation in biofiltration (Bressani-Ribeiro *et al.*, 2018; Chaudary *et al.*, 2003).

Contaminated air or water is placed on the biofilter substrate. Subsequently, contaminants undergo a phase transition from solid to liquid within the biofilter.

The subsequent section outlines the key microbial processes that facilitate pollutant phase transfer and degradation in biofiltration systems (Schroeder, 2002). These processes include

- biodegradation, which refers to the mechanism through which microorganisms decompose pollutants adsorbed onto organic media, such as a biofilter bed
- biofilm-mediated adsorption and biodegradation, whereby contaminants are captured and metabolised within microbial biofilms attached to the filter media
- pollutant dissolution and biodegradation in aqueous phases, where contaminants dissolve into water films and are subsequently biodegraded by microorganisms: following biodegradation, pollutants are effectively removed from the biofilter system (Adler, 2001).

Overall, the efficiency of biofilters is fundamentally dependent on microbial activity, as the effectiveness of all these processes relies on the metabolic capacity and viability of the microbial community (Baltrėnas and Baltrėnaitė, 2020).

Microorganisms play distinct roles in the modification of pollutant phases and the degradation of contaminants in polluted air and water. The vitality of these bacteria is essential for the effective elimination of toxic substances. To protect natural resources from harmful metals and pesticides, scientists have increasingly focused on the development and application of genetically modified microorganisms capable of enhanced bioremediation (Ogidi and Richard-Nwachikwu, 2024; Padhan *et al.*, 2021; Srivastava *et al.*, 2021).

This section provides a detailed explanation of how microorganisms contribute to biofiltration processes, including the mechanisms by which the biofilter bed confines, supports and immobilises the microbial community.

Two approaches exist for the immobilisation of microorganisms. Microbial adhesion or immobilisation to biofilter bed materials occurs through natural or artificial induction (Baltrėnas and Baltrėnaitė, 2020). Microbial immobilisation is designed to increase the survival rates of bacteria and the burden of cells within the filter bed material, thereby facilitating the rapid synthesis of the desired microbial product (Park and Chang, 2000; Rathore *et al.*, 2013).

3.2.2.1 Self-attachment of microorganisms

The biofilter bed material is adhered to by naturally occurring adherent bacteria in the absence of human intervention, without external forces. Maintaining a higher concentration of bacteria in the system compared to suspended microorganisms is essential, as the concentration of naturally attached microorganisms was found to be greater (Soccol *et al.*, 2003). In comparison to the suspended microorganism system, the microbial biofilm demonstrates enhanced metabolism and biodegradation (Cohen, 2001; Ogidi *et al.*, 2024d).

Microorganisms possess a polysaccharide layer on their cell membrane known as the glycocalyx. The structure of microorganisms and their surface attachment are primarily influenced by this (Soccol *et al.*, 2003). Recent studies indicate that biofilm formation in wastewater treatment is significantly influenced by rhamnolipids produced by bacteria (Peng *et al.*, 2020). Microorganisms utilise various forces for self-attachment, including covalent bond formation, electrostatic interactions and hydrophobic contacts. In addition, microorganisms are involved by way of a partial covalent connection with the surface hydroxyl group (Malhautier *et al.*, 2005; Soccol *et al.*, 2003). Factors such as the microorganisms used, the material of the filter bed's surface, the surrounding environment and the fluid's properties all had a role in the variation of these forces (Malhautier *et al.*, 2005; Soccol *et al.*, 2003).

The initial analysis demonstrated that the most potent force was electrostatic contact during the adsorption phase (Cohen, 2001). The majority of individuals concur that the nutrients provided by an organic filter bed material or packing are of great assistance to microorganisms (Malhautier *et al.*, 2005). Inorganic materials, such as glass or ceramics, can produce metal hydroxides on their surfaces. These metal hydroxides induced partial covalent bonding in the polymer matrix, enhancing the structural stability and adsorption capacity (Malhautier *et al.*, 2005; Soccol *et al.*, 2003).

3.2.2.2 Artificial immobilisation of microorganisms

There are five different methods for artificially immobilising microorganisms (Soccol *et al.*, 2003). Methods such as entrapment, membrane separation, microencapsulation, covalent bonding and covalent cross-linking are included.

3.2.2.2.1 COVALENT BONDING

This technique entails establishing a covalent bond between microbial cells and water-insoluble carriers. Carriers include proteins, synthetic polymers, inorganic compounds and water-insoluble polysaccharides (Cohen, 2001). The biofilter bed material comprises ligands that facilitate microbial interaction due to the presence of reactive groups within their cells (Malhautier *et al.*, 2005). Microbial cells possess reactive groups, some of which may exert detrimental effects. Consequently, a limited proportion of bacteria that adhere to biofilter media successfully attach, while cell division leakage occurs (Soccol *et al.*, 2003).

3.2.2.2.2 COVALENT CROSS-LINKING

The covalent cross-linking technique is a progression from the previous covalent bonding method, as it enables microorganisms to interact covalently to generate three-dimensional structures. According to what was said about the covalent bonding method, reactive group toxicity disrupts the immobilisation of microorganisms (Malhautier *et al.*, 2005).

3.2.2.2.3 ENTRAPMENT

Bacterial cells become enmeshed within a three-dimensional polymer matrix. This matrix is composed of various materials, including cellulose, polyurethane, polystyrene, agar and resins. The pores in the polymer matrix are smaller than the microbial cells and it effectively captures the cells of the microorganisms (Cohen, 2001).

Plasmid stability, high entrapped cell metabolism, gentle experimental conditions, cheap cost, physical separation and immobilisation of distinct bacteria, restriction to harmful substances, biodegradability and eco-friendliness are some of the advantages of entrapment immobilisation of cells. The major drawbacks are metabolic changes, a decrease in oxygen consumption within entrapped microbial cells, which damages those cells, and a severe diffusion limitation for some polymers (Cassidy *et al.*, 1996; Soccol *et al.*, 2003).

3.2.2.2.4 MEMBRANE SEPARATION

Microorganisms were extracted from a large liquid medium utilising the membrane separation technique. Ultrafiltration membranes with porosity between 0.002 and 0.1 μm are employed for this technique. Unique and non-porous membranes may also be applied in water purification (Shivajirao, 2012). Membrane fouling post treatment represents a significant concern (Cohen, 2001). The immobilised microbial cells of the biofiltration system are at risk due to conventional physical and chemical membrane cleansing techniques. Composite membranes and antifouling technologies have been extensively employed in the purification of water and the remediation of wastewater (Chen *et al.*, 2011; Gohari *et al.*, 2014).

3.2.2.2.5 MICROENCAPSULATION

Minute organisms are enclosed within a fragile membrane that resembles a droplet. The microorganism possesses unrestricted mobility within its capsule. The objective of microencapsulation is to protect the bacteria from mechanical and environmental challenges. The microorganisms exhibit an accelerated metabolism and enhanced growth potential (Rathore *et al.*, 2013). The diameters of the microcapsules may range from 10 to 100 μm, as indicated by Cohen (2001). The capsule's diminutive size enables the release of microbial metabolites and the distribution of substrates within it (Park and Chang, 2000; Rathore *et al.*, 2013).

Epoxy resins, nylon, cellulose nitrate and analogous materials are employed for encapsulation purposes (Cohen, 2001; Soccol *et al.*, 2003). Alternative immobilisation methods have limitations, including inadequate mechanical stability, contamination risks, reduced cell loading, diminished metabolism and cell leakage; microencapsulation addresses these challenges (Rathore *et al.*, 2013).

3.2.3 Parameters influencing the biofiltration process

The biofiltration process's efficacy is contingent on numerous physical, chemical and biological variables (Table 3.1).

3.2.3.1 Biological organisms

Biological organisms play a central role in biofiltration by driving the transformation and removal of contaminants through diverse metabolic and physical mechanisms.

The biofiltration process depends on a living organism functioning as the catalyst for the process. Biofilm is formed by microorganisms, including bacteria, protozoa, invertebrates and fungi. The predominant microorganisms likely present in the filter beds are bacteria and fungi. This breakdown of contaminants is conducted by immobilised heterotrophic microbial cells. The cells initially adhere to the surface of the reactor's biofilter bed. Subsequently, microbial cells colonise the surface, forming an active layer that captures the pollution. The formation of covalent bonds through multiple surface interactions and the secretion of polysaccharides, an extracellular component, accounted for this phenomenon (Cohen, 2001).

Another option is the artificial attachment of microbial cells to the biofilter bed. Methods employed include microentrapment, membrane utilisation, cross-linking, carrier bonding and artificial immobilisation of cells (Cohen, 2001). Beneficial microorganisms cultivated in the bioreactor bed decompose organic and inorganic contaminants in the influents, thereby mitigating odour. Fungal biofiltration demonstrated greater efficacy than bacterial methods in the removal of the volatile organic compound toluene (Maestre *et al.*, 2007; Zamir *et al.*, 2011). Fungi exhibit natural resistance to dry and acidic environments, explaining the observed phenomenon.

Biofiltration employing coconut fibre and compost achieved toluene removal capacities of 90 and 95 g/m^3/h, respectively, with the provision of nutrients (Maestre *et al.*, 2007). The saprophytic fungus *Phanerochaete chrysosporium* demonstrated an elimination capacity of 1913.7 mg/m^3/h, achieving clearance efficiencies between 50% and 92% (Zamir *et al.*, 2011). Another study indicated that the degrading species *Enterobacter* could eliminate nearly 90% of benzene from the waste gas stream at a rate of 21.46 g/m^3/h (Padhi and Gokhale, 2016). The biofilter exhibited microorganism concentrations between approximately 10% and 15% (Devinny *et al.*, 2017; Repeckiene *et al.*, 2015). Aerobic microorganisms, specifically actinobacteria, have recently shown the capability to degrade butyric acid and completely eliminate odours (100%) (Marquez *et al.*, 2021). Bacterial counts in biofilter compost beds vary from

Table 3.1 Parameters influencing the biofiltration process

Parameter	Description	Impact on biofiltration
Filter media	Type of material used in the filter, such as sand, gravel, act vated carbon or synthetic media.	Affects microbial attachment, pollutant adsorption and clogging potential.
Biofilm growth	Development of microbial layers on filter media that break down contaminants.	Thick biofilms enhance removal efficiency but may cause clogging over time.
Microbial population	Types and diversity of microorganisms (bacteria, fungi, protozoa) present in the filter.	More diverse microbial communities lead to better biodegradation performance.
Hydraulic retention time (HRT)	The duration for which water remains in the biofilter before exiting.	Longer HRT improves contaminant breakdown but may reduce flow efficiency.
Loading rate	The amount of contaminants introduced per unit time (e.g. organic matter, heavy metals, VOCs).	High loading rates can overload the system and reduce pollutant removal efficiency.
Oxygen availability	Presence of oxygen required for aerobic microbial activity.	Low oxygen levels can shift microbial processes to anaerobic degradation, affecting performance.
Temperature	Environmental temperature affects microbial metabolism and reaction rates.	Optimal temperatures (15–35°C) enhance microbial activity, while extremes slow down processes.
pH level	Acidity or alkalinity of water (typically optimal in the range of 6.5–8.5 for microbial processes).	Extreme pH levels can inhibit microbial growth and enzymatic reactions.
Nutrient availability	Presence of essential nutrients like nitrogen, phosphorus and carbon sources.	Sufficient nutrients support microbial growth, while deficiencies may slow down biodegradation.
Moisture content	Water saturation in the biofilter media.	Essential for microbial survival; too much can cause clogging, too little can dry out biofilms.
Flow rate	Speed at which water moves through the biofilter.	Affects contact time with biofilm and efficiency of pollutant removal.
Toxic substances	Presence of heavy metals, pesticides or other harmful chemicals.	Can inhibit microbial activity and biofilm formation, reducing filtration efficiency.
Biofilter design	Structure, depth and aeration system of the biofilter.	Proper design ensures optimal pollutant removal, prevents clogging and maintains efficiency.
Pre-treatment methods	Initial treatments such as sedimentation, aeration or chemical adjustments before filtration.	Helps remove large particulates and enhances biofilter lifespan and efficiency.

10^4 to 10^{10} CFU/g (Repeckiene *et al.*, 2015; Marquez *et al.*, 2021; Afrooz *et al.*, 2018). The efficacy of the biofilter in degrading pollutants is significantly influenced by the concentration and selection of microorganisms.

3.2.3.2 Biofilter bed

A biofiltration unit fundamentally consists of a bed of bacteria or packing material. The biofilter bed functions as a substrate for biofilm development, facilitating the immobilisation of microorganisms on its surface (Delhomenie and Heitz, 2005). Biofilter bed packing materials can be classified as either organic or inorganic. Common packing materials that are somewhat inexpensive include soil, compost, wood shavings, cocopeat, perlite, ceramics and polyurethane foam (Sakuma *et al.*, 2006). These packing materials are intended to have a high specific surface area (Andres *et al.*, 2006) and a high porosity (300–1000 m^2/m^3), which will facilitate the uniform distribution of influent.

Microbial development and biofilm formation are both facilitated by the availability of natural nutrients. The presence of numerous bacteria is significant. Enhanced microbial metabolism resulting from improved water retention capacity (40–60%) supports sustained microbial activity and improves pollutant degradation performance. Stability in both mechanical and thermal dimensions is essential to avoid filter bed packing. Soil acts as a natural containment system, retaining and degrading pollutants before they reach groundwater or surface water.The microbial diversity is extensive, characterised by high specific surface areas, yet it contains fewer intrinsic nutrients. In contrast, peat contains minimal inherent nutrients and microorganisms, despite its high specific surface area (Delhomenie and Heitz, 2005).

Composts are characterised by their nutrient content, water retention capacity, microbial abundance and air permeability (Delhomenie and Heitz, 2005). Compost decomposes, resulting in a reduction of pressure due to its significant water retention properties.

Wood chips demonstrate biodegradability following the attainment of adequate stability (Andres *et al.*, 2006). In contrast, symptoms indicative of inadequate nutrition absorption, reduced pressure and diminished pH buffering capacity are observed. By adjusting the loading concentration, the pollutant removal efficiency of wood chips can be assessed (Andres *et al.*, 2006).

Ceramics like biofoam and perlite exhibit favourable gas flow characteristics and possess superior thermal stability (Sakuma *et al.*, 2006). The performance of toluene removal was assessed using four different packing materials: perlite, open-pore polyurethane foam, horticulture porcelite and cattle-bone porcelite. Among these, open-pore polyurethane foam and cattle-bone porcelite showed the highest toluene removal efficiencies, highlighting the importance of material porosity and surface area in biofilter performance (Sakuma *et al.*, 2006).

3.2.3.3 Supply of nutrients

The efficacy of biofiltration is significantly influenced by diet. Microbial activity initiates the degradation process when pollutants penetrate the biofilter. These contaminants provide the microorganisms with energy by serving as carbon sources. Important macro- and micronutrients are abundant in the sustaining packing materials of the biofilter bed. Nitrogen, phosphorus, potassium and sulfur are the primary macronutrients, while vitamins and metals are considered micronutrients (Weckhuysen *et al.*, 1993).

Nutrients in solid or liquid form are introduced into the filter bed. The biofilter bed is supplied with a solution of mineral salts dissolved in water. Mineral salts such as $CaCl_2$, $FeSO_4$, KH_2PO_4, KNO_3, $(NH_4)_2SO_4$, $MgSO_4$, $MnSO_4$, NH_4Cl, NH_4HCO_3 and Na_2MoO_4 are commonly employed (Delhomenie and Heitz, 2005). Nutrient availability promotes microbial development, according to several studies (Wechhuysen *et al.*, 1993; Agu *et al.*, 2017).

Biofilters were used to remove butanal from contaminated air both with and without a nutrient source. In contrast to the 86% reduction achieved by a biofilter devoid of nutrients, the data demonstrate that the nutrient-containing biofilter achieved a removal efficiency of 97% (Weckhuysen *et al.*, 1993).

The selection of the packing material for the substantial elimination of pollutants has been the subject of multiple investigations. Organic packing material is superior to inorganic materials, according to these investigations (Anet *et al.*, 2013; Kim *et al.*, 2000).

3.2.3.4 Potential of hydrogen (pH)

When considering biofiltration performance, pH is another critical component. At an ideal pH, nutrients can play a role in the efficient breakdown of contaminants by microorganisms (Barzgar *et al.*, 2017; Han *et al.*, 2020). Because of the neutral pH of their living habitat, the heterotrophic bacteria that inhabit biofilters are known as neutrophils.

The importance of pH in the effectiveness of pollutant removal has been demonstrated in numerous research. Results showed that, at various pH levels, hydrogen sulfide oxidised methane. At an acidic pH of 4.5, the methane oxidation flux rate was 53 $g/m^2/day$, while at a neutral pH of 7.0, it was 146 $g/m^2/day$ (Barzgar *et al.*, 2017). During the biofiltration of reduced sulfur compounds and dimethyl sulfide, acidification takes place due to the generation of sulfuric acid. The acidification process alters pH levels and diminishes the efficiency of the biofilter. Therefore, methanol is added to the reactor to keep the pH stable and prevent the creation of sulfuric acid (Zhang *et al.*, 2007).

Biofilters for landfill odour treatment have recently been the subject of research into bioaerosol emission at neutral and low pH. When the inlet flow rate was large, heterotrophic bacteria had a better removal efficiency at low pH, while fungi had a better removal efficiency at neutral pH (Liu *et al.*, 2021a). Subsequently, the identical team of researchers has suggested using a biofilter to treat polluted gas that contains hydrogen sulfide, acetic acid, toluene and ammonia at neutral or low pH levels (Liu *et al.*, 2021b). Under neutral pH, microbial decomposition effectively removed ammonia and acetic acid with efficiencies of 99.90% and 99.92%, respectively. Low pH circumstances result in a high-pressure drop because fungi are present, but higher removal efficiencies of 99.24% and 99.90% for hydrogen sulfide and toluene, respectively (Liu *et al.*, 2021a).

The elimination of benzene and o-xylene was more effective at a neutral pH of 7, according to the fitted data, than at an acidic pH of 4. At an acidic pH of 4, the unidentified microbial culture exhibited a 22% increase in o-xylene and an 11% increase in benzene relative to neutral pH (Khoramfar *et al.*, 2020).

The studies emphasise the significance of sustaining an optimal pH level and its influence on the microbial degradation of pollutants.

3.2.3.5 Operating temperature

The biofilter's operational temperature controls the quantity of active microorganisms, which in turn degrade contaminants. A study found that the efficiency of a drinking water biofilter in removing organic contaminants varied with temperature (Moll *et al.*, 1999). Moll *et al.* (1999) applied operational temperatures of 5, 20 and 35°C to the water's surface. At a low temperature of 5°C, the clearance of organic materials was observed to decrease. This occurred due to a change in the metabolic rate of the substrate, which was influenced by the modified microbial structure. The effectiveness of biofilters in eliminating contaminants at temperatures of 20°C and 35°C was similar to that of disinfectants.

Biofilters were found to effectively remove VOCs from indoor air at both warm and cold temperatures (Darlington *et al.*, 2001).

Less microbial activity is one of the numerous benefits that biofilters exhibit when operating at lower temperatures, such as leading to a reduction in *Legionella* and mitigating the adverse effects of water vapour. Buildings experience diminished indoor air quality when biofilters are maintained at elevated temperatures (Darlington *et al.*, 2001). In the study, ethanol-containing waste air was treated using a biofilter that was inoculated with *Pseudomonas putida*. The ideal temperature for incubating *Pseudomonas putida*, according to earlier findings, was found to be approximately 26°C. Microbial growth activities were detected at incubation temperatures between 20 and 40°C. The optimal temperature for incubating biofilters with *Pseudomonas putida* is 30°C. The efficiency of total nitrogen (TN) and nitrate-nitrogen (NO_3-N) removal was demonstrated to be 81% and 85%, respectively, within the low temperature range of 6.4–9.8°C. The microbial community within the bioreactor exhibited a pattern in response to temperature variations.

Denitrification by heterotrophic bacteria (*Anaerolineae*) is enhanced, but denitrification by sulfur autotrophic bacteria (*Ferriticum*, *Sulfurimonas* and *Thiobacillus*) is diminished according to Li *et al.* (2020a).

3.2.3.6 Moisture contents

The function of microorganisms is fundamental to biofilter efficiency. Microorganisms will invariably grow uniformly as a biofilm on the biofilter bed if a high biodegradation rate is to be achieved (Badilla *et al.*, 2010). The presence of water content is essential for this achievement. Moisture in the material facilitates the transfer of nutrients to bacteria.

To address the pollutants in the influent, appropriate metabolic processes have been employed (Bohn and Bohn, 1999). The water absorption capacity of biofilter medium is contingent on its classification as either organic or inorganic. Organic materials exhibit hydrophobic properties, whereas inorganic materials demonstrate hydrophilic characteristics. Inorganic media retain moisture content more effectively than organic media. The moisture content of the biofilter can be assessed through volumetric analysis (Bohn and Bohn, 1999; Devinny *et al.*, 2017). It is essential to distinguish clearly between calculations based on dry weight and those based on wet weight. The optimal moisture content range for biofilters to achieve satisfactory degradation rates is between –0.2 and –3 bar (Bohn and Bohn, 1999).

Research is underway to improve contaminant removal through substrates that integrate organic and inorganic materials (Chang *et al.*, 2019; Kiedrzynska *et al.*, 2017). The optimal water content, porosity and surface areas of these filtration materials differ. Chang *et al.* (2019) constructed a biofilter utilising fibrous materials in conjunction with inorganic components, including ceramsite and lava rock. This hybrid biofilter effectively treated polluted river water. Denitrification was observed in organic materials at a rate approximately 44% greater than that of inorganic materials. Inorganic materials demonstrated the capacity to extract phosphorus at rates ranging from 76% to 94% and ammonium at rates between 87% and 97% (Chang *et al.*, 2019).

Key factors for maintaining moisture in a biofilter bed material include the temperature of the inlet gas, the duration required for the oxidation of pollutants through metabolic activity and the prehumidification of the inlet gas or air (Kiedrzynska *et al.*, 2017).

3.2.3.7 Pressure drop

Another method of evaluating pollution degradation is to measure the pressure decrease across the biofilter bed. The stability and efficiency of biofilters can be predicted by observing the relationship between pressure drop and biomass content (Morales *et al.*, 2003). A pressure reduction of 30–33 mm H_2O/m has been observed in polyurethane biomass filters for biomass concentrations ranging from 2.00 to 2.05 gDCW g/PU (Morales *et al.*, 2003).

The characteristics of the biofilter material, flow rate and moisture content are additional factors that affect pressure decrease (Morgan-Sagastume *et al.*, 2003; Ryu *et al.*, 2010). Ozone injection has been employed to investigate the efficacy of biofilters in eliminating VOCs from waste gas. The effectiveness of toluene removal was assessed by operating two biofilters, with one incorporating ozone injection and the other without (Malakar *et al.*, 2018). The pressure drop in an injected biofilter is less than that in an uninjected biofilter. The rate of toluene removal remained constant throughout the biofiltration process. Ozone exposure enhanced the microbial population, thereby increasing the biofilter removal rate and inhibiting biomass accumulation (Malakar *et al.*, 2018).

3.3. Biofiltration techniques in the removal of volatile organic compounds

The elimination of potentially hazardous and unpleasant VOCs from industrial effluent is a critical initial step prior to disposal. The operations and manufacturing processes of pharmaceutical companies, oil mills, petroleum refineries and other such industries release these compounds, necessitating the use of diverse biofiltration techniques for their removal (Ogidi, 2022). Air, water and land are contaminated by organic volatile substances. Toluene, benzene, ethylbenzene, xylene, styrene, α-pinene and various other compounds have been investigated in biofilter research and laboratory experiments (Darlington *et al.*, 2001; Khoramfar *et al.*, 2020; Malakar *et al.*, 2018; Padhi and Gokhale, 2016). Table 3.2 summarises experimental data of biofiltration techniques in the removal of VOCs.

Table 3.2 Experimental data of biofiltration techniques in the removal of volatile organic compounds (VOCS)

VOCs	Biofilter media	Inlet concentration (mg/m^3)	Removal efficiency (%)	Reference
Toluene	Compost and perlite	100–500	90–95	Rene *et al.*, 2018
Benzene	Activated carbon and bacteria	50–200	85–90	Li *et al.*, 2019a
Ethylbenzene	Peat moss and fungi	20–100	80–85	Singh *et al.*, 2020a
Xylene	Sawdust and microorganisms	30–150	75–80	Kumar *et al.*, 2018a
Styrene	Coconut coir and bacteria	40–200	70–75	Rajendran *et al.*, 2019a
Acetone	Compost and zeolite	50–250	65–70	Zhang *et al.*, 2020a
Methyl ethyl ketone (MEK)	Activated carbon and fungi	20–100	80–85	Singh *et al.*, 2020b
Methylene chloride	Peat moss and bacteria	10–50	75–80	Kumar *et al.*, 2018b
Trichloroethylene (TCE)	Sawdust and microorganisms	5–20	70–75	Rajendran *et al.*, 2019b
Formaldehyde	Coconut coir and bacteria	0.5–2	60–65	Zhang *et al.*, 2020b
n–Hexane	Compost and zeolite	10–50	55–60	Li *et al.*, 2019b

3.3.1 Biotrickling filter

In a biotrickling filtration system, an aqueous medium is continuously applied to the filter bed media, which supports immobilised microorganisms. The nutrient solution is required to facilitate microbial development. This aqueous medium facilitates the passage of polluted influent. On absorption into the biofilm, the contaminants experience degradation (Delhomenie and Heitz, 2005).

Biotrickling filters are utilised independently or in conjunction with existing methods. In comparison to other methods, such as regenerative catalytic oxidation and adsorption processes, this method demonstrates a high level of success in the removal of volatile organic compounds, with an elimination efficiency of approximately 90% (Yang *et al.*, 2019). To eliminate the volatile organic compounds produced by the chemical fibre effluent treatment system, biotrickling is implemented. The biotrickling filter's complete contaminant removal will be ensured by the homogeneity and specific degradation period of the microorganisms used (Yang *et al.*, 2019).

Biotrickling filters tend to allow contaminated gas to descend. The microbial count in the lower layer of the biofilter differs significantly from that of its intermediate and upper layers. The national standards for pollutant concentration are met when the empty bed residence time is between 32 and 59 s.

The concentration of pollutants emitted by biofilters is influenced by the types of microorganisms present and their distribution. The efficiency of the biotrickling filter diminishes with the accumulation of biomass. A complete fluidisation of the medium addresses this issue (Serial *et al.*, 1995).

Pelletised and monolithic channelised media were utilised by Serial *et al.* (1995) in a hybrid trickling bed biofilter for the removal of toluene. Activity was greater in the pelletised media than in the monolithic channelised medium. The pelletised medium was proven effective for biomass accumulation. Backwashing for 320 min effectively inhibited biomass formation in this context. No discernible back pressure was observed between successive backwashing cycles. In three operational runs, a toluene concentration of 0.725 kg COD/m^3/day resulted in a 99% efficiency rate.

Biotrickling filters with polyhedral spheres efficiently deodourised compost and concurrently eliminated ammonia and hydrogen sulfide. A theoretical examination was conducted to identify the microorganisms responsible for pollution degradation by Huan *et al.* (2021). However, advancing the method from pilot or laboratory settings to practical application necessitates a comprehensive understanding of the microbial metabolism pathway.

Biofilters have been employed to reduce odour-causing emissions from tanneries, successfully capturing hydrogen sulfide (H_2S) and ammonia. *Proteobacteria*, *Acinetobacter* and *Firmicutes* exhibited removal efficiencies exceeding 99% for hydrogen sulfide and ammonia (Gandu *et al.*, 2021). Biotrickling is also utilised in municipal wastewater treatment plants for the extraction of carbon, methane and other nutrients (Bezirgiannidis *et al.*, 2019; Khadiri *et al.*, 2021). The implementation of a biotrickling filter alongside chemically enhanced primary treatment (CEPT) has resulted in decreased operational costs and energy consumption.

The removal of solid wastes and organic loadings in municipal wastewater is achieved through coagulation-flocculation processes utilising cationic polyelectrolyte and polyaluminium chloride (PAC). Methane emissions globally were primarily generated by wastewater resulting directly from human activities. Biotrickling filters biooxidise methane found in the effluent gases of wastewater treatment plants (Khadiri *et al.*, 2021).

The hydrophilic nature of methane facilitates its dissolution in water. The methanol-degrading microorganisms in the filter readily absorb methanol as a nutrient source. The increase in microorganisms that can degrade methane correlates with a decrease in those capable of degrading α-pinene. The process is able to exclude either hydrophobic (α-pinene) or hydrophilic (methanol) volatile organic molecules by utilising biofilters for a limited period of time, as indicated by Mohseni and Allen (1999). The biofilter media used was composed of compost harvested from mushroom detritus and wood chips.

Methanol is initially absorbed and subsequently degraded by microbial activity. The degradation of α-pinene has been influenced by the concentration of incoming methanol. Methanol inhibits the growth of microorganisms responsible for α-pinene elimination. The degradation of methanol occurs within hours as a result of its hydrophilic properties. Conversely, α-pinene can be eliminated within 7–10 days due to its hydrophobic properties. The degradation of α-pinene remains unaffected by variations in methanol concentration, despite the influence of α-pinene concentration on methanol levels.

3.4. Biofiltration techniques in the removal of heavy metals

Heavy metals contaminate water and soil due to both contemporary and historical industrial production methods. Consequently, it is essential to eliminate these metal pollutants. Conventional methods for wastewater treatment are available. Conventional approaches to heavy metal accumulation, disposal and detoxification exhibit considerable limitations. Thus, concentrations of heavy metals in soil and water have been excessively elevated, historically. Relocating industries to rural areas was not a practical solution, as it merely shifted environmental impacts without addressing the root causes of pollution. The improper disposal of heavy metal ions exacerbates a severe issue of soil and water contamination (Bennicelli *et al.*, 2004). The primary motivation for researching green technology was that eliminating heavy metals would contribute to preservation of the environment. This section discusses phytoremediation, which is occasionally referred to as botanical biofilters, in conjunction with biofilters that utilise microorganisms and biomimetic membranes to eliminate heavy metals.

3.4.1 Macrophyte biofilters

Heavy metal pollution can be absorbed and stored by certain macrophytes, which are aquatic plants. While heavy metals are necessary for the growth of aquatic vegetation, an excessive amount of them can be detrimental to all aquatic life. Macrophytes function as biofilters that efficiently remove heavy metals from contaminated water. Macrophytes are capable of filtering contaminated water by bioaccumulating and biomagnifying heavy metal ions. The capacity to bioaccumulate and biosorb heavy metals has been associated with specific plant species. The capacity of aquatic plants to assimilate heavy metals is influenced by their affinity for such compounds. The metal ions that have been assimilated are accumulating in multiple locations. The operation is both straightforward and inexpensive – for example, Azolla fronds have the ability to passively absorb heavy metals from contaminated water, as demonstrated by Bennicelli *et al.* (2004).

Their significant affinity for bivalent metal ions has endowed the Azolla fern with enhanced storage and metal-binding capabilities. The ferns, subsequent to their use in the study (Bennicelli *et al.*, 2004), were disposed of securely to prevent any contact with water or air sources by heavy metals. The Azolla plants comprised (by dry mass) copper and cadmium at 3–5%, chromium at 5–11% and lead at 5–12%.

A biofilter's acid wash and ash are produced by way of incineration, which involves subjecting the heavy metal-laden biomass Azolla to extremely high temperatures. Reusable biofilters are made by the acid wash (Tel-Or *et al.*, 1997). The same holds true for the use of *Azolla caroliniana* in purifying municipal water systems from chromium and mercury ions.

In addition to removing heavy metal ions, plants may clean up water and soil polluted with chemicals, radioactive materials, organic solvents, chemical fertilisers, pesticides and more. A culture of *Azolla caroliniana* was introduced to a solution containing chromium ions and mercury. The *Azolla caroliniana* tissue exhibited the highest chromium (III) capture rates, ranging from 200 to 48 000 mg/dm^3, and the highest mercury absorption rates, reaching 578 mg/dm^3. Research indicates that ferns serve as effective bioaccumulators for the removal of heavy metals (Bennicelli *et al.*, 2004; Tel-Or *et al.*, 1997).

The phytoremediation procedure facilitates the effective capture of heavy metals in an environmentally acceptable manner. Toxic heavy metals are effectively degraded when they accumulate in plant tissues and roots. Phytoremediation methods can utilise plant organisms to effectively sequester substantial quantities of heavy metals. The biosorption technique has also been employed for the treatment of hard water and desalination utilising beetroot fibres (Rima *et al.*, 2004).

3.4.2 Stormwater biofilters

Biofilters exist that treat stormwater by eliminating heavy metals and other pollutants, alongside those designed for aquatic environments. The design and optimisation of stormwater biofiltration can be customised to address specific requirements. The effectiveness of the stormwater filter system in pollutant removal from wastewater is significantly enhanced by the synergistic interactions among plants, bacteria and filters. A stormwater biofiltration system was employed to investigate the removal of copper, cadmium, lead and zinc (Blecken *et al.*, 2011; Feng *et al.*, 2012). In both studies, conventional storm biofilters captured approximately 90% of heavy metals. A 100 mm-thick layer of soil was incorporated into the design, resulting in modifications. The metals were retained in the topsoil layer due to its enhanced sorption capacity. Experiments were conducted at 2, 8 and 20°C. No metal removals were affected, except for copper.

3.4.3 Microorganism-based biofilters

Microbiological bioremediation is the process by which heavy metal toxicity is reduced through the conversion of the metal into less hazardous forms. The bacteria used in the process may help manage odour while biofiltration is in operation (Liu *et al.*, 2016).

A variety of microorganisms, such as bacteria, fungi, algae and plants, are employed in bioremediation strategies. The primary method of degradation is metabolism, which is facilitated by an enzyme produced by the bacterium (Kang *et al.*, 2016). Numerous biological and environmental factors influence the maturation of bacteria. Enzyme activity, microbial composition, size, mutation and horizontal gene transfer are among the most notable phenomena in biology. Humidity, oxygen concentration, pH, nutritional quantities and temperature are all significant environmental factors (Kang *et al.*, 2016). Bacteria are able to surmount these obstacles as a result of the variability of their genes (Dhankar and Hooda, 2011). Bacteria are capable of developing resistance to the aforementioned components as a result of genetic differences. The bioremediation process was entirely successful due to genetically engineered microorganisms that were modified to enhance their pollutant-degrading capabilities.

Several genera of bacteria, including *Pseudomonas*, *Alcaligenes*, *Citrobacter*, *Klebsiella*, *Escherichia*, *Rhodococcus*, *Bacillus* and *Staphylococcus*, are utilised in bioremediation. *Rhizopus*, *Aspergillus*, *Pleurotus*, *Saccharomyces* and *Penicillium* are among the numerous fungi that are employed in bioremediation.

A multi-component system is able to eliminate heavy metals with the assistance of sulfate-reducing microorganisms. The removal efficiencies are relatively high, regardless of the metal concentration. The clearing efficacy decreased as the metal concentration increased. This can be attributed to a decrease in chemical oxygen demand and a suppression of sulfate (Kiran *et al.*, 2017).

Arsenates can be eliminated from *Penicillium chrysogenum* biomass that has been chemically modified, as demonstrated by Mycan (Loukidou *et al.*, 2003). Cationic polyelectrolytes and surfactants, including dodecylamine and hexadecyl-trimethylammonium bromide, are used to alter fungal biomass. This research demonstrated that the ionic content of biomass was elevated by 0.57 meq/g as a result of Mycan treatment.

The modification of biomass can be accomplished using inexpensive and straightforward methods. The biosorption values for heavy metals at a pH of 3 were 33.31 mg As/g for dodecylamine, 56.07 mg As/g for cationic polyelectrolyte and 37.85 mg As/g for Mycan-modified hexadecyltrimethylammonium bromide (Loukidou *et al.*, 2003). By altering the biomass, it is feasible to eradicate even the most minute quantities of arsenic.

Uranium can be bioprecipitated through the use of the radiation-resistant polyextremophile bacterium *Deinococcus radium*. By facilitating the process of uranium precipitation, lyophilisation contributes to the preservation of non-specific acid phosphatase activity. *Deinococcus radium* (Choi *et al.*, 2011; Giri *et al.*, 2020; Khairnar *et al.*, 2016; Misra *et al.*, 2012) is capable of efficiently collecting heavy metals, such as cadmium. These bacteriophages also assist in the resolution of wastewater-related issues, such as foaming, sediment thickening and colouration (Khan *et al.*, 2021; Rafida and Sallis, 2005; Rezooqi *et al.*, 2021).

Rafida and Sallis (2005) demonstrated that sulfate-reducing bacteria (SRB) can precipitate lead and zinc by colonising a laboratory-scale vertical-flow biofilter. The removal effectiveness was ascertained by examining the carbon-to-sulfate ratio, lead and zinc concentrations and zinc concentrations. Following 126 days of treatment with a carbon to sulfate ratio of 1:1, the efficiency of heavy metal removal achieved 80% (Rafida and Sallis, 2005).

Recent research has identified capacitance sensing of diverse contaminants in wastewater within microalgae diatoms characterised by silica-shelled eukaryotic cells (Khan *et al.*, 2021). Microalgae fulfil two functions in nanotechnology: they can detect contaminants and serve as intelligent nanocontainers. Microalgae effectively accumulate and transport various contaminants from wastewater, such as trace metals, pharmaceuticals, dyes, hydrocarbons and polymers. The advancement of nano-smart microalgae and their capacity to interact with specific ligands utilised by contaminants has rendered this possible. Natural nanosized microalgae are more economically viable for pollutant treatment compared to synthetic silica-based nanomaterials (Khan *et al.*, 2021).

Table 3.3 shows experimental data for biofiltration techniques in the removal of heavy metals.

Table 3.3 Experimental data of biofiltration technique in the removal of heavy metals

Heavy metal	Biofilter media	Removal efficiency (%)	Reference
Lead (Pb)	Peat moss and compost	95.6	Kumar *et al.*, 2018a
Cadmium (Cd)	Activated sludge and zeolite	92.1	Li *et al.*, 2019
Chromium (Cr)	Sawdust and bacteria	88.5	Singh *et al.*, 2020b
Copper (Cu)	Coconut coir and fungi	85.2	Rajendran *et al.*, 2019c
Zinc (Zn)	Compost and microorganisms	82.1	Khan *et al.*, 2017
Arsenic (As)	Iron oxide-coated sand and bacteria	78.5	Joshi *et al.*, 2019
Mercury (Hg)	Sulfur-reducing bacteria and limestone	75.6	Zhang *et al.*, 2020

3.5. Case studies and industrial applications of biofiltration of water pollutants

Biofiltration is an economical and eco-sensitive technology to treat polluted water. This approach leverages the naturally occurring microbial communities in ecosystems, using fungi and bacteria to degrade organic and inorganic contaminants.

Biofiltration is versatile because it may be used in a wide range of applications, including urban wastewater treatment, industrial effluent management and precipitation filtration. To reduce water pollution, several sectors – including mining, pharmaceutical and agriculture – are looking more and more to biofiltration. For example, while microbial biofilters are crucial in breaking down pharmaceutical residues found in industrial discharges, vegetative biofilters have shown success in removing heavy metals including lead and cadmium from mining effluent.

The capture of sediments, nutrients and hydrocarbons before they enter natural water bodies, systems like built wetlands and biofiltration swales become extremely important in urban environments for stormwater management (Ogidi, 2024). See Table 3.4 below for a thorough summary of case studies and industry-specific uses of biofiltration in water treatment.

3.6. Key challenges and future perspective

Biological filter systems have demonstrated a high degree of success in the removal of volatile organic compounds and heavy metals in wastewater remediation. The biofiltration system continues to have deficiencies and constraints that necessitate resolution. The effectiveness of biofilters is significantly influenced by the evolution of the microbial population over time. The focus of research has been on the role of microorganisms in the degradation of water pollutants. It is important to contemplate the potential for the deterioration process to be accelerated during the work process. Many laboratory-level pilot initiatives have failed to generate effective water remediation systems.

Table 3.4 Case studies and industrial applications of biofiltration of water pollutants

Industry/ application	Pollutant	Biofilter media	Removal efficiency (%)	Reference
Municipal wastewater	Nitrate, ammonia	Wood chips and denitrifying bacteria	90–95	Rajendran *et al.*, 2019
Industrial effluent	Heavy metals (Pb, Cd, Cr)	Activated carbon and metal-removing bacteria	75–80	Ogidi *et al.*, 2019a,b; Okereke *et al.*, 2016a,b; Okore *et al.*, 2021; Rajendran *et al.*, 2019b
Drinking water treatment	Turbidity, BOD	Sand and gravel, sawdust and microorganisms	60–70	Khan *et al.*, 2017; Zhang *et al.*, 2020a
Agricultural runoff	Pesticides (ATP, DDT)	Compost and pesticide-degrading bacteria	70–75	Singh *et al.*, 2020
Mining wastewater	Arsenic, fluoride	Iron oxide-coated sand and arsenic-removing bacteria, activated alumina and fluoride-removing bacteria	50–60	Joshi *et al.*, 2019; Rene *et al.*, 2018
Textile industry	COD, colour	Fungal-based biofilter	80–85	Kumar *et al.*, 2020
pharmaceutical industry	Organic compounds	Granular activated carbon and microorganisms	90–95	Li *et al.*, 2020b
Pulp and paper industry	Chlorinated compounds	Anaerobic granular sludge and microorganisms	85–90	Liu *et al.*, 2019
Food processing industry	Nutrients (N, P)	Zeolite and nutrient-removing bacteria	75–80	Kumar *et al.*, 2019
Oil and gas industry	Petroleum hydrocarbons	Activated carbon and petroleum hydrocarbon-degrading bacteria	80–85	Al-Mutairi *et al.*, 2019
Leather industry	Chromium, ammonia	Sawdust and chromium-reducing bacteria	70–75	Kumar *et al.*, 2019
Chemical industry	Volatile organic compounds (VOCs)	Compost and VOC-degrading bacteria	60–65	Rene *et al.*, 2018

The precise manner in which microorganisms degrade toxins through their metabolic activities remains unknown. It is imperative to monitor the metabolic pathways of microorganisms in order to ensure that pollutants are completely eliminated (Huan *et al.*, 2021). The results are contingent on the microorganism's characteristics, the genus in question and the population that resides in the biofilter substrate (Dhankhar and Hooda, 2011; Huan *et al.*, 2021; Li *et al.*, 2016).

To effectively implement biofiltration technology, it is imperative to possess a comprehensive understanding of the mechanisms of microbial development and disintegration. A thick biofilm layer is subsequently produced as a result of the corresponding increase in pollutant levels. The adsorption process takes place on the surface of the microbial biofilm. In addition, it penetrates the middle of the biofilm. The pollution ceases to migrate downward into the biofilm's depths after a period of time. The final result is a biofilm that is composed of inactive bacteria that are arranged in a variety of widths (Kumar *et al.*, 2011).

Accelerating degradation and enhancing pollutant clearance can be achieved through the utilisation of multi-microbial cultures (Shidhal *et al.*, 2021). Wastewater treatment plants, pharmaceutical residue, water treatment facilities, petroleum refineries and industrial and municipal sewage treatment plants are among the numerous sources of hazardous compounds. The type and severity of contaminants in the effluent needs to be considered in the design and construction of biofiltration systems or processes. Additionally, it is imperative to evaluate the disposal, recycling or application of the purified metals and water, particularly for potential value-added products.

In recent years, artificial intelligence (AI) has demonstrated intriguing applications in a variety of fields, such as water filtration (Al Aani *et al.*, 2019; Alam *et al.*, 2022; Faisal *et al.*, 2020; Zheng and Nguyen, 2022). It may predict the efficacy of various adsorbents (microorganisms) in relation to the quantities and types of certain contaminants in effluent (Briffa *et al.*, 2020; Engwa *et al.*, 2019; Rani and Shanker, 2020). Additionally, value-added products will be anticipated together with the simultaneous removal of impurities (Shindhal *et al.*, 2021; Yang *et al.*, 2019), free of secondary pollutants and fouling effects. Recent research indicates that contaminants can be effectively removed from wastewater through hybridisation techniques grounded in biological principles (Bello *et al.*, 2018; Kiedrzynska *et al.*, 2017; Majumder *et al.*, 2016). The most effective biobased method of water treatment utilising AI is likely to be realised in the near future.

3.7. Conclusions

Biofilters are indispensable for the long-term preservation of environmental purity and the purification of wastewater, as this chapter illustrates. Toxins in water can be absorbed or degraded by certain plants and bacteria. Microorganisms, filter bed media, temperature, pH, humidity, pressure and nutrients are critical factors that affect the effectiveness of biofiltration.

The most efficient elimination of the contaminants is ensured by fine-tuning the parameters. Plants and microorganisms are essential components of biological filtration, which are capable of eliminating a diverse array of pollutants. Several varieties of biofilters are effective in the treatment of wastewater, as indicated by the tested studies. In order to develop ubiquitous biofilters that are fully functional, additional research is imperative, as suggested by Ogidi and Izah (2024).

Biofilters are designed to degrade or absorb pollutants such as heavy metals, VOCs and a variety of other potentially hazardous and non-harmful compounds, thereby contributing to the preservation of a clean and safe environment. The research emphasises the absolute necessity of numerous factors for the complete degradation of contaminants. The initial ingestion concentration, loading rate, pH, retention period, temperature, microbial growth and flow rate are all of paramount importance. The advancement of a biofiltration system that is adaptable is consistent with the detection of pollution. By employing activated carbons and amyloid fibril membranes, recent advancements in biofiltration technology have enabled the implementation of a sustainable global water purification process. In recent years, machine learning has gained some popularity as a potential method for enhancing the efficacy and architecture of numerous technological systems. AI-trained biofilter models can predict the adsorption behaviour of different organisms across various filtering systems. Before commencing the real-time procedure, one can pre-plan the relevant experimental parameters and conditions.

In the near future, sustainable solutions to the challenges posed by various pollutants will likely emerge through the implementation of AI-controlled biological filters. These solutions will demonstrate broad adaptability and acceptance.

REFERENCES

Adler SF (2001) Biofiltration: a primer. *Chemical Engineering Progress* 7(4): 33–41.

Afrooz AN, Pitol AK and Kitt D (2018) Role of microbial cell properties on bacterial pathogen and coliphage removal in biochar-modified stormwater biofilters. *Environmental Science: Water Research & Technology* **4**(12): 2160–2169.

Agu IV, Ibiene AA and Okpokwasili GC (2017) Effect of micronutrients and macronutrients on the biodegradation of phenol in biological treatment of refinery effluent. *Microbiology Research Journal International* **18**(3): 1–2.

Al Aani S, Bonny T and Hasan SW (2019) Can machine language and artificial intelligence revolutionize process automation for water treatment and desalination? *Desalination* **458**: 84–96, 10.1016/j.desal.2019.02.005.

Alam G, Ihsanullah I and Naushad M (2022) Applications of artificial intelligence in water treatment for optimization and automation of adsorption processes: recent advances and prospects. *Chemical Engineering Journal* **427**: 130011, 10.1016/j.cej.2021.130011.

Al-Mutairi N, Al-Bloushi M and Saeed T (2019) Removal of petroleum hydrocarbons from contaminated water using activated carbon and petroleum hydrocarbon-degrading bacteria. *Journal of Environmental Science and Health, Part B* **54**(1): 58–65.

Amin MM, Rahimi A and Bina, B (2021) Performance evaluation of a scoria-compost biofilter treating xylene vapors. *Journal of Environmental Health Science and Engineering* **12**(1): 1–10.

Andres Y, Dumont E and Le Cloirec P (2006) Wood bark as packing material in a biofilter used for air treatment. *Environmental Technology* **27**(12): 1297–1301, 10.1080/09593332708618744.

Anet B, Couriol C and Lendormi T (2013) Characterization and selection of packing materials for biofiltration of rendering odourous emissions. *Water, Air, & Soil Pollution* **224**(7): 1–3.

Azimi A, Azari A and Rezakazemi M (2017) Removal of heavy metals from industrial wastewaters: a review. *ChemBioEng Reviews* **4**(1): 37–59.

Badilla DB, Gostomski PA and Dalida ML (2010) Influence of water content on biofiltration performance. *ASEAN Journal of Chemical Engineering* **10**(2): 31–39.

Baltrėnas P and Baltrėnaitė E (2020) Natural and inoculated microorganisms as important component for sustainability of biofiltration system. In Sustainable Environmental Protection Technologies 2020 (Baltrėnas P and Baltrėnaitė E (eds)). Springer, Cham, Switzerland, pp. 433–482.

Barzgar S, Hettiaratchi JP and Pearse L (2017) Inhibitory effects of acidic pH and confounding effects of moisture content on methane biofiltration. *Bioresource Technology* **245**: 633–640, 10.1016/j. biortech.2017.08.188.

Bello AO, Tawabini BS and Khalil AB (2018) Phytoremediation of cadmium-, lead-and nickel-contaminated water by Phragmites australis in hydroponic systems. *Ecological Engineering* **120**: 126–133, 10.1016/j.ecoleng.2018.05.035.

Bennicelli R, Stępniewska Z and Banach A (2004) The ability of azolla caroliniana to remove heavy metals (Hg (II), Cr (III), Cr (VI)) from municipal waste water. *Chemosphere* **55**(1): 141–146.

Bezirgiannidis A, Plesia-Efstathopoulou A and Ntougias S (2019) Combined chemically enhanced primary sedimentation and biofiltration process for low cost municipal wastewater treatment. *Journal of Environmental Science and Health* **54**(12): 1227–1232, 10.1080/10934529.2019.1633842.

Blecken GT, Marsalek J and Viklander M (2011) Laboratory study of stormwater biofiltration in low temperatures: total and dissolved metal removals and fates. *Water, Air, & Soil Pollution*. **Jul;219**(1–4): 303–317.

Boehm AB, Bell CD and Fitzgerald NJ (2020) Biocharaugmented biofilters to improve pollutant removal from stormwater–can they improve receiving water quality? *Environmental Science: Water Research & Technology* **6**(6): 1520–1537.

Bohn H (1992) Consider biofiltration for decontaminating gases. *Chemical Engineering Progress* **88**: 34–40.

Bohn HL and Bohn KH (1999) Moisture in biofilters. *Environmental Progress* **18**(3): 156–161.

Bressani-Ribeiro T, Almeida PG and Volcke EI (2018) Trickling filters following anaerobic sewage treatment: state of the art and perspectives. *Environmental Science: Water Research & Technology* **4**(11): 1721–1738.

Briffa J, Sinagra E and Blundell R (2020) Heavy metal pollution in the environment and their toxicological effects on humans. *Heliyon* **6**(9): e04691. DOI:10.1016/j.heliyon.2020.e04691.

Cassidy MB, Lee H and Trevors JT (1996) Environmental applications of immobilized microbial cells: a review. *Journal of Industrial Microbiology and Biotechnology* **16**(2): 79–101, 10.1007/BF01570068.

Chang J, Mei J and Jia W (2019) Treatment of heavily polluted river water by tidal-operated biofilters with organic/inorganic media: evaluation of performance and bacterial community. *Bioresource Technology* **279**: 34–42, 10.1016/j.biortech.2019.01.060.

Chen W, Su Y and Peng J (2011) Efficient wastewater treatment by membranes through constructing tunable antifouling membrane surfaces. *Environmental Science & Technology* **45**(15): 6545–6552, 10.1021/es200994n.

Choi J, Kotay SM and Goel R (2011) Bacteriophage-based biocontrol of biological sludge bulking in wastewater. *Bioengineered Bugs* **2**(4): 214–217, 10.4161/bbug.2.4.16211.

Cohen Y (2001) Biofiltration–the treatment of fluids by microorganisms immobilized into the filter bedding material: a review. *Bioresource Technology* **77**(3): 257–274, 10.1016/S0960-8524(00)00074-2.

Darlington AB, Dat JF and Dixon MA (2001) The biofiltration of indoor air: air flux and temperature influences the removal of toluene, ethylbenzene, and xylene. *Environmental Science & Technology* **35**(1): 240–246, 10.1021/es0010507.

Delhoménie MC and Heitz M (2005) Biofiltration of air: a review. *Critical Reviews in Biotechnology* **25**(1–2): 53–72, 10.1080/07388550590935814

Devda V, Chaudhary K and Varjani S (2021) Recovery of resources from industrial wastewater employing electrochemical technologies: status, advancements, and perspectives. *Bioengineered* **12**(1): 4697–4718, 10.1080/21655979.2021.1946631.

Devinny JS, Deshusses MA and Webster TS (2017) *Biofiltration for Air Pollution Control*. CRC Press, Boca Raton, FL, USA **5**: 22–35.

Dhankhar R and Hooda A (2011) Fungal biosorption–an alternative to meet the challenges of heavy metal pollution in aqueous solutions. *Environmental Technology* **32**(5): 467–491, 10.1080/09593330.2011.572922.

Engwa GA, Ferdinand PU and Nwalo FN (2019) Mechanism and health effects of heavy metal toxicity in humans. In *Poisoning in the Modern World: New Tricks for an Old Dog 10* (Karcioglu O and Arslan B (eds.)). IntechOpen, London, UK.

Faisal AA, Nassir ZS and Naji LA (2020) A sustainable approach to utilize olive pips for the sorption of lead ions: numerical modeling with aid of artificial neural network. *Sustainable Chemistry and Pharmacy* **15**:100220. 10.1016/j.scp.2020.100220

Feng W, Hatt BE and McCarthy DT (2012) Biofilters for stormwater harvesting: understanding the treatment performance of key metals that pose a risk for water use. *Environmental Science & Technology*. **May 1; 46**(9): 5100–5108, 10.1021/es203396f.

Fulazzaky MA, Talaiekhozani A and Ponraj M (2014) Biofiltration process as an ideal approach to remove pollutants from polluted air. *Desalination and Water Treatment* **52**(19–21): 3600–3615, 10.1080/19443994.2013.854102.

Gandu B, Palanivel S and Juntupally S (2021) Removal of NH3 and H2S from odor causing tannery emissions using biological filters: impact of operational strategy on the performance of a pilot-scale bio-filter. *Journal of Environmental Science and Health* **15**: 1.

Gautam RK, Sharma SK and Mahiya S (2014) Contamination of heavy metals in aquatic media: transport, toxicity and technologies for remediation. *Heavy Metals in Water: Presence, Removal and Safety* 1–24, 10.1039/9781782620174-00001.

Giri BS, Gun S and Pandey S (2020) Reusability of brilliant green dye contaminated wastewater using corncob biochar and brevibacillus parabrevis: hybrid treatment and kinetic studies. *Bioengineered* **11**(1): 743–758, 10.1080/21655979.2020.1788353.

Gohari RJ, Halakoo E and Lau WJ (2014) Novel polyethersulfone (PES)/hydrous manganese dioxide (HMO) mixed matrix membranes with improved anti-fouling properties for oily wastewater treatment process. *RSC Adv* **4**(34): 17587–17596.

Han MF, Wang C and Yang NY (2020) Performance enhancement of a biofilter with pH buffering and filter bed supporting material in removal of chlorobenzene. *Chemosphere* **251**: 126358, 10.1016/j.chemosphere.2020126358.

Huan C, Fang J and Tong X (2021) Simultaneous elimination of H2S and NH3 in a biotrickling filter packed with polyhedral spheres and best efficiency in compost deodorization. *Journal of Cleaner Production* **284**: 124708, 10.1016/j.jclepro.2020.124708.

Joshi DR, Sharma CM and Ghimire A (2019) Removal of arsenic from contaminated water using iron oxide-coated sand and bacteria. *Journal of Environmental Science and Health, Part B* **54**(1): 34–41, 10.1080/03601234.2018.1540894

Joshi DR, Sharma CM and Ghimire A (2019) Removal of fluoride from contaminated water using activated alumina and fluoride-removing bacteria. *Journal of Environmental Science and Health, Part B* **54**(1): 34–41.

Kang CH, Kwon YJ and So JS (2016) Bioremediation of heavy metals by using bacterial mixtures. *Ecological Engineering* **89**: 64–69. 10.1016/j.ecoleng.2016.01.023

Khairnar K, Chandekar R and Nair A (2016) Novel application of bacteriophage for controlling foaming in wastewater treatment plant-an eco-friendly approach. *Bioengineered* **7**(1): 46–49, 10.1080/21655979.2015.1134066.

Khan MA, Singh RP and Singh R (2017) Bioremediation of zinc-contaminated soil using compost and microorganisms. *Journal of Environmental Science and Health, Part B* **52**(1): 33–40, 10.1080/03601234.2016.1239294.

Khan MJ, Rai A and Ahirwar A (2021) Diatom microalgae as smart nanocontainers for biosensing wastewater pollutants: recent trends and innovations. *Bioengineered* **12**(2): 9531–9549, 10.1080/21655979.2021.1996748.

Khoramfar S, Jones KD and Ghobadi J (2020) Effect of surfactants at natural and acidic pH on microbial activity and biodegradation of mixture of benzene and oxylene. *Chemosphere* **260**: 127471, 10.1016/j.chemosphere.2020.127471.

Kiedrzyńska E, Urbaniak M and Kiedrzyński M (2017) The use of a hybrid sequential biofiltration system for the improvement of nutrient removal and PCB control in municipal wastewater. *Scientific Reports* **7**(1): 1–4, 10.1038/s41598-017-05555-y.

Kim NJ, Hirai M and Shoda M (2000) Comparison of organic and inorganic packing materials in the removal of ammonia gas in biofilters. *Journal of Hazardous Materials* **72**(1): 77–90, 10.1016/S0304-3894(99)00160-0.

Kiran MG, Pakshirajan K and Das G (2017) Heavy metal removal from multicomponent system by sulfate reducing bacteria: mechanism and cell surface characterization. *Journal of Hazardous Materials* **324**: 62–70. 10.1016/j.jhazmat.2015.12.042

Kumar P, Kumar V and Sharma S (2018a) Removal of lead from contaminated water using peat moss and compost-based biofilter. *Journal of Hazardous Materials* **358**: 242–248, 10.1016/j.jhazmat.2018.06.045

Kumar P, Kumar V and Sharma S (2018b) Removal of xylene from contaminated air using sawdust and microorganisms-based biofilter. *Journal of Environmental Science and Health, Part B* **53**(1): 34–41, 10.1080/03601234.2017.1393134

Kumar P, Kumar V and Sharma S (2019) Removal of chromium and ammonia from contaminated water using sawdust and chromium-reducing bacteria. *Journal of Environmental Science and Health, Part B* **54**(1): 66–73.

Kumar P, Kumar V and Sharma S (2020) Removal of COD and color from textile industry wastewater using fungal-based biofilter. *Journal of Environmental Science and Health, Part B* **55**(1): 42–49.

Kumar TP, Rahul MA, Chandrajit B (2011) Biofiltration of volatile organic compounds (VOCs): an overview. *Research Journal of Chemical Sciences*, **1**(8): 83–92.

Leson G and Winer AM (1991) Biofiltration: an innovative air pollution control technology for VOC emissions. *Journal of the Air & Waste Management Association* **41**(8): 1045–1054, 10.1080/10473289.1991.10466898.

Li X, Dong S and Yao Y (2016) Inoculation of bacteria for the bioremediation of heavy metals contaminated soil by agrocybe aegerita. *RSC Advances* **6**(70): 65816–65824.

Li X, Li Y and Li X (2019a) Removal of benzene from contaminated air using activated carbon and bacteria-based biofilter. *Environmental Science and Pollution Research* **26**(10): 10419–10426, 10.1007/s11356-019-04652-2

Li X, Li Y and Li X (2019b) Removal of cadmium from wastewater using activated sludge and zeolite-based biofilter. *Environmental Science and Pollution Research* **26**(10): 10411–10418, 10.1007/s11356-019-04651-3

Li X, Li Y and Li X (2020a) Removal of organic compounds from pharmaceutical industry wastewater using granular activated carbon and microorganisms. *Environmental Science and Pollution Research* **27**(10): 12385–12392.

Li Y, Wang Y and Wan D (2020b) Pilot-scale application of sulfur-limestone autotrophic denitrification biofilter for municipal tailwater treatment: performance and microbial community structure. *Bioresource Technology* **300**: 122682, 10.1016/j.biortech.2019.122682.

Liu J, Sun J and Lu C (2021a) Bioaerosol emissions of pilotscale low-pH and neutral-pH biofilters treating odors from landfill leachate: characteristics and impact factors. *Waste Management* **128**: 64–72, 10.1016/j.wasman.2021.04.040.

Liu J, Sun J and Lu C (2021b) Performance and substance transformation of low-pH and neutral-pH biofilters treating complex gases containing hydrogen sulfide, ammonia, acetic acid, and toluene. *Environmental Science and Pollution Research* **13**: 1–2.

Liu Y, Li X and Li X (2019) Removal of chlorinated compounds from pulp and paper industry wastewater using anaerobic granular sludge and microorganisms. *Environmental Science and Pollution Research* **26**(10): 10435–10442.

Loukidou MX, Matis KA and Zouboulis AI (2003) Removal of As (V) from wastewaters by chemically modified fungal biomass. *Water Research* **37**(18): 4544–4552, 10.1016/S0043-1354(03)00415-9.

Madkour LH (2020) Ecotoxicology of environmental heavy metal ions and free radicals on macromolecule cell organisms. *Bioengineering Applications of Carbon Nanostructures*, Springer Science and Business, Berlin/Heidelberg, Germany, pp. 1–46, 10.1007/978-3-030-37297-2_1.

Maestre JP, Gamisans X and Gabriel D (2007) Fungal biofilters for toluene biofiltration: evaluation of the performance with four packing materials under different operating conditions. *Chemosphere* **67**(4): 684–692, 10.1016/j.chemosphere.2006.11.004.

Majumder S, Raghuvanshi S and Gupta S (2016) Application of a hybrid biofilter column for the removal of Cr (VI) from aqueous solution using an indigenous bacterial strain Pseudomonas taiwanensis. *Bioremediation Journal* **20**(1): 10–23, 10.1080/10889868.2015.1113923.

Malakar S, Saha PD and Baskaran D (2018) Microbial biofilter for toluene removal: performance evaluation, transient operation and theoretical prediction of elimination capacity. *Sustainable Environment Research* **28**(3): 121–127, 10.1016/j.serj.2017.12.001.

Malhautier L, Khammar N and Bayle S (2005) Biofiltration of volatile organic compounds. *Applied Microbiology and Biotechnology* **68**(1): 16–22.

Márquez P, Herruzo-Ruiz AM and Siles JA (2021) Influence of packing material on the biofiltration of butyric acid: a comparative study from a physico-chemical, olfactometric and microbiological perspective. *Journal of Environmental Management* **294**: 113044, 10.1016/j.jenvman.2021.113044.

Misra CS, Appukuttan D and Kantamreddi VS (2012) Recombinant D. radiodurans cells for bioremediation of heavy metals from acidic/neutral aqueous wastes. *Bioengineered Bugs* **3**(1): 44–48, 10.4161/bbug.3.1.18878.

Mohseni M and Allen DG (1999) Transient performance of biofilters treating mixtures of hydrophilic and hydrophobic volatile organic compounds. *Journal of the Air & Waste Management Association (JA&WMA)* **49**(12): 1434–1441, 10.1080/10473289.1999.10463980.

Moll DM, Summers RS and Fonseca AC (1999) Impact of temperature on drinking water biofilter performance and microbial community structure. *Environmental Science & Technology* **33**(14): 2377–2382, 10.1021/es9900757.

Morales M, Hernández S and Cornabé T (2003) Effect of drying on biofilter performance: modeling and experimental approach. *Environmental Science & Technology* **37**(5): 985–992, 10.1021/es025970w.

Morgan-Sagastume JM, Noyola A and Revah S (2003) Changes in physical properties of a compost biofilter treating hydrogen sulfide. *Journal of the Air & Waste Management Association (JA&WMA)* **53**(8): 1011–1021, 10.1080/10473289.2003.10466249.

Ogidi OI (2022) Impact of pharmaceutical compounds on the microbial ecology of surface water resources. In *Biodiversity in Africa: Potentials, Threats and Conservation*. Springer Nature, Singapore, 10.1007/978-981-19-3326-4_12

Ogidi OI (2024) Strategies of sustainable management of water resources in the Global South. In *Water Crisis and Sustainable Management in the Global South*. Springer Nature, Singapore.

Ogidi OI and Akpan UM (2022) Aquatic biodiversity loss: Impacts of pollution and anthropogenic activities and strategies for conservation. In *Biodiversity in Africa: Potentials, Threats and Conservation*. Springer Nature, Singapore. https://doi.org/10.1007/978-981-19-3326-4_16

Ogidi OI and Izah SC (2024) Water contamination by municipal solid wastes and sustainable management strategies. In *Water Crisis and Sustainable Management in the Global South*. Springer Nature, Singapore.

Ogidi OI and Richard-Nwachukwu N (2024) Microbe assisted remediation of pesticide residues from soil and water. In *Eco-Restoration of the Environment*. CRC Press, Boca Raton, FL, USA.

Ogidi OI, Angaye TCN and Carbom HE (2024) Microorganisms in the monitoring of terrestrial and aquatic ecosystems: applications, challenges and future prospects. In *Biomonitoring of Pollutants in the Global South*. Springer Nature, Singapore.

Ogidi OI, Okereke JN and Anyalogbu EA (2019a) Potentials of consortium of banana, orange and potatos peels as bio-sorbents in the purification of paint industrial wastewater. *Elixir International Journal* **137**: 53979–53987.

Ogidi OI, Okereke JN and Anyalogbu EA (2019b) Effectiveness of composites of corn cobs, coconut husks and breadfruit peels in purifying selected paint effluents. *International Journal of Research and Scientific Innovation (IJRSI)* **6**(10): 236–248.

Okereke JN, Ogidi OI and Obasi KO (2016a) Biosorption of Cd and Pb in textile effluent using *Musa sapientum* (banana) peels. *The International Journal of Science & Technoledge* **4**(6): 31–37.

Okereke JN, Ogidi OI and Obasi KO (2016b) Removal of inorganic anions in brewery effluent using banana peels. *International Journal of Advanced Research in Biological Sciences* **3**(6): 192–197.

Okereke JN, Ogidi OI and Obasi KO (2016c) Environmental and health impact of industrial wastewater effluents in Nigeria – a review. *International Journal of Advanced Research in Biological Science* **3**(6): 55–67.

Okore CC, Ogbulie TE, Ejiogu C *et al.* (2021) Removal of Pb (II) from synthetic solution using activated carbons produced from agricultural wastes. *Global Scientific Journal* **9**(4): 2697–2707. www.globalscientificjournal.com

Padhan D, Rout PP and Kundu R (2021) Bioremediation of heavy metals and other toxic substances by microorganisms. *Soil Bioremediation and Sustainable Technology* **29**: 285–329.

Padhi SK and Gokhale S (2016) Benzene control from waste gas streams with a sponge-medium based rotating biological contactor. *International Biodeterioration and Biodegradation* **109**: 96–103, 10.1016/j.ibiod.2016.01.007.

Park JK and Chang HN (2000) Microencapsulation of microbial cells. *Biotechnology Advances* **18**(4): 303–319, 10.1016/S0734-9750(00)00040-9.

Peng C, Huang H and Gao Y (2019) A novel start-up strategy for mixotrophic denitrification biofilters by rhamnolipid and its performance on denitrification of low C/N wastewater. *Chemosphere* **239**: 124726, 10.1016/j.chemosphere.2019.124726.

Plaisance H, Vignau-Laulhere J and Mocho P (2017) Volatile organic compounds concentrations during the construction process in newly-built timber-frame houses: source identification and emission kinetics. *Environmental Science: Processes & Impacts* **19**(5): 696–710.

Rafida AI and Sallis PJ (2005) Removal of heavy metals from rainwater in vertical flow biofilters conditioned with sulphate reducing bacteria (SRB). In *XII International Rainwater Catchment Systems Conference 2005*, New Delhi, India.

Rajendran S, Khan MA and Singh RP (2019a) Bioremediation of copper-contaminated soil using coconut coir and fungi. *Journal of Environmental Science and Health, Part B* **54**(1): 42–49, 10.1080/03601234.2018.1540895.

Rajendran S, Khan MA and Singh RP (2019b) Bioremediation of heavy metals from contaminated water using activated carbon and metal-removing bacteria. *Journal of Environmental Science and Health, Part B* **54**(1): 50–57.

Rajendran S, Khan MA and Singh RP (2019c) Bioremediation of styrene-contaminated air using coconut coir and bacteria-based biofilter. *Journal of Environmental Science and Health, Part B* **54**(1): 50–57, 10.1080/03601234.2018.1540896

Ralebitso-Senior TK, Senior E and Di Felice R (2012) Waste gas biofiltration: advances and limitations of current approaches in microbiology. *Environmental Science & Technology* **46**(16): 8542–8573.

Rani M and Shanker U (2020) Environmental nanotechnology approaches for the remediation of contaminants. In *Bioremediation technology* (Fulekar MH and Pathak B (eds)). CRC Press, Boca Raton, FL, USA, pp. 311–348.

Rathore S, Desai PM and Liew CV (2013) Microencapsulation of microbial cells. *Journal of Food Engineering* **116**(2): 369–381, 10.1016/j.jfoodeng.2012.12.022.

Rene ER, Kumar P and Veiga MC (2018) Biodegradation of toluene using compost and perlite-based biofilter. *Journal of Hazardous Materials* **358**: 249–256, 10.1016/j.jhazmat.2018.06.046.

Repečkienė J, Švedienė J and Paškevičius A (2015) Succession of microorganisms in a plate-type air treatment biofilter during filtration of various volatile compounds. *Environmental Technology* **36**(7): 881–889, 10.1080/09593330.2014.965227.

Rezooqi AM, Atiyah AH, Mugheir HS (2021) Bioaccumulation of (Ni, Cd, Pb, Cr, Hg, and Co) from Al-rustomia wastewater using A. filiculoids. *Journal of Physics: Conference Series* **1853**(1): 012019.

Rima J, Aoun E and Hanna K (2004) Use of beetroot fibers to clean water contaminated with heavy metals, to soften hard water, and to desalinate seawater. *Toxicological and Environmental Chemistry* **86**(2): 105–116, 10.1080/02772240420002211113.

Ryu HW, Cho KS and Chung DJ (2010) Relationships between biomass, pressure drop, and performance in a polyurethane biofilter. *Bioresource Technology* **101**(6): 1745–1751, 10.1016/j.biortech.2009.10.018.

Sakuma T, Hattori T and Deshusses MA (2006) Comparison of different packing materials for the biofiltration of air toxics. *Journal of the Air & Waste Management Association* **56**(11): 1567–1575, 10.1080/10473289.2006.10464564.

Schroeder ED (2002) Trends in application of gas-phase bioreactors. *Reviews in Environmental Science and Biotechnology* **1**(1): 65–74.

Serial GA, Smith FL and Suidan MT (1995) Evaluation of trickle bed biofilter media for toluene removal. *Journal of the Air & Waste Management Association* **45**(10): 801–810, 10.1080/10473289.1995.10467410.

Shindhal T, Rakholiya P and Varjani S (2021) A critical review on advances in the practices and perspectives for the treatment of dye industry wastewater. *Bioengineered Bugs* **12**(1): 70–87, 10.1080/21655979.2020.1863034.

Shivajirao PA (2012) Treatment of distillery wastewater using membrane technologies. *International Journal of Advanced Engineering Research and Studies* **1**(3): 275–283.

Singh R, Kumar V and Singh RP (2020a) Removal of chromium from contaminated water using sawdust and bacteria-based biofilter. *Journal of Water Process Engineering* **33**: 101944, 10.1016/j.jwpe.2020.101944.

Singh R, Kumar V and Singh RP (2020b) Removal of ethylbenzene from contaminated air using peat moss and fungi-based biofilter. *Journal of Water Process Engineering* **33**: 101945, 10.1016/j.jwpe.2020.101945.

Soccol CR, Woiciechowski AL, Vandenberghe LP (2003) Biofiltration: an emerging technology. *Indian Journal of Biotechnology* **2**: 396–410.

Srivastava AK, Singh RK and Singh D (2021) Microbe-based bioreactor system for bioremediation of organic contaminants: present and future perspective. In *Microbe Mediated Remediation of Environmental Contaminants* (Kumar A, Singh VK, Singh, P and Sing D (eds)). Woodhead, Sawston, UK, pp. 241–253.

Swanson WJ and Loehr RC (1997) Biofiltration: fundamentals, design and operations principles, and applications. *Journal of Environmental Engineering* **123**(6): 538–546.

Tel-Or E, Sela M and Ravid S (1997) Biofiltration of heavy metals by the aquatic fern azolla. In *Modern Agriculture and the Environment*. Springer, Dordrecht, Germany, pp. 431–442.

Tisserand R and Young R (2014) 6–The respiratory system. In *Essential Oil Safety* (Tisserand R and Young R (eds.)) 2nd edn.), Churchill Livingstone, St. Louis, USA, pp. 99–110.

Wani AH, Branion RM and Lau AK (1997) Biofiltration: a promising and cost-effective control technology for odors, VOCs and air toxics. *Journal of Environmental Science & Health Part A* **32**(7): 2027–2055.

Weckhuysen B, Vriens L and Verachtert H (1993) The effect of nutrient supplementation on the biofiltration removal of butanal in contaminated air. *Applied Microbiology and Biotechnology* **39**(3): 395–399.

Yang Z, Li J and Liu J (2019) Evaluation of a pilot-scale biotrickling filter as a VOCs control technology for the chemical fibre wastewater treatment plant *Journal of Environmental Management* **246**: 71–76. 10.1016/j.jenvman.2019.05.102

Zamir SM, Halladj R and Nasernejad B (2011) Removal of toluene vapors using a fungal biofilter under intermittent loading. *Process Safety and Environmental Protection* **89**(1): 8–14. 10.1016/j.psep.2010.10.001

Zhang Y, Li X and Li X (2020a) Removal of acetone from contaminated air using compost and zeolite-based biofilter. *Environmental Science and Pollution Research* **27**(10): 12369–12376, 10.1007/s11356-020-08216-2.

Zhang Y, Li X and Li X (2020b) Removal of mercury from contaminated water using sulfur-reducing bacteria and limestone-based biofilter. *Environmental Science and Pollution Research* **27**(10): 12361–12368, 10.1007/s11356-020-08215-3.

Zhang Y, Liss SN and Allen DG (2007) Effect of methanol on pH and stability of inorganic biofilters treating dimethyl sulfide. *Environmental Science & Technology* **41** (10): 3752–3757, 10.1021/es062756p.

Zheng X and Nguyen H (2022) A novel artificial intelligent model for predicting water treatment efficiency of various biochar systems based on artificial neural network and queuing search algorithm. *Chemosphere* **287**: 132251, 10.1016/j.chemosphere.2021.132251.

Microbial Processes in Environmental Engineering: From Theory to Practice

Rajneesh Kumar, Ram Sharan Singh and Maulin P Shah
ISBN 978-1-83708-327-5
https://doi.org/10.1108/978-1-83708-326-820261004

Chapter 4
Heavy metal removal from contaminated water

Pooja Dhiman[1] and Anita Rani[2]

[1]*Department of Chemistry, School of Basic and Applied Sciences, Maharaja Agrasen University, Solan, (HP), India*
[2]*Department of Chemistry, School of Basic and Applied Sciences, IEC University, Solan, (HP), India*

Abstract

After an era of revolution in the manufacturing industry, contamination of water has posed a threat to the health of its consumers. The effluent discharged from textile, leather, mining, pharmaceutical, cosmetic, paint and personal care products industries contains chromium, lead, cadmium, copper, arsenic, nickel, zinc, mercury, iron and manganese. The consumption of heavy metal-contaminated water leads to acute and chronic toxicity, neurological disorders, reproductive ailments, birth defects, organ damage and cancer. The investigated heavy metal removal techniques from contaminated water include flotation, chemical precipitation, adsorption, membrane filtration, coagulation and flocculation, ion exchange and electrochemical separation. The working mechanisms, advantages and limitations of the removal techniques are discussed in this chapter.

Keywords: Heavy metals; removal; contaminated; adsorption.

4.1. Introduction

Water is fundamental to sustain life on this planet, and easy access to clean drinking water is essential, not only for humankind, but for the whole ecosystem. No doubt, during the last few decades, water quality has degraded due to the gradual increase in population, urbanisation, industrialisation and incautious use of naturally available resources (Talema, 2023).

Poly- and per-fluoro-alkyl substances, heavy metals, organic matter, nutrients, dyes, plastics, pharmaceutical and personal care products, insecticides and pesticides, radioactive residues, nanoparticles and pathogens are some of the most commonly detected pollutants of water, causing serious concern (Bak *et al.*, 2024).

Heavy metal ions are one of the most discharged pollutants and, for this reason, their detection, occurrence, distribution and toxicological effects are a matter for concern. Heavy metals are the elements expressing atomic density > 4 g/cm^3, including cadmium (Cd), copper (Cu), lead (Pb), zinc (Zn), arsenic (As), mercury (Hg), chromium (Cr), silver (Ag), platinum (Pt), manganese (Mn), cobalt (Co), nickel (Ni) and iron (Fe) (Roy *et al.*, 2024).

These metallic toxic elements get discharged into the water bodies from diverse natural and anthropogenic origins. In numerous regions across the globe, the mean concentrations of Mn, Cr, Co, Fe, As, Cd and Ni detected in surface water samples have been recorded at higher than maximum permitted limits for potable water. These matrices express a serious concern for public health (Hasan *et al.*, 2021). Heavy metals are non-biodegradable, tending to bioaccumulate, remaining longer and in higher concentrations in living organisms. They are persistent in nature and, directly and indirectly, impact various organisms due to their biomagnification (Ganesh *et al.*, 2022).

Numerous heavy metal ions have a proven record for being toxic and carcinogenic. Several heavy metals in small concentration, bioaccumulated over a duration of time, start impacting and damaging multiple organs, affecting kidneys, lungs, prostate, liver, stomach, skin and esophagus. There are published documents that provide evidence of neuro-degenerative disorders as well as diseases including Parkinson's and Alzheimer's diseases (Doroszkiewicz *et al.*, 2023). Heavy metals can impact zooplankton, phytoplankton and aquatic organisms such as fish, bioaccumulating in numerous organs consequently leading to oxidative damage, endocrine disruption and weakening of the immune system, which can negatively impact the survival and growth of the living organism (Anetor *et al.*, 2022).

Because of revolutionary developments in the discharged effluent from numerous sources (batteries, pesticides, plating and electroplating industry, rayon industry, mining industry, tanning industry, metal rinse processes, textile industry, metal smelting, paper manufacturing, fluidised bed bioreactors, electrolysis applications and petrochemicals), there is a significant presence of heavy metal contamination in wastewater (Zhang *et al.*, 2023b). Wastewater contaminating the environment with heavy metals threatens the health of human beings and the all the biotic components of affected ecosystems. Being non-biodegradable in nature, heavy metals can potentially be carcinogenic; therefore, the presence of carcinogenic heavy metals in waste water in concentrations exceeding acceptable limits results in severe health problems to all encountered biotic components of the biosphere (Jagaba *et al.*, 2024). The most commonly detected heavy metals are zinc (Zn), lead (Pb), nickel (Ni), mercury (Hg), copper (Cu), cadmium (Cd), arsenic (As) and chromium (Cr). The presence of these heavy metals even in traces is hazardous (Singh *et al.*, 2023a).

Several practical strategies have been designed and explored in order to remove hazardous heavy metals from polluted water. Conventional water treatment techniques including electrochemical removal, ion exchange and chemical precipitation have demonstrated several limitations, such as generation of hazardous sludge, inefficient removal and high energy requirements (Ahmed *et al.*, 2022). Other disadvantages pertain to impacts of the development of industrial and infrastructure assets, including their operation and associated maintenance, which can result in inefficient deployments of such technological development, particularly in developing nations. That is why there is the utmost requirement for the development of efficient, effective and eco-friendly outcomes. This may include biological or nanotechnology approaches and adsorption-based strategies deploying polymeric materials, magnetic adsorbents or composite adsorbents for the removal of heavy metals from contaminated water (Aziz *et al.*, 2023).

This chapter describes the primary sources of heavy metal-contaminated water, potential health hazards caused by their presence, and the conventional and advanced water treatment methods for their effective removal. It also elaborates on the factors affecting the removal of these toxic heavy metals.

4.2. Sources of heavy metal contamination

According to previously published reports, around 40% of the water bodies in terms of rivers and lakes of the planet have been contaminated by the presence of hazardous heavy metals. The origin of contamination by toxic heavy metals can either be natural or human-made (anthropogenic) (Madhav *et al.*, 2020).

The reported natural cause of contamination of water by heavy metals includes volcanic eruptions as well as weathering of metal-rich rocks naturally occurring in the environment. Eruptions from volcanoes cause giant, sporadic discharge, explosive volcanic outflow or gradual prolonged outrush at slower rates, with inclusion of geothermal pursuits and degassing (Yang *et al.*, 2022a).

The studied origins of release of heavy metals from anthropogenic activities includes domestic (discharge of liquid detergents and garbage), industry (combustion of fossil fuels and metal processing) and the release of pesticides into natural water bodies. Sources of heavy metals are depicted in Figure 4.1.

The mining industry, among other industrial sectors, has made remarkable contributions to the economies of both developed and developing nations (Yang *et al.*, 2022a). The extraction of metals and their transportation may lead to discharge of heavy metals, often after their dissolution in water bodies in ionic form or as sediment (Miranda *et al.*, 2022). Dissolved metallic salts accumulate through seepage action, entering groundwater and causing a scarcity of usable water, which leads to the inability to grow agricultural crops and subsequent erosion of soil with consequential health impacts on humans and animals (Mukherjee *et al.*, 2020).

In the last decade, industrialisation, urbanisation and other anthropogenic activities have been factors causing elevated pollution of water by heavy metals. The documented concentration of arsenic (As) in drinking water was 107 μg/L in India (Panda *et al.*, 2023). The water samples collected from surroundings of lead–zinc mining regions of Abakaliki, Nigeria, when analysed exhibited surprisingly greater concentrations of Pb^{2+}, Mn^{2+} Ni^{2+}, Cr^{3+}, Ag^{+}, Cd^{2+}, As^{2+}, Hg^{2+}, Zn^{2+}, Co^{2+} and selenium Se^{2+} (Zhang *et al.*, 2023b). A further study identified that around 4.5 million people in Latin America have been exposed to water critically

Figure 4.1 Sources of heavy metals

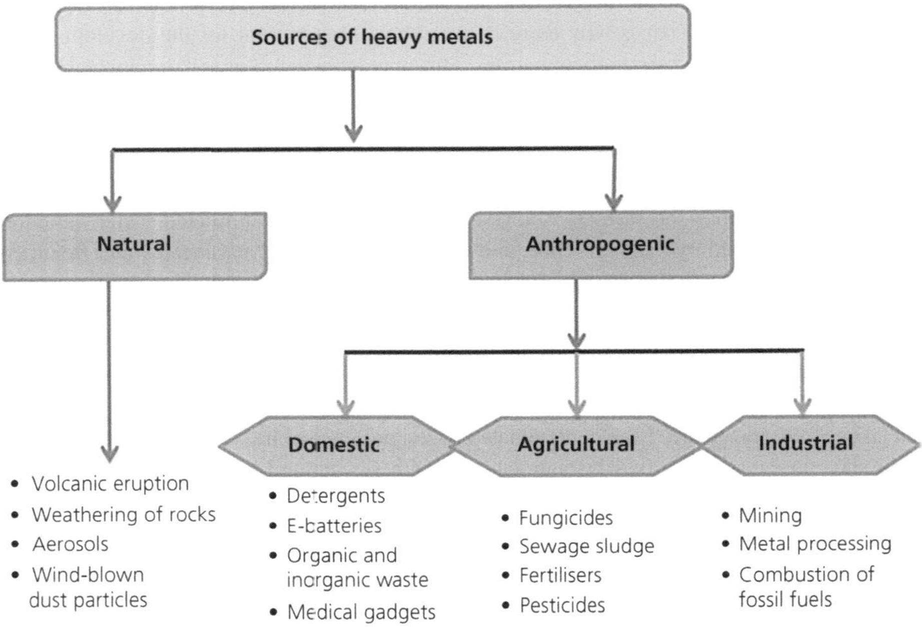

contaminated with arsenic (As) > 50 μg/L, and in some other regions up to 2000 μg/L (Xu *et al.*, 2024b).

The analysed concentrations were observed to be critically higher than the reference value (10 μg/L) as established by the World Health Organization (WHO).

Additionally, approximate 140 million consumers in 50 nations have been drinking water contaminated with arsenic (As) in concentrations greater than the WHO reference value of 10 μg/L (Rahman *et al.*, 2021). From China's Laizhou Bay basin, water samples were analysed after their collection from nine different coastal rivers and expressed elevated quantities of Zn^{2+}, Cu^{2+} and Cd^{2+} (Wang *et al.*, 2020). The highest reported concentrations were greater than the WHO's recommended limit values for potable water.

In the Coyuni river basin, Venezuela, a remarkably high contamination with mercury was recorded due to amalgamation deployed in the mining of artisan gold in the region. From two main rivers of Antalya, Turkey, several heavy metals were detected (Öktüren ASRİ *et al.*, 2024). Particularly in reagent water, strontium (Sr) and aluminium (Al) concentrations were also found exceeding the permissible value of standard specifications.

In order to decrease the toxicological effects of these hazardous heavy metals on living beings, interventions are needed. However, insufficient economic capacity is one of the restricting factors for the development and application of strategies and technologies for the removal of heavy metals (Wang *et al.*, 2022).

4.3. Toxicological effects of heavy metals

The bioaccumulation of heavy metals in their compound molecular form makes their removal from water more challenging. However, elements such as copper (Cu), chromium (Cr), iron (Fe), manganese (Mn), molybdenum (Mo), nickel (Ni), zinc (Zn) and cobalt (Co) in trace quantities are required by the living body for proper metabolic activity (Teschke, 2022). However, excessive amounts of such heavy metals can also cause severe health issues including neurological degeneration, muscular and physical deformities, and alterations in genetic composition (Ohiagu *et al.*, 2022). Prolonged exposure of the body to toxic heavy metals has been associated with Alzheimer's disease, muscular dystrophy and diverse forms of cancer (Althobaiti, 2024).

Other documented toxicological effects of heavy metals include abnormal mental and central nervous system responses, degradation of liver, lungs and blood, along with other impacts on other key organs, including kidneys (Jomova *et al.*, 2024). The type and extent of alteration is greatly governed by the concentration of heavy metals with the potential to interfere with enzymatic processes.

4.4. Policies and regulations regarding heavy metals in water

The cleanliness parameters and the maximum permissible limit for several other contaminating pollutants have been established by environmental regulations. The limits set within environmental regulations are influenced by the nature and characteristic features of a pollutant, including the extent to which they affect human beings geographically, available infrastructure and the technology available for their treatment or removal (Briffa *et al.*, 2020).

Specific policies and regulations have been established by environmentally concerned regulatory bodies for controlling the maximum permissible quantities of heavy metals (Singh *et al.*, 2024). Safety criteria have also been established by advisory and regulatory bodies for a wide variety of water pollutants in different types of ecosystems, including aquatic (Singh *et al.*, 2023b).

As a legal reference framework, in some nations, these policies utilise international standards for protection of the environment. Such regulatory policies are often accompanied by multiple stakeholder and cross-disciplinary committees.

Depending on the geographical region under study, the regional standards are utilised (Haswell *et al.*, 2024). For instance, in the USA, it is quite usual that United States Environmental Protection Agency (USEPA) environmental regulations are used. In Europe, the European Environmental Agency (EEA) has proposed the categorisation of 'water and marine environment', from which two subcategories have been developed: seas and coasts, and freshwater water bodies. Limits for water pollution and heavy metals have been discussed for both of these subcategories.

Other established sets of standards regarding environmental policies and regulations have been proposed by the Organization for Economic Cooperation and Development (OECD). In Asia, the three main aspects of the water cycle (Canton, 2021) have been addressed by the Asian Water Council (AWC), namely (*a*) general utilisation of water, (*b*) water for natural environmental protection and (*c*) water for mitigation of natural hazards (Poudel *et al.*, 2024). Environment-related conservation and protection-related policies and strategies for several water resources have been established by the AWC.

The majority of the policies and regulations have been established on the basis of conclusions of previous analysis, statistical, mathematical and geographical studies (Liu *et al.*, 2023b). However, it has been observed that many nations do not have the necessary human capacity, infrastructure or economic and technological development for their implementation. Such nations have developed national policies and regulations by adopting random criteria copied often arbitrarily from adjacent nations (Srivastav *et al.*, 2023). This action can result in severe issues in implementation of regulations. In some cases, these regulations do not appropriately consider the climatic, ecosystem, socio-economic and geographical nuances of their specific region (Yang *et al.*, 2022b). This situation makes the development of standards, regulations, policies, procedures and best practices to control and regulate the concentrations of aquatic heavy metals challenging at a global scale (Sarker *et al.*, 2022).

A considered approach should include: (*a*) keeping the specific features of the anthropogenic source in consideration and its contribution in the reduction of the concentration of the potential contaminants discharge; (*b*) in situ facilitation to deploy the appropriate approaches permitting the precise identification and quantification of the contaminants in various ecosystems and (*c*) supporting the deployment of economic eco-friendly materials and technical approaches deployed for contaminated water treatment (Singh *et al.*, 2023b).

4.5. Techniques for removal of heavy metals

There are three stages of techniques for water treatment, namely primary, secondary and tertiary water treatment.

Organic matter and suspended solids are removed from polluted water by physical and chemical processes in the primary water treatment process (Pachaiappan *et al.*, 2022). For the removal of heavy metals, primary water treatment techniques include chemical filtration, microfiltration, flocculation and coagulation.

Secondary water treatment processes introduce a variety of microorganisms that efficiently convert organic and inorganic pollutants into non-hazardous substances (Saleh *et al.*, 2022). The efficiency of removal of heavy metals through microbial processes is under investigation, and current research has identified increasing contaminant removal potential.

Electrochemical precipitation (EP), chemical oxidation (CO), crystallisation, distillation, membrane technologies and ion exchange processes are available tertiary water treatment processes (Ko, 2024). However, adsorption and photo-catalysis are advanced water treatment processes available for either removal or mineralisation of pollutants.

Keeping the nature of the pollutants in consideration, after application of the appropriate water treatment processes, polluted water is transformed into treated water. Symbiotic application of the advanced water treatment processes with conventional and non-conventional processes can demonstrate removal efficiencies of up to 99% (Ayach *et al.*, 2024).

4.6. Factors affecting removal of heavy metals

Taking the basic chemical nature of the element into consideration, each heavy metal has a unique set of characteristic features and, hence, behaviour. There is a wide variety of parameters which have potential to influence removal rate and efficiency from the polluted matrix. Temperature, pH, inorganic impurities and organic matter in groundwater, lakes, rivers and domestic wastewater are the principal factors affecting the removal of heavy metals (Akhtar *et al.*, 2021).

4.6.1 Effect of temperature

The different treatment techniques carried out at higher temperature have given better removal efficiencies – for example, at 55°C, treatment plants had removal efficiencies for Cu^{2+}, Ni^{2+}, Cd^{2+} and Pb^{2+} greater than at relatively lower temperatures (Hamoudi *et al.*, 2023).

The pattern of adsorptive removal efficiency of Pb^{2+} and Cu^{2+} by hydrogels has been investigated at 22°C, 30°C, 35°C and 40°C by Zou (2024). The optimum removal of Pb^{2+} was observed at 30°C while that of Cu^{2+} was recorded at 35°C. A slight increase in removal efficiencies was also observed by Sanad *et al.* (2024) at a further increase in temperature up to 40°C. An increase in temperature from 5 to 40°C demonstrated a boosted rise in removal efficiencies of Cr^{6+} by residues of pistachios (Suter *et al.*, 2025). The rationale behind this is the rise in the number of adsorption sites on the residue surface.

An increase in temperature from 10 to 60°C also brought improvements in the biosorptive removal efficiencies of heavy metals on the surface of biomass derived from fungi because of the greater attraction of the active sites available on the biomass-derived substrates (Priya *et al.*, 2022). Contrary to this, a minor rise in temperature also led to a decrease in adsorptive removal of heavy metals and could decrease the removal efficacy of heavy metals. As evidence, in the case of magnetic oxides of iron nanoparticles as adsorbents, a decrease in the removal efficiencies of Cu^{2+}, Ni^{2+} and Al^{3+} was observed (Khan *et al.*, 2020).

A higher temperature results in a greater rate of chemisorption, which subsequently inhibits the adsorptive removal of heavy metals due to the reduced ability of heavy metals to approach new active sites on the adsorbent surface. In some cases it has also been observed that temperature has brought alterations in the dimensions of the pores of nanoparticles resulting in a decrease in adsorptive removal efficiency.

4.6.2 Effect of pH

Alteration in pH conditions of water has the potential to imbalance the ratio of organic as well as inorganic content, later impacting the mitigation efficiencies of the target pollutant. These alterations can be introduced by the addition of chemicals while remediation treatment is ongoing, but they can also result in instabilities from a biological point of view (Ahmed *et al.*, 2021). It has been observed that at both neutral values of pH=7 and under acidic conditions, heavy metals prefer to exist in cationic form and tend to be more actively moving through the water because of their greater solubility. Evidence has been observed that pH values affect not only adsorptive capacity but also the removal efficiency of Ni^{2+} as a target pollutant within natural clay. At higher pH values (i.e. under basic conditions), the adsorptive removal efficiency of Ni^{2+} was observed to be enhanced (Sosun *et al.*, 2022).

4.6.3 Effect of inorganic impurities

The ionic strength of an aqueous solution is calculated from the total concentration of ionic species and the chemical charge borne by these chemical species. This strength has been reported to impact the mechanism of ion exchange between adsorbent surface and the target contaminant, and their potential to eliminate pollutants (Qasem *et al.*, 2021).

With a rise in concentration of NaCl, a little decrease in the adsorptive removal of Pb^{2+}, Zn^{2+}, Cu^{2+} and Cr^{3+} ions on a porous surface of magnetic composite has been recorded (Ribeiro *et al.*, 2019). In yet another investigation, at pH 5.5, the sorption of Ni^{2+} and Cu^{2+} on the surface of bentonite has been observed to be enhanced, with a reported decrease in ionic strength, achieving optimum adsorption at concentration lower than 0.05M NaCl (Merrikhpour *et al.*, 2022). Higher ionic strength has been reported to reduce the electrostatic repulsion between molecules, thereby leading to an increase in the aggregation and agglomeration of surrounding particles, which potentially decreases the availability of the attaching sites in the composite system, ultimately reducing the adsorption (Shrestha *et al.*, 2020). Besides this, chloride ions (Cl^-) can potentially favour the generation of soluble chlorides of heavy metals which are more challenging to eliminate from an aqueous matrix (Zhou *et al.*, 2023). The rationale behind this is that such soluble complexes were observed to have poor affinity for adsorption (Shan *et al.*, 2020)

4.6.4 Effect of organic content

Natural organic matter (NOM) has been observed to be a part of sources of water, particularly in surface waters, due to the natural interactions between the geosphere and biosphere with dead or degraded plant and animal matter coming into contact with a water source (Upadhyay *et al.*, 2023).

Consequently, NOM can potentially interact with heavy metals in a water source thereby impacting mechanisms, including chelation, adsorption and ion exchange. However, it is quite challenging to find out the exact impact on the mitigation of heavy metals because NOM has been observed to impact their bioavailability, mobility, toxicity and reactivity. As evidence, lead (Pb^{2+}) ions have been reported to form complexes with NOM in both natural and wastewater, enabling greater mobility of Pb^{2+}. Thereby, the removal of lead ions became more challenging in conventional water treatment processes including coagulation/flocculation and chemical precipitation (Wang *et al.*, 2021).

Research carried out on nanotubes of titanate embedded on powdered form of activated charcoal demonstrated that the adsorptive removal potential of lead (Pb^{2+}) ions was impacted by weak electrostatic interactions between nanotubes of titanates and dissolved organic matter (Bi *et al.*, 2020). Large quantities of NOM can potentially limit the adsorptive mitigation of heavy metal contamination of water because of the generation of their complex during ongoing ionic competition and simultaneous chelation between NOM and heavy metals (Zhai *et al.*, 2023).

These factors and their interactions have been reported to offer remarkable challenges for the mitigation of heavy metals as water contaminants, consequently affecting their reproducibility which ought to be considered when investigating their other mitigation strategies (Zaimee *et al.*, 2021).

Different conventional and advanced heavy metal removal techniques are illustrated in Figure 4.2.

4.7. Conventional methods of removal

There are several conventional strategies for the removal of heavy metals from contaminated water. The conventional strategies include membrane technologies, coagulation/flocculation, chemical precipitation, ion exchange technologies and electrochemical technologies (Kumar *et al.*, 2023).

4.7.1 Membrane technologies

The past few years have seen significant development in membrane technology. The improvement in membrane technology has not only facilitated the deployment of membranes for filtration but also for the elimination of water contaminants (Vo *et al.*, 2020).

Figure 4.2 Conventional and advanced heavy metal removal techniques

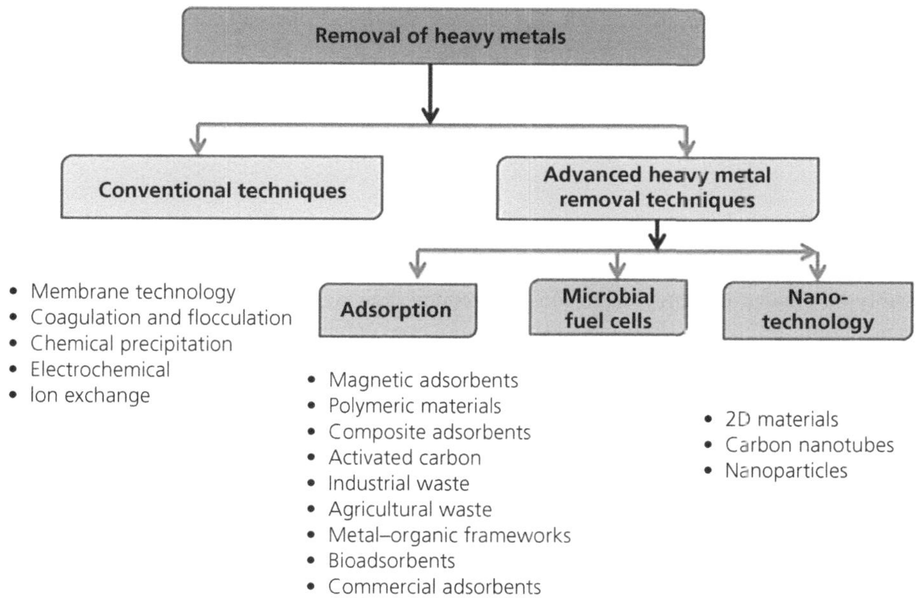

Membrane technologies include microfiltration, nanofiltration, ultrafiltration, reverse osmosis, electrodialysis and other membrane-based technologies (Siddique, 2023).

4.7.1.1 Microfiltration

Microfiltration (MF) utilises microporous membranes to eliminate micron-dimensioned entities, viruses, bacteria, protozoa, metallic contaminants as pollutants and so on from a liquid matrix. The MF process has been described as a low pressure run membrane-based filtration process, with the membrane pores being within a range of 0.1–10 μm (Mondal and Majumder, 2020). MF membranes generally comprise ceramics, silica, zirconia, PVC, alumina, PTFE, polysulfone, polypropylene, polyamides, PVDF, polycarbonate, cellulose esters, cellulose acetate or composite materials.

The industrial uses of MF have broadly been in medicine- and biology-based commercial industries (Yantasee *et al.*, 2010). Uses of MF technology include the removal of suspended particles from rinsing water in the semiconductor industries, the separation and purification of wine and other beverages, and for wastewater treatment. However, for the elimination of heavy metals, deployment of MF is not yet fully effective due to its poor mitigation potential (Cai *et al.*, 2024).

MF has two different modes of operation, namely dead-end and cross flow.

4.7.1.2 Nanofiltration

Nanofiltration (NF) is deployed to filter individual particles with a molecular weight >1000 Da and for the mitigation of solid particles with dimensions of 0.0005–0.007 μm and molecular weights >200 Da (Siddique, 2023). Therefore, the operating range of NF is in between that of ultrafiltration and the reverse osmosis (RO) processes. The membranes of NFs consist of polymer composites of multi-layered thin films of negatively charged chemical species. NF membranes with anti-fouling properties constituted of Ce_7O_{12}/CeO_2 as well as polyethersulfonate (PES) have been reported to be originated by way of phase inversion. NF has also been used to extract Al^{3+}, Fe^{3+}, Co^{2+}, Cu^{2+}, humic acid and Cd^{2+} from polluted water with optimum extraction efficiency in the range of 94–98% (Mansor *et al.*, 2021).

4.7.1.3 Ultrafiltration

Ultrafiltration (UF) operates at relatively lower trans-membrane pressure (TMP). As UF membrane pores can be greater than the sizes of heavy metal ions, addendums can also possibly be bonded to heavy metal ions to broaden their dimensions. That is why micellar enhanced ultrafiltration (MEUF) in addition to polymer enhanced ultrafiltration (PEUF) have been put forward.

MEUF has been recorded to be formed by bonding of UF with surfactant (Lin, 2020). MEUF is facilitated with relatively larger flux as well as with higher selectivity, subsequently resulting in the consumption of less energy and achieving a greater elimination efficiency with limited infrastructural requirements. MEUF has been deployed for wastewater remediation when the concentration of heavy metals is low. In MEUF, a surfactant is blended with wastewater in concentrations greater than critical micellar concentration (CMC) (Al-Soufi and Novo, 2021).

Surfactant monomers gather and generate micelles in the matrix. The surfactant molecule consists of a hydrophilic head with hydrophobic tail. The inner hydrophobic centre of micelles can solubilise organic matter (with lower molecular weight) as solubilisate, simultaneously surface adsorbing available metal ions on their exposed surface due to electrostatic interactions. Surfactants with electric charge opposite to that of metal ions generally require longer retention time. Cationic surfactants, polyelectrolytes (PE) and anionic surfactants (such as sodium dodecyl sulfate (SDS)) are deployed for efficient extraction of heavy metals (Bediako *et al.*, 2023).

The workability of MEUF is governed by numerous factors, namely the inclusion of addendum solutes, the nature of surfactant, working parameters and the type and nature of membrane deployed.

PEUF has been generated through the combination of UF with binding polymers. The active functional groups of bonding polymers can be phosphonic, sulfonate, amine or carboxylated because these are united through electrostatic or chelating bonds. PEUF are also renowned as polymer-based, complexation, polymer-assisted dimensions as well as complexation-improved ultrafiltrations (Qasem *et al.*, 2021). During permeation of water and simplified (un-complexed) counterparts through membrane pores, PEUF has been reported to block and extract polymer-united metal ions. PEUF has demonstrated effective polymer bonding, efficient extraction, potential to recover and reutilise complexation polymer of retentate and a lesser requirement for energy, in addition to being economic from an operational cost perspective. However, selection of suitable water-soluble polymer macro-ligands exists as a prime challenge in the development of filtration technology (Can *et al.*, 2022).

4.7.1.4 Reverse osmosis

Reverse osmosis (RO) is a separation process which is run under pressure by employing a semipermeable membrane with pore size 0.5–1.5 nm, permitting only small dimensioned molecules to permeate through. The RO process is the reversed form of the ordinary osmosis process by the application of pressure in the range of 20–70 bar greater than the osmotic pressure of the input matrix (Li and Lin, 2024).

Solutes which cannot pass through the semipermeable membrane lie in the range of 0.00025–0.003 μm. The RO process has potential to extract about 95–99% dissolved inorganic salts in addition to organic ions. However, the RO process has a smaller filter size with greater blockage potential (Ramasamy, 2019). Additionally, degradation and fouling of the semipermeable membrane are reported weaknesses of RO systems.

The RO segregation process is used for extraction of heavy metal ions, such as Ni^{2+}, Cu^{2+} and Cr^{6+}, from electroplating of wastewater, with an elimination potential greater than 98.75%. Additionally, RO has been deployed in order to purify industrially polluted waters from mining operations in Victoria, Australia, with an average extraction efficiency of 95%, 82%, 66%, 48% and 10% for Sb^{3+}, Ni_2^{+}, As^{3+}, Zn^{2+} and Fe^{3+} respectively (Scrase and Brereton, 2024).

4.7.1.5 Forward osmosis

Forward osmosis (FO) is a version of the normal osmosis process that demands filler media for maintaining selectivity in association with permeated water flux (Field *et al.*, 2021). Semi-permeable membranes deployed in forward osmosis segregate matrix solution from refined solution. The refined solution has generally been at greater osmotic pressure in comparison to the feed solution (Law and Mohammad, 2018). Because the osmotic pressure of matrix solutions differs from refined solutions, the transportation of water from matrix solution to refined solution retains non-permeated solutes on the matrix solution side and, thereafter, treated water on the refined solution side.

Hydraulic pressure is not required in the case of forward osmosis, hence the process is economic from an energy point of view (Kook *et al.*, 2018). The forward osmosis process is environmentally friendly, convenient to clean and experiences less fouling; these features make FO one of the most commonly deployed wastewater treatment processes. However, there are certain disadvantages with the FO process, including re-concentration of refined solution, challenges in the selection of the membrane, and internal and external concentration polarisation (Liao *et al.*, 2018).

4.7.1.6 Electrodialysis

Electrodialysis (ED) is deployed to segregate ions of different electric potential. ED generally follows a sequence of cation exchange membranes (CEM) in addition to anion exchange membranes (AEM), alternatively set up parallel to each other in order to segregate ionic (cationic and anionic) solutes (Ahmad and Ahmed, 2023).

In the ED process, the anions permeate by way of anion exchange membranes (AEM) and cations permeate by way of cation exchange membranes (CEM). In the treatment scenario, treated diluate is generated from half of the ED stack channels, while a concentrated portion is drained from the other half (Pawlowski *et al.*, 2019). The characteristic features of ED include greater recovery of water and no changes in phases, without any intermediate reaction or chemical interactions. ED additionally covers a wide range of pH values for its operation. However, it has been found to experience fouling of its membranes, requires the deployment of expensive membranes and has electric potential requirements.

ED has previously been deployed to segregate Pb^{2+}, K^{+} (potassium) as well as Ni^{2+} from artificial anthropogenic solutions by way of novel ED heterogeneous CEM (containing 2-acrylamido-2-methyl propane sulfonic acid-supported hydrogel as well as PVC) to achieve elimination efficiencies of 99.9% for both Pb^{2+} and Ni^{2+} and 96.9% for K^{+}. A batch ED process has been used to eliminate Pb^{2+} with an optimum segregation efficiency of 100%.

Pilot ED assemblies have also been deployed to eliminate Ni^{2+}, Cu^{2+} in addition to traces of Fe^{3+}, Cd^{2+}, Zn^{2+} and Cr^{6+} with removal rates exceeding 90%. As^{5+} and As^{3+} were eliminated from effluent discharges from the metallurgical industry by ED at an efficiency of 91.38% (Kuldeep *et al.*, 2021).

4.7.1.7 Other membrane-based methods

Membrane distillation (MD) and liquid membrane (LM) can be used for treatment of wastewater.

MD is a hybrid thermally driven membrane segregation process consisting of cold and hot chambers segregated by a microporous, water-repelling hydrophobic membrane (Safi and Waisi, 2024). MD permits only vapours to pass through the pores of the membrane, restricting larger molecules. MD operates in four modes: sweeping gas membrane distillation, direct contact membrane distillation, vacuum membrane distillation and air gap membrane distillation (Mesquita *et al.*, 2025). MD processes have been observed to achieve more than 96% elimination of Mg^{2+}, Ca^{2+}, Fe^{2+} and Fe^{3+}, and eliminations exceeding 99% for As^{5+} and As^{3+}.

On the other hand, LM is made of a liquid phase or thin-layer organic phase which acts as a barrier between two aqueous phases. With LM the feed solution and retentate solution are immiscible, and the stripping and extraction processes are combined in a single stage (Das *et al.*, 2022). LM has reportedly been greatly efficient, relatively selective and can acquire particular molecular individuality. However, according to the previously published literature, the membrane is not durable.

The LM process is available as polymer inclusion membrane (PIM), supported liquid membrane (SLM), bulk liquid membrane (BLM) and emulsion liquid membrane (ELM). From all available types of membrane-based separation techniques, the SLM process is one of the most promising choices over conventional techniques such as solvent extraction to eliminate heavy metals from raw feed (Kaczorowska, 2023). In the elimination of Cd^{2+}, Zn^{2+}, Fe^{3+} and Cu^{2+}, SLM has acquired elimination efficiency up to 89%.

The different conventional technologies for removal of heavy metals from contaminated water are summarised in Figure 4.3.

4.7.2 Chemical-based separation

In comparison to other separation techniques, chemical strategies for elimination of heavy metals from wastewater have proven effective over a long period (Khalidi-Idrissi *et al.*, 2024). In this section, chemical-based strategies are discussed, including of flocculation, precipitation, coagulation as well as flotation.

4.7.2.1 Precipitation

Chemical precipitation (popularly known as coagulation precipitation) has been widely deployed in industries, and is considered one of the most efficient and durable methods of treatment. It modifies the form of dissolved metal ions into solid particles to facilitate their effective sedimentation (Bao *et al.*, 2023).

The coagulant-based process precipitates metal ions by altering electro-oxidising potential, pH and co-precipitation. Coagulants are used to increase the rate of settlement of metallic sediments. The most commonly deployed coagulant is hydroxide (OH^-) because of its easy operation, performance under various pH conditions and it is cost effective. Precipitation takes place by addition of hydroxides to agitated wastewater which causes the formation of

Figure 4.3 Different conventional technologies for removal of heavy metals from contaminated water

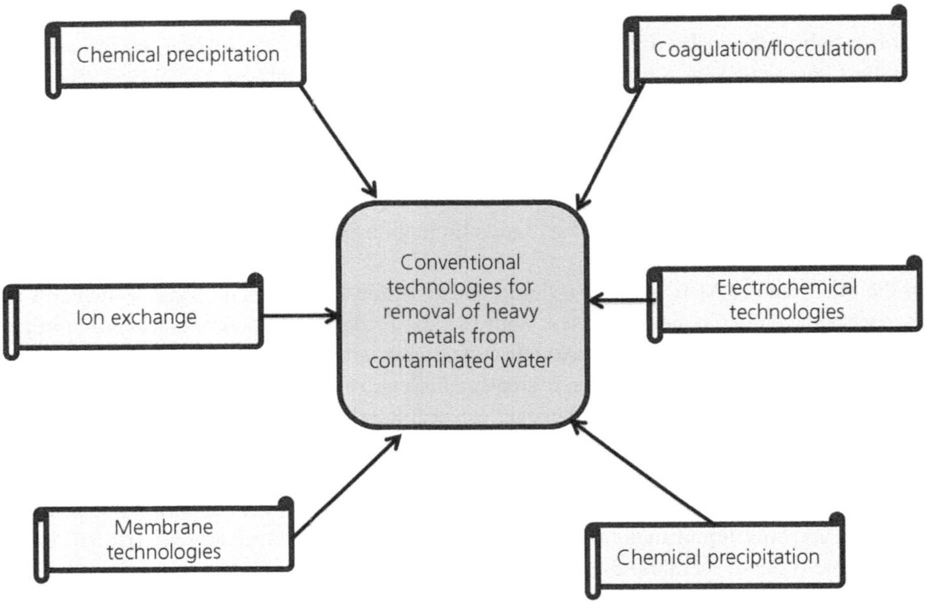

insoluble hydroxides of metal (Khaki *et al.*, 2017) – for example, a metal ion can interact with calcium hydroxide ($Ca(OH)_2$) to settle hydroxide of metal and calcium ions.

However, elevated pH levels have proven a disadvantage in the present method because it demands relatively greater amounts of precipitates. Lime (either as CaO or as $Ca(OH)_2$) is among the most effective hydroxide precipitates for the treatment of inorganic effluents of heavy metal concentrations of 1000 mg/L (Ma *et al.*, 2024). According to the previously conducted research, it has been observed that metal ions eliminated by this precipitation method are Cu^{2+}, Zn^{2+}, Ni^{2+}, Cr^{3+} and Pb^{2+}. The amphoteric nature of metal hydroxide precipitation, with the availability of complexing agents, creates relatively larger volumes of settled sludge. This subsequently leads to dewatering requirements and disposal-related concerns, which are the associated limitations pertaining to higher dosages used to get optimal pH (Hamraoui *et al.*, 2024).

The precipitation by sulfide method, in comparison to the hydroxide method, offers relatively higher removal efficiencies and lower dissolved solids quantities. The precipitation method has been proven to remove toxic heavy metal ions from the feed matrix. It is noted that a lower sulfide input results in a higher concentration of zinc ions in the effluent, and a higher sulfide input results in the creation of foul odour due to elevated residual sulfide (Hamraoui *et al.*, 2024), potentially generating toxic hydrogen sulfide (H_2S) gas. To balance these issues, precipitation by the sulfide method is suggested to be carried out at neutral pH.

The toxicity of sulfide and high costs are the two principal limitations of the precipitation by sulfide method.

Precipitation by carbonate, similarly to hydroxide precipitation, has demonstrated considerable effectiveness with optimum precipitation under acidic conditions by the deployment of either calcium carbonate ($CaCO_3$) or sodium carbonate (Na_2CO_3) (Wang and Wang, 2023). For effective precipitation, the process demands greater quantities of coagulants but produces less sludge with evolution of carbon dioxide (CO_2) as bubbles.

Published literature is available that describes methodologies for the efficient elimination of manganese (Mn), copper (Cu), lead (Pb) and zinc (Zn) by the general precipitation process (Xu *et al.*, 2023).

The efficiency of chemical precipitation methods can be upgraded with the help of the Fenton reaction. In order to remove complexes of heavy metals, either the Fenton reaction or Fenton-like oxidation is deployed for de-complexation. In order to adjust the pH of the real scenario elimination process for workability of the chemical precipitation mechanism, sodium hydroxide (NaOH) can be deployed (Romano *et al.*, 2023). However, involvement of the Fenton mechanism in the chemical precipitation method involves the use of several active intermediates, including free hydroxyl radicals and $[Fe(IV)O]^{2+}$ in several sequential steps.

4.7.2.2 Coagulation/flocculation

Coagulation is the practical de-stabilisation of colloids by neutralisation of the electrical charges that prevent their agglomeration and settlement. Flocculation is the agglomeration of such de-stabilised matter.

Examples of traditionally deployed coagulants include sulfates of ferrous $[Fe(SO_4)]$ and aluminium $[Al_2(SO_4)_3]$, and ferric chloride ($FeCl_3$) (Ottink, 2024). Flocculation joins neutral particles causing them to form agglomerates of bigger size by the addition of flocculants. Some examples of flocculants are polyferric sulfate (PFS), polyaluminium chloride (PAC) and polyacrylamide (PAM) (Hu *et al.*, 2024). Polyelectrolyte is considered the most commonly deployed flocculant, but the generated sludge may possibly be toxic. The majority of deployed flocculants are not only naturally occurring but also biodegradable.

The recorded drawbacks of the process include the toxic nature of its products, health concerns around inorganic coagulants, quantity of sludge produced, each flocculant has selectivity to a narrow range of metals, it is ineffective for emerging pollutants, it results in a darkened colour of effluent, and unsatisfactory results are achieved if natural coagulants are deployed in real practical situations (Badawi *et al.*, 2023). Over the last few years, heavy metals eliminated either by coagulation or flocculation include lead (Pb^{2-}), nickel (Ni^{2+}) and copper (Cu^{2+}). Other metals which can also be eliminated by this method include selenium (Se^{2+}), arsenic (As^{2+}), chromium (Cr^{2+}), antimony (Sb^{5+}, Sb^{3+}), in addition to silver (Ag^{2+}) (Wang *et al.*, 2021).

4.7.2.3 Flotation

For some time, flotation has been deployed to eliminate several undesirable metal ions. Dissolved air flotation (DAF), ion flotation (IF) and precipitation flotation (PF) have been investigated in detail.

In the case of dissolved air flotation, air (or gas) is made to pass through wastewater in order to generate micro-bubbles that can encapsulate metal ions, generating agglomerates of lesser density, and subsequently developing flocs in the wastewater (Haleem *et al.*, 2023). Sludge collected from the upper layer can conveniently be removed. The ion flotation process, to a great extent, depends on the hydrophobicity of metallic moieties by deploying surfactants; these hydrophobic moieties are eliminated through air bubbles. Added surfactants act as accumulators and, at the same time, frothers regulate the indexes of ion flotation.

Ion flotation seems ineffective in the case of concentrations of metal ions recorded and is inferior for bulk volumes of wastewater (Sun *et al.*, 2024). The specific features of the ion flotation process include treatment volume limitations, lower consumption of energy, selective treatment and lower volumes of sludge generation.

The precipitation flotation process is basically a chemical precipitate process that uses micro-bubbles. Precipitation flotation occurs over a shorter span of time in an efficient way. The reported benefits of PF include the speed of the operation, its compact assembly with standard operational procedures and its average operational cost (Yuan *et al.*, 2025).

Ion flotation is another promising flotation process. IF, to a great extent, depends on surfactants to accumulate contaminants and it is essential that non-toxic, efficient surfactants are used in its processes. Chemical artificial surfactants are available on the market, showing appreciable selectivity toward a range of contaminants as well as potential accumulation capability. However, it has been observed that toxicity and cost concerns are restricting their use.

Coagulation and flocculation processes can either be a sequence of two processes or combined into a single process (Zhang *et al.*, 2023a). The sedimentation step can be followed by either a filtration process or another process specific to the particular conditions encountered.

Chemicals including accumulators are generally incorporated, as well as micro-bubbles, to encapsulate metal ions, sometimes resulting in an inferior density of floating agglomerates that causes difficulty in their removal from the upper surface of the treatment chamber. Consequently, nanoparticles have been introduced as novel accumulators, demonstrating both merits as biosurfactants and as artificially synthesised surfactants (Fernandes *et al.*, 2023).

4.7.2.4 Chemical precipitation

Chemical precipitation is an efficient strategy to eliminate heavy metals, particularly for effluents released from electroplating industries and paper mills. In this process, insoluble precipitates are made to settle out when ferrous salts [$Fe(OH_3)$], lime (CaO), alum [$K_2SO_4.Al_2(SO_4)_3.24H_2O$] and some polymers are added as chemical precipitants, later interacting with heavy metals available in the wastewater (Gómez-García and Dobrosz-Gómez, 2024).

Metallic ionic moieties can be eliminated more conveniently by chemical precipitation. Elimination potential as well as efficiency can be upgraded by modifying operational parameters including temperature, pH, ion charge and initial pollutant concentration (Periyasamy, 2024).

The practical mechanism of the elimination of heavy metals by chemical precipitation is governed by Equation 4.1.

$$M^{2+} + 2(OH)^- \leftrightarrow M(OH)_2 \tag{4.1}$$

in which M^{2+} and OH^- are metal ions and precipitant, respectively, and $M(OH)_2$ is metal hydroxide.

The pH of the real practical reaction mixture needs to be adjusted to alkaline conditions in the range of pH 9–11 to optimise the reaction effectiveness (Liang *et al.*, 2024). Chemical precipitation results in metallic pollutant ions taking the form of sulfides and hydroxides. Deployment of coagulants in hydroxide precipitation can improve the elimination of heavy metals either by sedimentation or filtration. Additionally, sludge settled as precipitates of metallic sulfides is eliminated either by filtration or gravity settling (Reshma, 2023). This process demands both pre- as well as post-treatment. Additionally, the precise application of reagents is necessary because of the toxicity of both the resultant metal sulfide ion and H_2S.

4.7.2.5 Ion exchange technologies

The ion exchange process is a reversible chemical reaction deployed to transform unwanted metallic ions into compounds that are eco-friendly and non-toxic. Heavy metal ion pollutants are eliminated from a wastewater matrix by binding them to particles loaded on an immobile solid surface instead of solid cation particles (Althomali *et al.*, 2023).

The material of solid ion exchange particles can either be natural, such as inorganic zeolites, or artificially fabricated, such as organic resins. The ion exchange process can efficiently eliminate targeted heavy metal ion contaminants, including Hg^{2+}, Pb^{2+}, Cd^{2+}, V^{4+}, Ni^{2+}, V^{5+}, Cr^{4+}, Cr^{3+}, Zn^{2+} and Cu^{2+}, from wastewater.

The mechanism for the elimination of heavy metal ion contaminants from wastewater uses the ion exchange process (Cai *et al.*, 2024). In the case of cation exchange resins, the anions remain immobilised but cationic (Na^+/H^+) ions are exchanged with cations of wastewater – for example, DIAION CR11 and amberlite have been explored for the elimination of cations from wastewater (Pandit and Sharma, 2024).

Recent research has demonstrated that zeolite displays outstanding ion exchange potential because of its negative charge, originating from Si^{4+}, fits in the centre of a tetrahedron and gets isomorphously replaced with Al^{3+} cations (Zhu *et al.*, 2024).

Very recently, it has been observed that metal–organic frameworks (MOFs) are promising candidates for the elimination of metal pollutants through an ion exchange process. According to previously published literature, MOFs deployed for ion exchange reactions were AMOF-1 (for elimination of Hg^{2+}, Cd^{2+} and Pb^{2+}), ZIF-8 (to remove Cu^{2+}) and ZIF-67 (to remove Cr^{6+}). Similar to adsorption, removal of metallic contaminants through ion exchange strategies demands more in-depth research for their stability and reusability (Barman *et al.*, 2023).

4.7.2.6 Electrochemical technologies

This section describes the various electrochemical (EC) technologies that are available, including electrochemical reduction (ER), electroflotation (EF) and electrooxidation (EO).

In electrochemical assemblies, oxidation takes place at the positive electrode (i.e. anode), where electrons flow towards the negative electrode (i.e. cathode), during which a gain of electrons (i.e. reduction) takes place (Onoh *et al.*, 2024). The simultaneous occurrence of these two chemical reactions is known as redox (reduction-oxidation), resulting in purification of water by way of removal of metals. It is the nature of the electrochemical process that determines which metals get deployed as the anode and which as the cathode, impacting the removal efficiencies of certain specific metals.

Different arrangements of electrodes are possible: bipolar electrodes in series (BP-S), monopolar electrodes in parallel (MP-P) and monopolar electrodes in series (MP-S) (Nippatlapalli and Selvaraj, 2022). In the majority of cases, MP-P as well as MP-S are preferred because of their high yield and associated lower cost.

ER is also known as electrodeposition or electroplating, where either targeted atoms or molecules get deposited at the surface of a cathode. Unlike other processes, sludge, demanding further management, does not form in this treatment. Cathodes with high over potential with respect to hydrogen evolution have been reported to be more efficient either in the reduction or elimination of contaminants. Cathodes, either consisting of carbon-based or sulfur mixture, in various ratios under acidic conditions are favourable for eliminating Cd^{2+}, Hg^{2+}, Cu^{2+} and Pb^{2+} from polluted water (Dinu *et al.*, 2024). Titanium with iridium oxide as the anode coating have been observed to be ideal for the removal of Cd^{2+}, with 100% effectiveness irrespective of its initial concentration.

In order to improve the efficiency of wastewater treatment, reactor design and operational working parameters need to be evaluated (Singh *et al.*, 2022). Consumption of energy is a known restriction that needs to be resolved in order to commercialise treatment by this process in industrial applications. Either aluminium (Al) or steel (Fe) electrodes, being more reliable as well as non-toxic, are most commonly deployed in EC. The EC method follows the order: dissolved cations of anodic metal, followed by formation of hydroxo complex (coagulants) stability of agglomerate in addition to phase separation, precipitation and flotation. Cations are accountable for the destabilisation of the anode, colloidal particles as well as constitute complex hydroxides in polymeric form with metal ions (coagulants), which interact with negatively charged contaminants prevailing in wastewater. Therefore, larger-sized metal hydroxide, whose density is greater than water, forms and settles as a precipitate, or flocs and floats to the surface if their density is less than that of water.

Coagulants developing in situ, because of oxidation of the anode, are generally accepted as an additional advantage of EC methods over other competent technologies in the market (Mourdikoudis and Dominguez-Benetton, 2025). With alternating current (AC), a lower consumption of energy with greater mitigation efficiency has been practically possible compared to direct current (DC).

It has also been observed that if the process is carried out at an elevated temperature, pH and voltage decrease the duration of treatment. The disadvantages of EC methods include passivation of electrodes and a greater consumption of energy with particular challenges for energy consumption for large-scale deployments (Torrinha *et al.*, 2022).

Numerous practical approaches have previously been proposed in order to passivate electrodes, including operation by alternating current, excessive addition of ions, polarity reversal, cleaning of electrodes mechanically, ultra-sonication, chemical cleaning of electrodes and hydrodynamic scouring. Each EC option has its own disadvantages, including high cost involved, auxiliary treatment requirement, infrastructure needs, increasing volumes of deposit and the generation of hazardous byproducts (Cetinkaya *et al.*, 2024). EC processes are not yet satisfactorily mature treatment applications.

The mechanism of EF is primarily the electrolysis of water using insoluble electrodes, although flotation effects have been introduced in order to facilitate the treatment process (Fobang, 2023). The efficiency of the process relies on the generation of uniform and tiny bubbles (c. 0.15 mm). It has also been observed that Ti-based anodes have widely been deployed in EF processes. These anodes have been observed to be not only insoluble but non-corrosive giving high catalytic potential with O_2 as well as Cl_2 in the electrolyte. Optimal contaminant concentration has been recorded in the range of 10–100 mg/L, although the maximum could be 200 mg/L (Priyadarshini *et al.*, 2022). The mitigation efficiency of EF processes for heavy metal removal has been restricted because of poor evolution of O_2 over the potentials (Wu *et al.*, 2024). This is why a hybridised system using EC, membrane technology and EF has the potential to improve the overall performance of a system for removal of heavy metals (Moneer, 2023).

Deploying EO mechanisms enables contaminants to be removed from the wastewater both directly and indirectly. The performance of indirect oxidation by chlorine (Cl_2) is governed by the NaCl concentration and is independent of current intensity (Shmychkova *et al.*, 2022). The contaminants exchange electrons (e-) directly with the anode surface, depositing a polymeric layer at the surface, subsequently leading to deactivation of the electrodes and decreasing efficiency (Ren *et al.*, 2023).

Oxidised forms of pollutants appear in the matrix when organic contaminants intermingle with prevailing oxidants. Research has shown that gold (Au), platinum (Pt), manganese (Mn), boron-doped diamond (PbO_2) and hybrid metal oxide (HMO) (including PbO_2, SnO_2, IrO_2, Ti/TiO_2, RuO_2, Sb_2O_5,) have been commonly deployed as anodes (Xu *et al.*, 2024a). The most capable, effective and efficient anodic materials are costly. The presence of other metallic impurities in practical scenarios interfere with the efficiency of the water treatment process. To overcome this situation, more anodic materials need to be explored to increase the efficiency of the water treatment process.

Additionally, the presence of several types of metallic ions in wastewater matrices has affected the efficiency of water treatment (Agasti, 2021). The need for more effective and efficient anodic materials is identified for such situations. More research in this domain needs to be carried out in order to investigate the efficiency of integrating EO with competent water technologies to overcome the operational challenges.

The advantages and disadvantages of conventional water treatment techniques for the removal of heavy metals are summarised Table 4.1.

4.8. Advanced removal strategies

Advanced heavy metal removal strategies include adsorption, magnetic adsorbents, polymeric materials, activated carbon, industrial waste, agricultural waste, composite adsorbents, and metal–organic frameworks.

4.8.1 Adsorption

The mechanism of adsorption is influenced by the physicochemical features of deployed adsorbents, the heavy metals present and the process operational parameters (including pH, temperature, contact time, adsorbent dose, as well as initial quantities of metal ions).

In general, heavy metal ions need to be adsorbed onto the surface of an adsorbent (Chakraborty *et al.*, 2022). Adsorption has been reported to have had relatively lower operational cost, greater mitigation potential, convenient implementation and ease of treatment by way of regeneration of adsorbed heavy metals as ions.

A variety of adsorbents has been explored for their potential for heavy metal remediation – for example, lactone phenyl and carboxyl groups. From various treatment strategies, oxidation, nitrogenation and sulfuration have generally been deployed to improve pore structure, specific surface area, adsorption capacity, mechanical strength and thermal stability (Liu and Zhang, 2022). However, all these matrices are greatly governed by the nature and composition of the adsorbent, which in some cases may not be cost effective. Consequently, the cost of the adsorbent is a crucial parameter for consideration when selecting the most appropriate adsorbent for a particular treatment scenario.

Previous research has shown that a modification in surface features has generally resulted in a reduction of surface area but an increase in the number of surface functionalities (Sultana *et al.*, 2022), eventually resulting in greater adsorption of metal ions from the matrix. The extent of adsorption increases by raising adsorbent dose, adsorbent surface area, contact time and initial concentration of metal ions.

Research on this matter has concluded that multiwalled carbon nanotubes (MWCNTs) offer potential to mitigate heavy metals in real-world situations. MWCNTs are reasonably hydrophobic in nature and, therefore, display instant aggregation in an aqueous matrix, due to the presence of considerable Van der Waals forces, resulting in decreased adsorption capacity (Bao *et al.*, 2024). At present, quantitative assessments of potential functional combinations of adsorbents in the adsorptive removal of heavy metal ions is unavailable.

Table 4.1 Advantages and disadvantages of conventional water treatment techniques for the removal of heavy metals

Conventional water treatment techniques	Advantages	Disadvantages	Source
Ion exchange	Rapid kinetic, faster removal capacity, efficient in removing inorganic contaminants, ease of operation during field applications, no disposal of sludge, economic material, faster operation, as well as exhausted resins can be reactivated.	Higher capital and operation cost, greater sensitivity to pH of matrix, non-selective, offensive odour from metallic contaminants, favourable for lesser concentration of metals.	Qasem *et al.*, 2021
Flotation/ coagulation	Relatively easy operability, more simple and economic process.	Disposal of sludge is a problem, inefficient removal of heavy metals, high operational cost due to the quantity of chemicals used.	Kumar *et al.*, 2023; Akhil and Raj, 2020
Electrochemical technology	Less laborious, process is simple and environmentally friendly, and can conserve considerable power.	Significant capital cost, high power consumption and associated cost, and inefficient for less contaminated aqueous matrices.	Javed and Mushtaq, 2023; Amin and Elsayed, 2024
Membrane-based technologies	Removes both organic and inorganic contaminants, greater mitigation efficiency, lower power consumption, there is no need for additional chemicals, no changes in existing phase, no secondary pollution, convenience of synthesis, and environmentally friendly.	Poor flux permeation, complex operation process, greater involved cost, fouling of the membrane.	Ayach *et al.*, 2024; Algieri *et al.*, 2021
Chemical precipitation	Lower capital investment, easy operation as well as convenient, automated treatment strategy.	Generates larger amount of sludge enriched with hazardous contaminants, requiring many chemicals in both the process and to achieve acceptable disposal conditions, slower metal precipitation, longer rate of settlement, and environmental issues.	Yadav *et al.*, 2021; Zhu *et al.*, 2019

Additionally, present surface modification strategies require strong acid/base and greater heat/pressure, in addition to intensive oxidation/reduction interactions (Khezerlou *et al.*, 2025). The complicated preparatory process makes carbon-supported adsorbents (adsorbents supported on carbon base like biochar, hydrochar, activated carbon and so on) more expensive, but they offer utility in industrial applications. There is a clear need for research into innovative, environmentally friendly and cost-effective surface modification strategies.

4.8.2 Magnetic adsorbents

Magnetic adsorbents are specific materials containing iron particles (generally, magnetic nanoparticles including Fe_3O_4) (Singh *et al.*, 2021). These magnetic nanoparticles can be supported on chitosan (CS), carbon, polymers, biomass or starch. According to previously conducted research it has been observed that the adsorption process is greatly impacted by surface charge, magnetic field strength and redox activity characteristics.

Cost-effectiveness, characteristic surface charge, convenient synthesis and reusability are the key features associated with magnetic adsorbents. Several magnetic adsorbents have been cited and investigated in previous literature, including iron oxides: hematite (α-Fe_2O_3), zero-valent iron nanoparticles (ZVI NPs), maghemite (γ-Fe_2O_3), spinel ferrites and magnetite (Fe_3O_4) (Phouthavong *et al.*, 2022). In such cases, the behaviour of magnetic adsorbents as well as surface morphological features are the key factors determining not only the mechanism of operation but the kinetics of the adsorption process also. Adsorption by magnetic adsorbents is also impacted, in practical experimental situations, by irradiation time, pH, dose of magnetic adsorbent, initial amount of heavy metal ions and ambient temperature (Mubarak *et al.*, 2022).

It is the presence of iron nanoparticles in an adsorbent that efficiently contributes to the removal of metal ions from industrial effluent. Some previous studies have emphasised that the coating of Fe_3O_4 particles on adsorbents provides efficient mitigation of heavy metal ions from the matrix. High-gravity technology, co-precipitation and grafting are among the most generally deployed strategies (Chauhan and Punia, 2023). Simplicity and flexibility are the key features that make the grafting method most preferable. However, it greatly depends on the concentration of the generated active hydroxyl on the surface of Fe_3O_4 particles in addition to the number of active functional groups on the surface. However, the synthesised adsorbents are not sufficiently cyclically stable, which is a barrier in the commercialisation of the present method.

4.8.3 Polymeric materials

Polymeric materials offer not only high proccessibility but stability also, additionally providing specific binding sites for specific contaminants of the matrices (Wu *et al.*, 2022). Polymeric materials have been reported to have features that give easy functionality, and their malleability and characteristic structure can also be tuned to significantly boost their adsorption abilities (Mokoena and Mofokeng, 2023).

Some polymers have been associated with pollution, which is potentially problematic as both governments and industry demand that environmentally friendly polymeric materials are used as adsorbents (Lazar *et al.*, 2023). Consequently, polymer adsorbents are required to be non-toxic in nature, economic, biodegradable, reusable, biocompatible and convenient to synthesise (Elella *et al.*, 2022).

Several prime characteristics have been observed to be generally present in numerous natural polymers, including alginate, silk, lignin, cellulose and chitosan. Alginate composites have been reported to exhibit remarkable mitigation potential for heavy metals, dyes, as well as other water pollutants (Doyo *et al.*, 2023). Lignin-based materials also possess a mitigation capacity exceeding 90% from water (Barman *et al.*, 2022). Similarly, cellulose-based materials and chitosan have been extensively deployed for the remediation of contaminants, notably for the mitigation of grease, oils and heavy metals from aqueous matrices (Gul Zaman *et al.*, 2021). Specifically, cellulose has been noted as one of the most abundantly available naturally occurring polymer materials, demonstrating a vast variety of unique biological and physicochemical features, fulfilling numerous expectations of an appropriate agent for the remediation of contaminants from water (Ravikumar and Udayakumar, 2020). Consequently, biopolymers are among the most favourable materials for remediation applications for contaminated aqueous matrices.

4.8.4 Activated carbon

Carbon-supported nanoporous adsorbents, specifically carbon nanotubes (CNTs), activated carbons (ACs) and graphene (GN), have been extensively deployed in remediation applications of heavy metals from aqueous matrices because of their enormous exposed surface area (500–1500 m^2/g) (Mokoena and Mofokeng, 2023).

AC has mostly been used as adsorbents for remediation of heavy metals which can be synthesised by carbonisation/pyrolysis of a wide range of carbon-containing materials, followed by an activation process. An adsorption potential as great as 900 mg/g has been reportedly achieved in the case of Cr^{3+} remediation from an aqueous matrix deploying agro-industrial residue as a precursor for AC (Maneechakr and Mongkollertlop, 2020).

Several raw materials (including agricultural residues/petroleum pitch), processing parameters and mitigation conditions have been evaluated for treatment by AC. Research has focused on minimising the cost, while raising its adsorption potential, innovating environment friendly activation strategies, enhancing its working life and improving recyclability. Carbon surface charges can potentially be raised by surface functional groups (including phenyl, carboxyl as well as lactone groups) to treat heavy metals (Zhang *et al.*, 2024).

Several modification strategies exist, with oxidation, nitrogenation and sulfuration being those most generally deployed to enlargen specific surface area, increase adsorption capacity, enhance pore structure, increase mechanical strength and achieve thermal stability (Kong *et al.*, 2021). However, these are dependent primarily on the specific adsorbent materials used, which in some cases may not be cost effective. Hence, the cost of adsorbent needs to be taken into consideration while selecting the most appropriate adsorbent.

Surface modification generally decreases the surface area of an adsorbent, which later increases the quantity of surface functional groups. This subsequently results in the adsorption of a larger quantity of heavy metal ions (Sultana *et al.*, 2022). The extent of adsorption also reportedly improves by increasing the exposed surface area of adsorbent, contact time, initial quantities of heavy metal ions and adsorbent dose (Sultana *et al.*, 2022). However, MWCNTs have so far attracted particular attention from researchers for the removal of heavy metal ions from wastewater matrices (Rodríguez *et al.*, 2020). MWCNTs are highly hydrophobic in nature and suffer rapid aggregation in aqueous matrix because of large Van der Waals forces, subsequently reducing adsorption capacity. Literature in support of the contribution of functional groups in the quantitative assessment of heavy metal ions through adsorption is not accessible.

Additionally, current surface modification strategies require strong acid/base and greater heat/ pressure, in addition to intensive oxidation/reduction interactions (Khezerlou *et al.*, 2025). The complicated preparatory process makes carbon-supported adsorbents more expensive, but they offer utility in industrial applications. There is a clear need for research into innovative, environmentally friendly and cost-effective surface modification strategies.

4.8.5 Industrial waste

Industrial wastes have been used to mitigate heavy metal ions from wastewater as they are abundantly available, high performing and low cost. Industrial waste includes materials such as blast furnace slag, sludge, flue dust, lignin, red mud and fly ash (Wang *et al.*, 2023a) – for example, sludge has been effective in mitigating Cd^{2+}, Pb^{2+} and Cr^{3+} (Sultana *et al.*, 2023), red mud can mitigate Cu^{2+} from effluent, and organic acids can be used to extract Zn^{2+} and Cu^{2+} from sewage sludge (Radovanovic *et al.*, 2023). Researchers generally tune operation parameters (including dose of adsorbent, contact time, working temperature, ionic strength, initial pollutant concentration as well as pH) so that sludge can be recycled, upgrading mitigation efficiency (Rajendran *et al.*, 2022).

4.8.6 Agricultural waste

Agricultural residues primarily consist of cellulose, hemicellulose, lignin and starch, which are all biopolymers demonstrating strong extraordinary adsorption potential. They possess a wide diversity of functional groups which facilitate the complexation of metals (Sabir *et al.*, 2021). Additionally, one of the prime merits of these biosorbents is that they can reportedly be fine-tuned to boost their adsorption potential with adaptability at an industrial scale (Tsade *et al.*, 2020). Such residues and their byproducts are highly efficient, less expensive and they minimise biological or chemical sludge compared with traditional mitigation technologies (Tsade *et al.*, 2020). The effectiveness of these residues is greatly dependent on parameters including temperature, pH, concentration of heavy metals and adsorbent dose.

4.8.7 Composite adsorbents

In several investigations, composite adsorbents have been reported as offering promising treatment options. They are formed by combining two or more materials in order to obtain an adsorbent with characteristic features that are far improved from those of its constituents – for example, ligno-cellulosic sawdust residues (LR) have been impregnated with magnetite nanoparticles (MNP) in order to obtain composites with LR: MNP ratios of 1:1 as well as 3:1 (Janani *et al.*, 2022).

Composite adsorbents require limited contact time to achieve similar elevated removal efficacies obtained by nanoparticles or residues alone. Additionally, since adsorptive mitigation performance is similar for both LR and MNP, it has been observed that relatively large amounts of low-cost LR could be combined with small quantities of the more costly MNP to obtain a highly efficient composite adsorbent (Soliman *et al.*, 2020). Alginates and chitosan have been used in order to functionalise AC to create composites with improved durability, greater mechanical strength, a high affinity for diverse contaminants and improved hydraulic properties (Rana *et al.*, 2021). AC has also been impregnated with magnetic nanoparticles to adsorb various metallic ions, including Cd^{2+}, Pb^{2+}, Cu^{2+}, Ni^{2+}, Zn^{2+} and Cr^{6+} (Igberase *et al.*, 2018). Magnetic biosorbents can effectively mitigate heavy metal ions in addition to other organic and inorganic pollutants (Liu *et al.*, 2023a) – for example, magnetite/magnemite nanoparticles embedded in chitosan matrix demonstrates relatively greater adsorption potential (221 mg/g in the case of Pb^{2+}).

4.8.8 Metal–organic frameworks

Metal–organic frameworks (MOFs) have been developed through reticular synthesis during which heavy metal ions get strongly bonded to organic linkers. According to research conducted over the last few years, researchers have investigated thousands of MOFs. It has also been noticed that the majority of the organic ligands deployed to fabricate several MOFs are considerably costly and hazardous (Abdollahi *et al.*, 2022).

The family of zirconium MOFs (including UiO-66) has been reported as promising, where nanomaterials for adsorption, due to deployments of functional groups, give stable hydrolytic-thermal compounds, including carboxylic, hydroxyl, oxygen and amine, while demonstrating beneficial cross-linking strategies (Olawade *et al.*, 2024). Composite supported MOF adsorbents can achieve advanced adsorption potential. However, while MOFs have the capability to mitigate many heavy metal ions, they are enriched with tiny micropores which are not accessible by some target metals. It has also been observed that the majority of them have not proven stable in aqueous matrices.

Iron (Fe), manganese (Mn) and copper (Cu) have been deployed to form MOFs, although the majority of them had outcomes with low chemical stability (Olawade *et al.*, 2024). Hence, more research needs to be carried out with respect to the structural and morphological features of MOFs, in addition to their scale-up for industrial wastewater applications. Not only this, diverse functionalisation strategies need to be developed and implemented to boost their adsorption kinetics stability.

4.8.9 Bioadsorbents

The placement of several active functional groups (including amino, carboxyl, hydroxyl, thiol and phosphate) on a bioadsorbent's surface accelerates the biosorption process (Baby *et al.*, 2023).

It has generally been noticed that interaction between contaminants and the bioadsorbent surface takes place by way of aggregation, electrostatic interaction, complexation/coordination, ion exchange, microprecipitation and/or reduction/oxidation (Wang and Wang, 2023b).

The pH of the solution impacts the bioadsorbent's surface charge density and the ionisation of functional groups situated on the bioadsorbent surface (Adewuyi, 2020). Under acidic conditions, when pH value is low, cations are approximately stable and can be attached to the surface of the bioadsorbent. Under alkaline conditions, when pH value is high, dissolution of metallic cations reduces the potential for precipitation to occur.

Bioadsorbent dose has been a crucial factor impacting mitigation potential by facilitating a greater quantity of empty biosorption sites. The capacity of bioadsorbents is raised at elevated temperatures because of reduced viscosity of the matrix, decreased Gibb's free energy and the rupturing of bonds. Such rationale elevates collision frequency in terms of both kinetic energy as well as mobility between the surface of both the bioadsorbent and the heavy metal ions, increasing the amount of active sites of the bioadsorbent and resulting in greater affinity (Jun *et al.*, 2020). However, bonding force between vacant sites of bioadsorbent and contaminants can decrease at elevated temperatures in certain scenarios, when adsorption on the bioadsorbent surface decreases.

An increase in the rate of agitation enhances mitigation efficiency (Hussain *et al.*, 2024).

4.8.10 Microbial fuel cells

Microbial fuel cells (MFCs) have been identified as a potential remediation technology as they deploy organic matter prevalent in wastewater to generate electricity through biocatalysts including microorganisms. In MFCs, microorganisms generate not only protons but electrons also at the anode electrode. This action is anaerobic and transfers protons by way of a proton exchange membrane. Electrons are generated by way of an external circuit towards the cathode cell, which is aerobic.

At the cathode, oxygen completes a circuit and reduces to water because of its higher redox potential (Yaqoob *et al.*, 2022). Treatment of heavy metals within MFCs with a greater redox potential, in comparison to anode electrodes, enables improved mitigation and their recovery in cathode chambers (Das *et al.*, 2024). MFCs have been deployed for in situ remediation in rivers, in which outcomes have demonstrated mitigation outcomes of 87.7% Cu^{2+}, 97.3% Hg^{2+} and 98.5% Ag^{1+} after 60 days of MFC operation (Shi *et al.*, 2023). Additionally, MFCs effectively boost the biodegradation of organic matter, generating electricity and offering an alternative approach for efficiently remediating multiple heavy metal pollutants with simultaneous recovery of bioenergy.

4.8.11 Nanotechnology

Nanotechnology-supported treatments deploy nanomaterials, which have been studied for the past few decades, resulting in greater surface-to-volume ratios as well as extraordinary optical, electronic and magnetic features (Gong *et al.*, 2021).

Nanofiltration has been one of the most deployed nanotechnology processes for the removal of heavy metals. Systems supported on alumina nanofibres have been considerably capable of mitigating soluble heavy metal pollutants through chemisorption. Weaker dimensional structures, including nanocarbon, single or multi-metal oxides, magnetic nanoparticles, nanoclays and non-metal oxides, have been deployed for mitigation of heavy metals from matrices, as well as for disinfection and purification (Suhan *et al.*, 2023). Such nanostructures have demonstrated great reactivity as well as having enormous exposed surface areas. Additionally, several of them either are very economical or are plentiful in nature. Likewise, nano-assemblies, microspheres with nanosheets, nanoplates and hierarchical ZnO nano-rods have been deployed for the treatment of wastewater (Kumar *et al.*, 2021). Although information regarding the environmental issues, toxicity and health impacts of nanomaterial use is still being studied, they demonstrate potential for practical use.

Nanocarbon, either in the form of graphene and its derivatives or carbon nanotubes (CNTs), have been deployed as a highly promising adsorbent (Chandran *et al.*, 2023). The carbonaceous portion of CNTs as well as their highly reactive surface area permits interaction among water contaminants by way of covalent, hydrophobic, π-π, electrostatic interactions and hydrogen bonding types of interactions (Ethaib *et al.*, 2022). Additionally, their rod-like, hollow shapes facilitate outstanding water remediation, with numerous opportunities for adsorption to occur. However, challenging characteristic features of CNTs have to be transferred to the macroscopic horizon through preparation of composite materials such as reinforced polymers, films, fibres, aerogels and gels (Yang *et al.*, 2024).

The CNT surface has easily been modified through metal impregnation, acid treatment as well as the grafting of functional molecules/groups. The latter permit regulation of nanotube surface charge, surface area, hydrophobicity, tunable adsorption capacity as well as dispersion. In order to mitigate heavy metal pollutants, nitric, sulfuric or hydrochloric acid functionalisation creates oxy-functional groups that have reportedly been highly effective in mitigating contaminants from wastewater, including copper, cadmium, mercury and lead (Salek Maghsoudi *et al.*, 2021).

Graphene is another promising material for water treatment. Its morphologically flat and two-dimensional (2D) honeycomb atomic arrangement, accompanied by its hybridised sp^2-structure and associated electronic band structure, can easily be reengineered for customised functionalities (Khalid *et al.*, 2020). Chemical oxidation of graphene can be achieved by way of acid functionalisation, generating carboxylic acids that are hydrophilic in nature and speed up the adsorptive removal of heavy metals from matrices (Marimuthu *et al.*, 2023).

Graphene oxide (GO)-supported materials, both chemically engineered and composite, have been deployed in the mitigation of heavy metal applications. Other graphene derivatives in their reduced form (RGO) are currently being deployed successfully for treatment of water (Iqbal *et al.*, 2023) – for example, 3D graphene-supported macrostructures generally present a larger surface area and porous network, accompanied by excellent chemically active sites, high electronic mobility and ultra-lightness. All such structures have the potential to serve as scaffolds that fix nanomaterials, consequently broadening their potential for tunable applications in water treatment (Inobeme *et al.*, 2023). Such technologies have demonstrated synthesis of a mesoporous silica 3D scaffold doped by flakes of graphene oxide (GOFs) and shown potential for the mitigation of As^{3+} and Pb^{2+} ions from groundwater samples. Such GO-supported mesoporous material has demonstrated the capacity to be recycled four times without any effective loss in activity.

4.8.12 2D material

2D materials are also referred to as single-layer materials, and they include molybdenum disulphide (MoS_2) nanosheets, borophene, silicene, germanene, 2D metal carbides (MXenes) and graphene.

MXenes oxidise functional groups, have active binding sites and unique layered structures that make them unique adsorbents for the mitigatation of heavy metals from aqueous matrices (Nibe and Gaikwad, 2025). The most investigated MXene is Ti_3C_2Tx, which has been deployed in several ways, ranging from nanosheets to fabricated composites. 2D MXene-supported material has demonstrated elevated adsorption potential, very high and rapid kinetics and thus mitigation potential. It is competitive against industrial adsorbents, including AC, under a broader range of situations (Nibe and Gaikwad, 2025).

MoS_2-supported materials have been very commonly used for their performance as better adsorbents because the sulfur portion of their composition is reported to possess improved affinity for adsorptive removal of heavy metals, although molybdenum (Mo) offers improved catalytic-based adsorptive and mineralisation ability for organic contaminants (Shawky *et al.*, 2022). MoS_2 nanosheets have also previously been dispersed in aqueous matrices, coated as thin-film membranes or imbibed within fabricated composites. Under such situations, these materials have demonstrated improved adsorption potential, rapid kinetics and upgraded removal efficiencies.

4.8.13 Nanoparticles

Iron (Fe_2O_3), silicon (SiO_2), titanium (TiO_2), magnesium (MgO), manganese (MnO) and zinc (ZnO) oxides have been very popular for their deployment for removal of heavy metal contaminants from aqueous matrices.

Silica (SiO_2) nanoparticles and silanol groups have a greater porosity at their exposed surface that can easily be reengineered through silane chemistry. Silica (SiO_2) nanoparticles, either singly or when associated with materials such as GO or functionalised GO, could conveniently mitigate heavy metal ions, including Pb^{2+}, Hg^{2+} and Ni^{2+} (Ali *et al.*, 2017).

Previous research shows that magnetic iron oxide nanoparticles (IONPs) have been effective in the segregation and mitigation of target heavy metal contaminants in aqueous matrices. Those that have been deployed most are supported on ferric oxide nanoparticles, primarily because of their enormous exposed surface charge, elevated redox potential and their deployability. Hydroxyl groups (OH^-) on the surface of IONPs facilitate their potential for reengineering their surface configuration. In the case of silica (SiO_2), IONPs have shown the potential to mitigate several heavy metals, and their adsorption potential can potentially be upgraded through functionalisation (Sharma *et al.*, 2022).

TiO_2 nanoparticles have catalytic features that are a significant improvement compared with their bulk counterparts. Such nanoparticles could potentially adsorb heavy metals as well as metalloids including Cd^{2+}, Zn^{2+}, As^{5+}, La^{3+}, Nd^{3+} and Cr^{6+} (Rajendran *et al.*, 2022).

Both chemical and biological non-conventional strategies to mitigate heavy metals have their own particular advantages and disadvantages (Ayach *et al.*, 2024). However, a lot of effort has already been made by researchers to develop nanotechnologies and their bespoke products in order to mitigate heavy metals. Their deployment has been very challenging in the majority of cases because of their associated cost. Biological strategies have had more acceptance because of their cost effectiveness and eco-friendly nature.

REFERENCES

Abdollahi N, Moussavi G and Giannakis S (2022) A review of heavy metals' removal from aqueous matrices by metal-organic frameworks (MOFs): state-of-the art and recent advances. *Journal of Environmental Chemical Engineering* **10**: 107394.

Abiodun O-AO, Oluwaseun O, Oladayo OK *et al.* (2023) Remediation of heavy metals using biomass-based adsorbents: adsorption kinetics and isotherm models. *Clean Technologies* **5**: 934–960.

Adewuyi A (2020) Chemically modified biosorbents and their role in the removal of emerging pharmaceutical waste in the water system. *Water* **12**: 1551.

Agasti N (2021) Decontamination of heavy metal ions from water by composites prepared from waste. *Current Research in Green and Sustainable Chemistry* **4**: 100088.

Ahmad M and Ahmed M (2023) Characterization and applications of ion-exchange membranes and selective ion transport through them: a review. *Journal of Applied Electrochemistry* **53**: 1537–1562.

Ahmed M, Mavukkandy MO, Giwa A *et al.* (2022) Recent developments in hazardous pollutants removal from wastewater and water reuse within a circular economy. *NPJ Clean Water* **5**: 1–25.

Ahmed SF, Mofijur M, Nuzhat S *et al.* (2021) Recent developments in physical, biological, chemical, and hybrid treatment techniques for removing emerging contaminants from wastewater. *Journal of Hazardous Materials* **416**: 125912.

Akhil NV and Raj LA (2020) Heavy metal sequestration from wastewater: current trends and technologies – a review.

Akhtar N, Syakir Ishak MI, Bhawani SA and Umar K (2021) Various natural and anthropogenic factors responsible for water quality degradation: a review. *Water* **13**: 2660.

Algieri C, Chakraborty S and Candamano S (2021) A way to membrane-based environmental remediation for heavy metal removal. *Environments* **8**: 52.

Ali A, Naz S, Mannan A *et al.* (2017) Adsorption of Ni^{2+}, Hg^{2+}, Pb^{2+}, Cr^{3-}, and Co^{2+} on iron oxide nanoparticles immobilized on cellulose fiber: equilibrium, kinetic, thermodynamic, mechanisms, and statistical supposition. *Desalination and Water Treatment* **95**: 234–246.

Al-Soufi W and Novo M (2021) A surfactant concentration model for the systematic determination of the critical micellar concentration and the transition width. *Molecules* **26**: 5339.

Althobaiti NA (2024) Heavy metals exposure and Alzheimer's disease: Underlying mechanisms and advancing therapeutic approaches. *Behavioural Brain Research* **115212**.

Althomali RH, Jabbar HS, Kareem AT *et al.* (2023) Various methods for the synthesis of $NiTiO_3$ and $ZnTiO_3$ nanomaterials and their optical, sensor and photocatalyst potentials: a review. *Inorganic Chemistry Communications* **158**: 111493.

Amin M and Elsayed MM (2024) Electrochemical techniques applied for industrial wastewater treatment: a review. *Egyptian Journal of Chemistry* **67**: 7–33.

Anetor GO, Nwobi NL, Igharo GO *et al.* (2022) Environmental pollutants and oxidative stress in terrestrial and aquatic organisms: examination of the total picture and implications for human health. *Frontiers in Physiology* **13**: 931386.

Ayach J, El Malti W, Duma L *et al.* (2024) Comparing conventional and advanced approaches for heavy metal removal in wastewater treatment: an in-depth review emphasizing filter-based strategies. *Polymers* **16**: 1959.

Aziz KHH, Mustafa FS, Omer KM *et al.* (2023) Heavy metal pollution in the aquatic environment: efficient and low-cost removal approaches to eliminate their toxicity: a review. *RSC Advances* **13**: 17595–17610.

Baby R, Hussein MZ, Zainal Z and Abdullah AH (2023) Preparation of functionalized palm kernel shell bio-adsorbent for the treatment of heavy metal-contaminated water. *Journal of Hazardous Materials Advances* **10**: 100253.

Badawi AK, Salama RS and Mostafa MMM (2023) Natural-based coagulants/flocculants as sustainable market-valued products for industrial wastewater treatment: a review of recent developments. *RSC Advances* **13**: 19335–19355.

Bak MP, Kroeze C, Janssen ABG and Strokal M (2024) Modelling future coastal water pollution: impacts of point sources, socio-economic developments and multiple pollutants. *Environmental Research Letters* **19**: 64028.

Bao T, Wang P, Hu B *et al.* (2023) Mobilization of colloids during sediment resuspension and its effect on the release of heavy metals and dissolved organic matter. *Science of the Total Environment* **861**: 160678.

Bao T, Wang P, Hu B *et al.* (2024) Adsorption and distribution of heavy metals in aquatic environments: The role of colloids and effects of environmental factors. *Journal of Hazardous Materials* **474**: 134725.

Barman MK, Bhattarai A and Saha B (2023) Applications of ion exchange resins in environmental remediation. *Vietnam Journal of Chemistry* **61**: 533–550.

Barman P, Kadam R and Kumar A (2022) Lignin-based adsorbent for effective removal of toxic heavy metals from wastewater. *Emergent Materials* **5**: 923–943.

Bediako JK, El Ouardi Y, Mouele ESM *et al.* (2023) Polyelectrolyte and polyelectrolyte complex-incorporated adsorbents in water and wastewater remediation–a review of recent advances. *Chemosphere* **325**: 138418.

Bi J, Huang X, Wang J *et al.* (2020) Titanate for water remediation: synthesis, application, mechanism and optimization. *Journal of Materials Chemistry A* **8**: 14415–14440.

Briffa J, Sinagra E and Blundell R (2020) Heavy metal pollution in the environment and their toxicological effects on humans. *Heliyon* **6**.

Cai Y, Fang M, Tan X *et al.* (2024) Highly efficient selective elimination of heavy metals from solutions by different strategies. *Separation and Purification Technology* **127975**.

Can MF, Arslan F and Çelik MS (2022) Modelling of selective retention of Cd-Ni ions from aqueous solutions by polymer enhanced ultrafiltration. *Physicochemical Problems of Mineral Processing* **58**.

Canton H (2021) Organisation for economic co-operation and development—OECD. In *The Europa Directory of International Organizations 2021*. Routledge, Abdingdon, UK, pp. 677–687.

Cetinkaya A, Budak F, Ozkan E and Ozkan SA (2024) The importance of electrochemistry in environmental monitoring: advantages and limitations. In *Recent Trends and Perspectives on Electrochemical Sensors for Environmental Monitoring*. Elsevier, Amsterdam, the Netherlands, pp. 3–38.

Chakraborty R, Asthana A, Singh AK *et al.* (2022) Adsorption of heavy metal ions by various low-cost adsorbents: a review. *International Journal of Environmental Analytical Chemistry* **102**: 342–379.

Chandran DG, Muruganandam L and Biswas R (2023) A review on adsorption of heavy metals from wastewater using carbon nanotube and graphene-based nanomaterials. *Environmental Science sand Pollution Research* **30**: 110010–110046.

Chauhan NS and Punia A (2023) Heavy metal removal and recovery: sustainable and efficient approaches. In *Wastewater Resource Recovery and Biological Methods*. Springer, Cham, Switzerland, pp. 87–124.

Das P, Sharma A, Singh Y *et al.* (2022) MEUF for removal and recovery of valuable organic components present in effluents: a process intensified technology. *Water Environment Research* **94**: e10761.

Das S, Kumar S, Mehta AK and Ghangrekar MM (2024) Heavy metals removal by algae and usage of activated metal-enriched biomass as cathode catalyst for improving performance of photosynthetic microbial fuel cell. *Bioresource Technology* **406**: 131038.

Dinu A, Bounegru AV, Iticescu C *et al.* (2024) Electrochemical detection of Cd^{2+}, Pb^{2+}, Cu^{2+} and Hg^{2+} with sensors based on carbonaceous nanomaterials and Fe_3O_4 nanoparticles. *Nanomaterials* **14**: 702.

Doroszkiewicz J, Farhan JA, Mroczko J *et al.* (2023) Common and trace metals in Alzheimer's and Parkinson's diseases. *International Journal of Molecular Sciences* **24**: 15721.

Doyo AN, Kumar R and Barakat MA (2023) Recent advances in cellulose, chitosan, and alginate based biopolymeric composites for adsorption of heavy metals from wastewater. *Journal of the Taiwan Institute of Chemical Engineers* **151**: 105095

Elella MHA, Goda ES, Gab-Allah MA *et al.* (2022) Biodegradable Polymeric nanocomposites for wastewater treatment. In *Advances in Nanocomposite Materials for Environmental and Energy Harvesting Applications*. Springer, Cham, Switzerland, pp. 245–298.

Ethaib S, Al-Qutaifia S, Al-Ansari N and Zubaidi SL (2022) Function of nanomaterials in removing heavy metals for water and wastewater remediation: a review. *Environments* **9**: 123.

Fang K, Deng L, Yin J *et al.* (2022) Recent advances in starch-based magnetic adsorbents for the removal of contaminants from wastewater: a review. *International Journal of Biological Macromolecules* **218**: 909–929.

Fernandes NAT, Simões LA and Dias DR (2023) Comparison of biodegradability, and toxicity effect of biosurfactants with synthetic surfactants. In *Advancements in Biosurfactants Research*. Springer, Cham, Switzerland, pp. 117–136.

Field RW, She Q, Siddiqui FA and Fane AG (2021) Reverse osmosis and forward osmosis fouling: a comparison. *Discover Chemical Engineering* **1**(1) p 6.

Fobang EY (2023) *Application of Electro-flotation Method to Agricultural and Livestock Wastewater Treatment.* Doctoral dissertation, Saitama University, Saitama City, Japan.

Ganesh G, Philominal P, Naga Mahesh LV and Sravya P (2022) Impact of heavy metal pollution on fish health and its consequent effect on human health. *Journal of Experimental Zoology India* **25**.

Gómez-García M-Á and Dobrosz-Gómez I (2024) The scope of alum coagulation-flocculation assisted by slaked lime for the treatment of industrial wastewater containing highly concentrated Acid Black 194 dye. Optimization, molecular weight distribution and toxicity analysis. *Results in Engineering* **23**: 102676.

Gong Z, Chan HT, Chen Q and Chen H (2021) Application of nanotechnology in analysis and removal of heavy metals in food and water resources. *Nanomaterials* **11**: 1792.

Gouma V, Tziasiou C, Pournara AD and Giokas DL (2022) A novel approach to sorbent-based remediation of soil impacted by organic micropollutants and heavy metals using granular biochar amendment and magnetic separation. *Journal of Environmental Chemical Engineering* **10**: 107316.

Gul Zaman H, Baloo L, Pendyala R *et al.* (2021) Produced water treatment with conventional adsorbents and MOF as an alternative: a review. *Materials* **14**: 7607.

Haleem N, Osabutey A, Albert K *et al.* (2023) Flocculation of livestock wastewater using cationic starch prepared from potato peels. *Environmental Science: Water Research and Technology* **9**: 1690–1700.

Hamoudi SA, Khelifa N, Nouri L *et al.* (2023) Removal of Pb^{2+} and Cd^{2+} by adsorption onto Y zeolite and its selectivity of retention in an actual contaminated effluent. *Colloid and Polymer Science* **301**: 631–645.

Hamraoui L, Bergani A, Ettoumi M *et al.* (2024) Towards a circular economy in the mining industry: possible solutions for water recovery through advanced mineral tailings dewatering. *Minerals* **14**: 319.

Hasan MF, Nur-E-Alam M, Salam MA *et al.* (2021) Health risk and water quality assessment of surface water in an urban river of Bangladesh. *Sustainability* **13**: 6832.

Haswell F, Edelenbosch OY, Piscicelli L and van Vuuren DP (2024) The geography of circularity missions: a cross-country comparison of circular economy policy approaches in the Global North and Global South. *Environmental Innovation and Societal Transitions* **52**: 100883.

Hu Z, Wu K, Wang Z *et al.* (2024) Research progress of magnetic flocculation in water treatment. *Magnetochemistry* **10**: 56.

Hussain MK, Khatoon S, Nizami G *et al.* (2024) Unleashing the power of bio-adsorbents: efficient heavy metal removal for sustainable water purification. *Journal of Water Process Engineering* **64**: 105705.

Igberase E, Osifo P and Ofomaja A (2018) Mathematical modelling of Pb^{2+}, Cu^{2+}, Ni^{2+}, Zn^{2+}, Cr^{6+} and Cd^{2+} ions adsorption from a synthetic acid mine drainage onto chitosan derivative in a packed bed column. *Environmental Technology* **39**: 3203–3220.

Inobeme A, Mathew JT, Adetunji CO *et al.* (2023) Recent advances in nanotechnology for remediation of heavy metals. *Environmental Monitoring and Assessment* **195**: 111.

Iqbal MS, Aslam AA, Iftikhar R *et al.* (2023) The potential of functionalized graphene-based composites for removing heavy metals and organic pollutants. *Journal of Water Process Engineering* **53**: 103809.

Jagaba AH, Lawal IM, Birniwa AH *et al.* (2024) Sources of water contamination by heavy metals. In *Membrane Technologies for Heavy Metal Removal from Water*. CRC Press, Boca Raton, FL, USA, pp. 3–27.

Janani R, Gurunathan B, Sivakumar K *et al.* (2022) Advancements in heavy metals removal from effluents employing nano-adsorbents: way towards cleaner production. *Environmental Research* **203**: 111815.

Javed A and Mushtaq A (2023) A critical review of electrocoagulation and other electrochemical methods. *International Journal of Chemical and Biochemical Sciences* **23**: 98–110.

Jomova K, Alomar SY, Nepovimova E *et al.* (2024) Heavy metals: toxicity and human health effects. *Archives of Toxicology* 1–57.

Jun KC, Raman AAA and Buthiyappan A (2020) Treatment of oil refinery effluent using bio-adsorbent developed from activated palm kernel shell and zeolite. *RSC Advances* **10**: 24079–24094.

Kaczorowska MA (2023) The latest achievements of liquid membranes for rare earth elements recovery from aqueous solutions – a mini review. *Membranes* **13**: 839.

Khaki MRD, Shafeeyan MS, Raman AAA and Daud WMAW (2017) Application of doped photocatalysts for organic pollutant degradation – a review. *Journal of Environmental Management* **198**: 78–94, 10.1016/j.jenvman.2017.04.099.

Khalid Z Bin, Nasrullah M, Nayeem A *et al.* (2020) Application of 2D graphene-based nanomaterials for pollutant removal from advanced water and wastewater treatment processes. In *Adapting 2D Nanomaterials for Advanced Applications*. ACS Publications, Washington, DC, USA, pp. 191–217.

Khalidi-Idrissi A, Hartal O, Madinzi A *et al.* (2024) Natural flotation and coagulation–flocculation: a dual approach to refinery wastewater treatment. *Euro-Mediterranean Journal for Environmental Integration* **10**(3):1425–1443.

Khan FSA, Mubarak NM, Khalid M *et al.* (2020) Magnetic nanoadsorbents' potential route for heavy metals removal—a review. *Environmental Science and Pollution Research* **27**: 24342–24356.

Khezerlou A, Tavassoli M, Abedi-Firoozjah R *et al.* (2025) MOFs-based adsorbents for the removal of tetracycline from water and food samples. *Scientific Reports* **15**: 502.

Ko YG (2024) Hybrid method integrating adsorption and chemical precipitation of heavy metal ions on polymeric fiber surfaces for highly efficient water purification. *Chemosphere* **142909**.

Kong Q, Shi X, Ma W *et al.* (2021) Strategies to improve the adsorption properties of graphene-based adsorbent towards heavy metal ions and their compound pollutants: a review. *Journal of Hazardous Materials* **415**: 125690.

Kook S, Lee C, Nguyen TT *et al.* (2018) Serially connected forward osmosis membrane elements of pressure-assisted forward osmosis-reverse osmosis hybrid system: process performance and economic analysis. *Desalination* **448**: 1–12.

Kuldeep, Badenhorst WD, Kauranen P *et al.* (2021) Bipolar membrane electrodialysis for sulfate recycling in the metallurgical industries. *Membranes* **11**: 718.

Kumar P, Kumar A and Kumar R (2021) Phytoremediation and nanoremediation. In *New Frontiers of Nanomaterials in Environmental Science*. Springer, Cham, Switzerland, pp. 281–297.

Kumar S, Mondal P and Purkait MK (2023) Hazardous effects of heavy metals from industrial wastewaters and their remediation through green technology. In *Green Technologies for Industrial Waste Remediation*. Springer Nature, Cham, Switzerland, pp. 29–55.

Law JY and Mohammad AW (2018) Osmotic concentration of succinic acid by forward osmosis: Influence of feed solution pH and evaluation of seawater as draw solution. *Chinese Journal of Chemical Engineering* **26**: 976–983.

Lazar MM, Ghiorghita C-A, Dragan ES *et al.* (2023) Ion-imprinted polymeric materials for selective adsorption of heavy metal ions from aqueous solution. *Molecules* **28**: 2798.

Li M and Lin S (2024) Elucidating the use of pressure-recovery diagrams for analyzing energy consumption in reverse osmosis desalination. *Desalination* **591**: 118033.

Li X, Zhang Q and Yang B (2020) Co-precipitation with $CaCO_3$ to remove heavy metals and significantly reduce the moisture content of filter residue. *Chemosphere* **239**: 124660.

Liang L, Deng Y, Wang W *et al.* (2024) Influences of lower pH on phytoplankton growth in alkaline lakes after water transfer: insights from a coupled hydrodynamic–algal ecological model and experimental analysis. *Environmental Research* **257**: 119201.

Liao Y, Tian M, Goh S *et al.* (2018) Effects of internal concentration polarization and membrane roughness on phenol removal in extractive membrane bioreactor. *Journal of Membrane Science* **563**: 309–319.

Lin W (2020) *Micellar- and Polymer-enhanced Ultrafiltration for Heavy Metal and Sulfate Removal from Aqueous Solutionss.* Doctoral dissertation, Memorial University of Newfoundland, Canada.

Liu C and Zhang H-X (2022) Modified-biochar adsorbents (MBAs) for heavy-metal ions adsorption: A critical review. *Journal of Environmental Chemical Engineering* **10**: 107393.

Liu K-G, Bigdeli F, Sharifzadeh Z *et al.* (2023a) Role of metal-organic framework composites in removal of inorganic toxic contaminants. *Journal of Cleaner Production* **404**: 136709.

Liu Z, Xu J, Liu M *et al.* (2023b) Remote sensing and geostatistics in urban water-resource monitoring: a review. *Marine and Freshwater Research* **74**: 747–765.

Ma Y, Li C, Yan J *et al.* (2024) Application and mechanism of carbonate material in the treatment of heavy metal pollution: a review. *Environmental Science and Pollution Research* **31**: 36551–36576.

Madhav S, Ahamad A, Singh AK *et al.* (2020) Water pollutants: sources and impact on the environment and human health. In *Sensors in Water Pollutants Monitoring: Role of Material.* Springer, Cham, Switzerland, pp. 43–62.

Maneechakr P and Mongkollertlop S (2020) Investigation on adsorption behaviors of heavy metal ions (Cd^{2+}, Cr^{3+}, Hg^{2+} and Pb^{2+}) through low-cost/active manganese dioxide-modified magnetic biochar derived from palm kernel cake residue. *Journal of Environmental Chemical Engineering* **8**: 104467.

Mansor ES, Ali EA and Shaban AM (2021) Tight ultrafiltration polyethersulfone membrane for cheese whey wastewater treatment. *Chemical Engineering Journal* **407**: 127175.

Marimuthu T, Chee CY and Sulaiman NMN (2023) A review on the use of cellulose nanomaterials for wastewater remediation of heavy metal ions. *International Journal of Environmental Science and Technology* **20**: 3421–3436.

Merrikhpour H, Mobarakpour S and Azimi SB (2022) Adsorption of Cd^{2+}, Cu^{2+}, and Ni^{2+} onto surfactant modified bentonite. *Desalination and Water Treatment* **271**: 157–165.

Mesquita CRS, Gómez AOC, Cotta CPN and Cotta RM (2025) Comparison of different polymeric membranes in direct contact membrane distillation and air gap membrane distillation configurations. *Membranes* **15**: 91.

Miranda LS, Ayoko GA, Egodawatta P and Goonetilleke A (2022) Adsorption-desorption behavior of heavy metals in aquatic environments: influence of sediment, water and metal ionic properties. *Journal of Hazardous Materials* **421**: 126743.

Mokoena LS and Mofokeng JP (2023) A review on graphene (GN) and graphene oxide (GO) based biodegradable polymer composites and their usage as selective adsorbents for heavy metals in water. *Materials* **16**: 2527.

Mondal S and Majumder SK (2020) Fabrication of the polysulfone-based composite ultrafiltration membranes for the adsorptive removal of heavy metal ions from their contaminated aqueous solutions. *Chemical Engineering Journal* **401**: 126036.

Moneer AA (2023) The potential of hybrid electrocoagulation-membrane separation processes for performance enhancement and membrane fouling mitigation: a review. *Egyptian Journal of Aquatic Research* **49**: 269–282.

Mourdikoudis S and Dominguez-Benetton X (2025) Physicochemical vs electrochemical technologies for metal recovery–main insights, comparison, complementarity and challenges. chemistry-methods. *Chemistry-Methods* **5**(3): e202400046.

Mubarak MF, Selim H and Elshypany R (2022) Hybrid magnetic core–shell TiO_2@ $CoFe_3O_4$ composite towards visible light-driven photodegradation of Methylene blue dye and the heavy metal adsorption: isotherm and kinetic study. *Journal of Environmental Health Science and Engineering* **20**: 265–280.

Mukherjee S, Patel AK and Kumar M (2020) Water scarcity and land degradation nexus in the anthropocene: reformations for advanced water management as per the sustainable development goals. In *Emerging Issues in the Water Environment during Anthropocene: A South East Asian Perspective*. Springer, Cham, Switzerland, pp. 317–336.

Nibe RL and Gaikwad RW (2025) A comprehensive review on application of different natural and chemically modified nanomaterials adsorbents for heavy metals removal. *Journal of Inorganic and Organometallic Polymers and Materials*. Springer, Singapore, pp. 1–21.

Nippatlapalli N and Selvaraj A (2022) A study on the removal of long chain perflourodecanoic acid in simulated aqueous solution using enhanced electrochemical technique: Metabolites, kinetic and isotherm model analysis. *Journal of Water Process Engineering* **49**: 103045

Ohiagu FO, Chikezie PC, Ahaneku CC and Chikezie CM (2022) Human exposure to heavy metals: toxicity mechanisms and health implications. *Materials Science and Engineering* **6**: 78–87.

Öktüren Asri F, Demirtaş EI, Şimşek M *et al.* (2024) Heavy metal contamination and human health risk assessment in the soils grown field crops in Antalya, Türkiye. *International Journal of Environmental Analytical Chemistry* 1–20.

Olawade DB, Wada OZ, Egbewole BI *et al.* (2024) Metal and metal oxide nanomaterials for heavy metal remediation: novel approaches for selective, regenerative, and scalable water treatment. *Frontiers in Nanotechnology* **6**: 1466721.

Onoh EU, Nwanya AC, Shinde NM *et al.* (2024) Understanding the chemistry of graphene oxide on redox flow lithium-ion batteries with a view to enhancing the battery's high-density storage. *Asia-Pacific Journal of Chemical Engineering* **19**: e2995.

Ottink T (2024) *Recycling of Steel Grinding Swarf by way of Production of Iron Chloride Coagulants for Water Treatment*. Chalmers Tekniska Hogskola, Sweden.

Pachaiappan R, Cornejo-Ponce L, Rajendran R *et al.* (2022) A review on biofiltration techniques: recent advancements in the removal of volatile organic compounds and heavy metals in the treatment of polluted water. *Bioengineered* **13**: 8432–8477.

Panda AP, Jha U, Kumar SA and Swain SK (2023) A Simplified and affordable arsenic filter to prevent arsenic poisoning: lab-scale study and pilot experiment. *ACS ES & T Water* **3**: 4092–4102.

Pandit DJ and Sharma AK (2024) Advanced techniques in wastewater treatment: a comprehensive review. *Asian Journal of Environment and Ecology* **23**(10):1–26

Pawlowski S, Crespo JG and Velizarov S (2019) Profiled ion exchange membranes: a comprehensible review. *International Journal of Molecular Sciences* **20**: 165.

Periyasamy AP (2024) Recent advances in the remediation of textile-dye-containing wastewater: prioritizing human health and sustainable wastewater treatment. *Sustainability* **16**: 495.

Phouthavong V, Yan R, Nijpanich S *et al.* (2022) Magnetic adsorbents for wastewater treatment: advancements in their synthesis methods. *Materials* **15**: 1053.

Poudel S, Wickramasinghe D, Mukherjee M *et al.* (2024) The status and prospect on nature-based solution in South Asia: a policy-based analysis. *PLOS Climate* **3**: e0000289.

Priya AK, Gnanasekaran L, Dutta K *et al.* (2022) Biosorption of heavy metals by microorganisms: evaluation of different underlying mechanisms. *Chemosphere* **307**: 135957.

Priyadarshini M, Ahmad A, Das S and Ghangrekar MM (2022) Application of innovative electrochemical and microbial electrochemical technologies for the efficacious removal of emerging contaminants from wastewater: a review. *Journal of Environmental Chemical Engineering* **10**: 108230.

Qasem NAA, Mohammed RH and Lawal DU (2021) Removal of heavy metal ions from wastewater: a comprehensive and critical review. *npj Clean Water* **4**: 1–15.

Radovanovic D, Dikić J, Štulović M *et al.* (2023) Sorption of Pb^{2+}, Zn^{2+}, Cu^{2+} and Ni^{2+} ions on na-enriched natural zeolite for wastewater treatment process: a kinetic approach. *Metallurgical and Materials Engineering* **29**: 20–35.

Rahman MA, Kumar S, Lamb D and Rahman MM (2021) Health risk assessment of arsenic, manganese, and iron from drinking water for high school children. *Water, Air, and Soil Pollution* **232**: 269.

Rajendran S, Priya AK, Kumar PS *et al.* (2022) A critical and recent developments on adsorption technique for removal of heavy metals from wastewater – a review. *Chemosphere* **303**: 135146.

Ramasamy B (2019) Short review of salt recovery from reverse osmosis rejects. *Salt in the Earth* **64**: 80.

Rana AK, Mishra YK, Gupta VK and Thakur VK (2021) Sustainable materials in the removal of pesticides from contaminated water: perspective on macro to nanoscale cellulose. *Science of The Total Environment* **797**: 149129.

Ravikumar K and Udayakumar J (2020) Preparation and characterisation of green clay-polymer nanocomposite for heavy metals removal. *Chemistry and Ecology* **36**: 270–291.

Ren G, Li R, Zhao M *et al.* (2023) Membrane electrodes for electrochemical advanced oxidation processes: preparation, self-cleaning mechanisms and prospects. *Chemical Engineering Journal* **451**: 138907.

Reshma FR (2023) *Iron Settling and Sludge Dewatering for Pigment Production from Acid Mine Drainage Remediation at low pH*. Master's thesis, Ohio University, Athens, OH, USA.

Ribeiro IHS, Reis DT and Pereira DH (2019) A DFT-based analysis of adsorption of Cd^{2+}, Cr^{3+}, Cu^{2+}, Hg^{2+}, Pb^{2+}, and Zn^{2+}, on vanillin monomer: a study of the removal of metal ions from effluents. *Journal of Molecular Modeling* **25**: 267.

Rodríguez C, Briano S and Leiva E (2020) Increased adsorption of heavy metal ions in multi-walled carbon nanotubes with improved dispersion stability. *Molecules* **25**: 3106.

Romano S, Trespi S, Achermann R *et al.* (2023) The role of operating conditions in the precipitation of magnesium hydroxide hexagonal platelets using NaOH solutions. *Crystal Growth and Design* **23**: 6491–6505.

Roy R, Samanta S, Pandit S *et al.* (2024) An overview of bacteria-mediated heavy metal bioremediation strategies. *Applied Biochemistry and Biotechnology* **196**: 1712–1751.

Sabir A, Altaf F, Batool R *et al.* (2021) Agricultural waste absorbents for heavy metal removal. In *Green Adsorbents to Remove Metals, Dyes and Boron from Polluted Water*. Springer, Cham, Switzerland, pp. 195–228.

Safi NN and Waisi BI (2024) Enhanced hydrophobic double-layer nanofibers membranes for direct contact membrane distillation. *Ecological Engineering and Environmental Technology* **25**.

Saleh TA, Mustaqeem M and Khaled M (2022) Water treatment technologies in removing heavy metal ions from wastewater: A review. *Environmental Nanotechnology, Monitoring and Management* **17**: 100617.

Salek Maghsoudi A, Hassani S, Mirnia K and Abdollahi M (2021) Recent advances in nanotechnology-based biosensors development for detection of arsenic, lead, mercury, and cadmium. *International Journal of Nanomedicine* 803–832.

Sanad MN, Okil M, Mostafa AM and ElFaham MM (2024) Cost-effective Cu^{2+} and Pb^{2+} remediation using peanut shell-derived biosorbents: a physicochemical investigation. *International Journal of Environmental Analytical Chemistry*, pp. 1–21.

Sarker A, Kim J-E, Islam ARMT *et al.* (2022) Heavy metals contamination and associated health risks in food webs—a review focuses on food safety and environmental sustainability in Bangladesh. *Environmental Science and Pollution Research* **29**: 3230–3245.

Scrase A and Brereton J (2024) The perfect storm: mine closure in the Latrobe Valley, Victoria. In: *Mine Closure 2024: Proceedings of the 17th International Conference on Mine Closure. Australian Centre for Geomechanics*, pp. 1297–1310.

Shan R, Shi Y, Gu J *et al.* (2020) Single and competitive adsorption affinity of heavy metals toward peanut shell-derived biochar and its mechanisms in aqueous systems. *Chinese Journal of Chemical Engineering* **28**: 1375–1383.

Sharma V, Singh H, Guleria S *et al.* (2022) Application of superparamagnetic iron oxide nanoparticles (SPIONs) for heavy metal adsorption: A 10-year meta-analysis. *Environmental Nanotechnology, Monitoring and Management* **18**: 100716.

Shawky A, Mohamed RM, Alahmadi N and Zaki ZI (2022) Enhanced photocatalytic reduction of hexavalent chromium ions over S-Scheme based 2D MoS_2-supported TiO_2 heterojunctions under visible light. *Colloids and Surfaces A: Physicochemical and Engineering Aspects* **641**: 128564.

Shi K, Yao Y, Gao Y *et al.* (2023) Microbial electrolysis cell as an effective tool for mitigating the toxic effects of Cu^{2+} on microorganisms in the acid mine drainage treatment: Efficacy and mechanisms. *Chemical Engineering Journal* **473**: 145190.

Shmychkova OB, Girenko DV and Velichenko AB (2022) Review on influence of electrode material nature on anodic processes in chloride solutions. *Journal of Chemistry and Technologies* **30**: 312–332.

Shrestha S, Wang B and Dutta P (2020) Nanoparticle processing: understanding and controlling aggregation. *Advances in Colloid and Interface Science* **279**: 102162.

Siddique JA (2023) Nanofiltration membrane use for separation of heavy metals from wastewater. In *Emerging Techniques for Treatment of Toxic Metals from Wastewater*. Elsevier, Amsterdam, The Netherlands, pp. 523–549.

Singh A, Chaudhary S and Dehiya BS (2021) Fast removal of heavy metals from water and soil samples using magnetic Fe_3O_4 nanoparticles. *Environmental Science and Pollution Research* **28**: 3942–3952.

Singh A, Singh BP, Raghav Y and Mishra V (2024) Regulatory frameworks and guidelines aimed at controlling and minimizing heavy metals exposure. In *Heavy Metal Contamination in the Environment*. CRC Press, Boca Raton, FL, USA, pp. 171–179.

Singh A, Srivastava A, Saidulu D and Gupta AK (2022) Advancements of sequencing batch reactor for industrial wastewater treatment: major focus on modifications, critical operational parameters, and future perspectives. *Journal of Environmental Management* **317**: 115305.

Singh S, Paswan SK, Kumar P *et al.* (2023a) Heavy metal water pollution: an overview about remediation, removal and recovery of metals from contaminated water. *Metals in Water* 263–284.

Singh V, Singh N, Rai SN *et al.* (2023b) Heavy metal contamination in the aquatic ecosystem: toxicity and its remediation using eco-friendly approaches. *Toxics* **11**: 147.

Soliman EM, Ahmed SA and Fadl AA (2020) Microwave-enforced green synthesis of novel magnetic nano composite adsorbents based on functionalization of wood sawdust for fast removal of calcium hardness from water samples. *Water Environment Research* **92**: 2112–2128.

Sosun, Ali A, Mannan A *et al.* (2022) Removal of of toxic metal ions (Ni^{2+} and Cd^{2+}) from wastewater by using TOPO decorated iron oxide nanoparticles. *Applied Water Science* **12**: 86.

Srivastav AL, Markandeya, Patel N *et al.* (2023) Concepts of circular economy for sustainable management of electronic wastes: challenges and management options. *Environmental Science and Pollution Research* **30**: 48654–48675.

Suhan MBK, Shuchi SB, Al-Mamun MR *et al.* (2023) Enhanced UV light-driven photocatalytic degradation of methyl orange using MoO_3/WO_3-fluorinated TiO_2 nanocomposites. *Environmental Nanotechnology, Monitoring and Management* **19**, 10.1016/j.enmm.2022.100768.

Sultana M, Ahmmed MR, Ul Hoque MI *et al.* (2023) Biodegradable composite of gelatin blend microcrystalline cellulose for Cd^{2+}, Pb^{2+}, and Cr^{3+} adsorption from an aqueous solution. *Advances in Polymer Technology* **2023**: 1893660.

Sultana M, Rownok MH, Sabrin M *et al.* (2022) A review on experimental chemically modified activated carbon to enhance dye and heavy metals adsorption. *Cleaner Engineering and Technology* **6**: 100382.

Sun K, Nguyen NN and Nguyen A V (2024) Unique selective activating effect of lead ions on the flotation of NaCl crystals against KCl crystals with fatty acid collector colloids. *Minerals Engineering* **216**: 108908.

Suter E, Rutto H and Omwoyo W (2025) Continuous packed bed column removal of Cr^{6+}, Cd^{2+}, and Pb^{2+} ions from synthetic wastewater using polymeric ultra-permeable and biodegradable ferromagnetic nanocomposite membrane. *The Canadian Journal of Chemical Engineering* **103**(6): 2642–2658.

Talema A (2023) Causes, negative effects, and preventive methods of water pollution in Ethiopia. *Quality Assurance and Safety of Crops and Foods* **15**: 129–139.

Teschke R (2022) Aluminum, arsenic, beryllium, cadmium, chromium, cobalt, copper, iron, lead, mercury, molybdenum, nickel, platinum, thallium, titanium, vanadium, and zinc: molecular aspects in experimental liver injury. *International Journal of Molecular Sciences* **23**: 12213.

Torrinha Á, Oliveira TMBF, Ribeiro FWP *et al.* (2022) Advantages and limitations of functionalized graphene-based electrochemical sensors for environmental monitoring. In *Functionalized Nanomaterial-Based Electrochemical Sensors*. Elsevier, Amsterdam, The Netherlands, pp. 487–520.

Tsade H, Murthy HA and Muniswamy D (2020) Bio-sorbents from agricultural wastes for eradication of heavy metals: a review. *Journal of Materials and Environmental Science* **11**: 1719–1735.

Upadhyay SK, Rani N, Kumar V *et al.* (2023) A review on simultaneous heavy metal removal and organo-contaminants degradation by potential microorganisms: current findings and future outlook. *Microbiological Research* **273**: 127419.

Vo TS, Hossain MM, Jeong HM and Kim K (2020) Heavy metal removal applications using adsorptive membranes. *Nano Convergence* **7**: 36.

Wang G, Xiang J, Liang G *et al.* (2023a) Application of common industrial solid waste in water treatment: a review. *Environmental Science and Pollution Research* **30**: 111766–111801.

Wang LK and Wang M-HS (2023) Sustainable water and wastewater treatment systems consisting of magnesium coagulation-precipitation, dissolved air flotation, recarbonation, and filtration. In *Control of Heavy Metals in the Environment*, Volume **1**. CRC Press, Boca Raton, FL, USA, pp. 207–231.

Wang LK, Wang M-HS, Shammas NK and Hahn HH (2021) Physicochemical treatment consisting of chemical coagulation, precipitation, sedimentation, and flotation. In *Integrated Natural Resources Research*. Springer, Cham, Switzerland, pp. 265–397.

Wang R, Liu Y, Lu Y *et al.* (2023b) Fabrication of a corn stalk derived cellulose-based bioadsorbent to remove Congo red from wastewater: Investigation on its ultra-high adsorption performance and mechanism. *International Journal of Biological Macromolecules* **241**: 124545.

Wang X, Fu R, Li H *et al.* (2020) Heavy metal contamination in surface sediments: a comprehensive, large-scale evaluation for the Bohai Sea, China. *Environmental Pollution* **260**: 113986.

Wang Z, Luo P, Zha X *et al.* (2022) Overview assessment of risk evaluation and treatment technologies for heavy metal pollution of water and soil. *Journal of Cleaner Production* **379**: 134043.

Wu H, Lin G, Liu C *et al.* (2022) Progress and challenges in molecularly imprinted polymers for adsorption of heavy metal ions from wastewater. *Trends in Environmental Analytical Chemistry* **36**: e00178.

Wu L, Garg S and Waite TD (2024) Progress and challenges in the use of electrochemical oxidation and reduction processes for heavy metals removal and recovery from wastewaters. *Journal of Hazardous Materials* **135581**.

Xu J, Li Y, Wang S *et al.* (2023) Sources, transfers and the fate of heavy metals in soil-wheat systems: the case of lead (Pb)/zinc (Zn) smelting region. *Journal of Hazardous Materials* **441**: 129863.

Xu S, Wan L, Liu J *et al.* (2024a) Confined PbO_2 nanoparticles in carbon nanotubes on boron-doped diamond substrate for highly efficient chlortetracycline degradation. *Applied Surface Science* **648**: 159085.

Xu Z, Jaiswal A, Liu X *et al.* (2024b) Microfluidics Evolution and Surface Functionalization: A Pathway to Enhanced Heavy Metal Ion Detection. *Advanced Sensor Research* **3**: 2400008.

Yadav M, Singh G and Jadeja RN (2021) Physical and chemical methods for heavy metal removal. In *Pollutants and Water Management: Resources, Strategies and Scarcity*. Wiley, Hoboken, NJ, USA, pp. 377–397.

Yang J, Sun Y, Wang Z *et al.* (2022a) Heavy metal pollution in agricultural soils of a typical volcanic area: risk assessment and source appointment. *Chemosphere* **304**: 135340.

Yang L, Meng F, Ma C and Hou D (2022b) Elucidating the spatial determinants of heavy metals pollution in different agricultural soils using geographically weighted regression. *Science of The Total Environment* **853**: 158628.

Yang Y, Li X, Wan C *et al.* (2024) A comprehensive review of cellulose nanomaterials for adsorption of wastewater pollutants: focus on dye and heavy metal Cr adsorption and oil/water separation. *Collagen and Leather* **6**: 35.

Yantasee W, Rutledge RD, Chouyyok W *et al.* (2010) Functionalized nanoporous silica for the removal of heavy metals from biological systems: adsorption and application. *ACS Applied Materials and Interfaces* **2**: 2749–2758.

Yaqoob AA, Idris MO, Ahmad A *et al.* (2022) Removal of toxic metal ions from wastewater through microbial fuel cells. In *Microbial Fuel Cells for Environmental Remediation*. Springer Nature, Singapore, pp. 299–325.

Yuan H, Miao Z, Wan C *et al.* (2025) Recent advances in centrifugal microfluidics for point-of-care testing. *Lab on a Chip*. Royal Society of Chemistry, London, UK.

Zaimee MZA, Sarjadi MS and Rahman ML (2021) Heavy metals removal from water by efficient adsorbents. *Water* **13**: 2659.

Zeng X, Zhang G, Zhu J and Wu Z (2022) Adsorption of heavy metal ions in water by surface functionalized magnetic composites: a review. *Environmental Science: Water Research and Technology* **8**: 907–925.

Zhai M, Fu B, Zhai Y *et al.* (2023) Simultaneous removal of pharmaceuticals and heavy metals from aqueous phase by way of adsorptive strategy: a critical review. *Water Research* **236**: 119924.

Zhang B, Peng C, Zhang S *et al.* (2023a) Comprehensive analysis of the combined flocculation and filtration process for microalgae harvesting at various operating parameters. *Science of the Total Environment* **857**: 159658.

Zhang P, Yang M, Lan J *et al.* (2023b) Water quality degradation due to heavy metal contamination: health impacts and eco-friendly approaches for heavy metal remediation. *Toxics* **11**: 828.

Zhang Q, Ma L, Peng Y and Yan X (2024) Sustainable bioleaching of heavy metals from coal tailings using Bacillus inaquosorum B. 4: Mechanistic insights and environmental implications. *Journal of Environmental Chemical Engineering* **12**: 113400.

Zhou S, Ogawa Y and Kawai K (2023) Influence of chloride solutions on the leaching of heavy metals from cement hydrates. *Construction and Building Materials* **378**: 131108.

Zhu H, Zhang W, Yang Z *et al.* (2024) Enhanced blue emission of $KBaYSi_2O_7$: Bi^{3+} phosphors through partial substitution of Si^{4+} with Al^{3+} and Ge^{4+} for white light-emitting diodes. *Journal of Alloys and Compounds* **1004**: 175908.

Zhu Y, Fan W, Zhou T and Li X (2019) Removal of chelated heavy metals from aqueous solution: A review of current methods and mechanisms. *Science of the Total Environment* **678**: 253–266.

Zou Y (2024) Cu^{2+}, Cd^{2+}, and Pb^{2+} ions adsorption from wastewater using polysaccharide hydrogels made of oxidized carboxymethyl cellulose and chitosan grafted with catechol groups. *Iranian Polymer Journal* **33**: 57–66.

Microbial Processes in Environmental Engineering: From Theory to Practice
Rajneesh Kumar, Ram Sharan Singh and Maulin P Shah
ISBN 978-1-83708-327-5
https://doi.org/10.1108/978-1-83708-326-820261005

Chapter 5
Sustainable biohydrogen through microbial fuel cells: a green revolution

Ashwini S Thakre, Diwakar Z Shende and Kailas L Wasewar

Advanced Separation and Analytical Laboratory (ASAL), Department of Chemical Engineering, Visvesvaraya National Institute of Technology (VNIT), Nagpur, India

Abstract

Growing environmental concerns worldwide and the quick depletion of finite fossil resources like coal and petroleum underscore the urgent need for sustainable energy technologies. Microbial fuel cells (MFCs) offer a potentially effective carbon-neutral solution for biohydrogen generation by leveraging the microorganism's metabolic processes to convert organic waste into hydrogen energy. This environmentally sustainable technology employs electrogenic bacteria that generate electricity through biochemical reactions involving organic materials, especially wastewater. An MFC functions as a bioreactor that uses anaerobic microbial catalytic processes to convert the chemical energy of organic substances into electrical energy. This innovative approach not only facilitates sustainable hydrogen production but also effectively manages waste. Microorganisms in the MFC produce biohydrogen by oxidising organic compounds at the anode, which releases electrons that move to the cathode by way of an external circuit. Additionally, the MFC process aids pollutant removal, serving a dual-purpose role in hydrogen production and the purification of wastewater. Advances in electrode materials, microbial community optimisation and reactor designs have significantly enhanced MFC efficiency and scalability. Despite ongoing challenges such as low power outputs, operational costs and technical complexities, continuous research efforts are expanding the applicability and effectiveness of MFC technology. Consequently, MFCs hold substantial promise as a critical innovation for integrating renewable energy production with environmental sustainability.

Graphical abstract

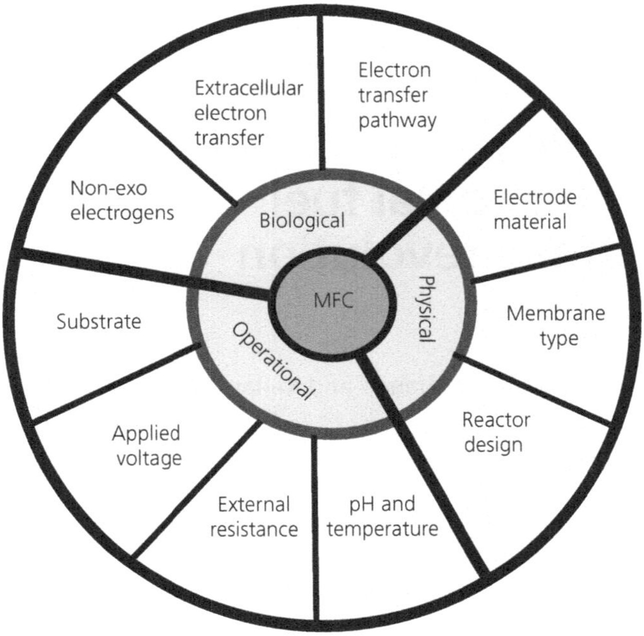

Keywords: Microbial fuel cells; biohydrogen; microorganisms; sustainability.

5.1. Introduction

Global population growth has rapidly accelerated economic growth and industrialisation, consequently elevating global energy demands. Meeting this rising energy requirement remains a vital challenge, influencing both economic advancement and environmental sustainability (Sharma *et al.*, 2023). Historically, fossil fuels have dominated the energy landscape, constituting nearly 88% of energy consumption within industries (Le *et al.*, 2024). Nonetheless, prolonged dependence on fossil fuels raises serious concerns related to resource scarcity and environmental harm (Wang *et al.*, 2024). Furthermore, combustion of fossil fuels significantly impacts public health by contributing to air pollution, linked to various health conditions such as respiratory disorders, cardiovascular diseases and developmental issues, especially affecting children (Li *et al.*, 2024). As reported by Jiang *et al.*, fossil fuel combustion contributes around 21.6 billion tons of carbon dioxide (CO_2) per year, with transportation being a prominent source. Addressing these issues necessitates an urgent transition to sustainable energy alternatives such as solar power, wind energy and hydrogen fuel (Yusaf *et al.*, 2024).

Among clean energy alternatives such as solar, wind, hydropower, geothermal, biogas, solid biomass and biogas, hydrogen has emerged prominently due to its capability to substitute fossil fuels across diverse sectors. Hydrogen is increasingly recognised as the optimal zero-carbon fuel source capable of fulfilling international energy policy commitments towards sustainability (Ramos *et al.*, 2020). Notably, hydrogen possesses an exceptional energy density of approximately 142 MJ/kg, significantly surpassing conventional fuels such as methane (57 MJ/kg), natural gas (58 MJ/kg) and gasoline (49 MJ/kg) (Ishaq *et al.*, 2022).

Biohydrogen's attractiveness stems from its excellent energy-carrying capacity, superior energy density, exceptional conversion efficiency (reaching up to 99% in fuel cells), and the absence of greenhouse gas emissions during combustion, positioning it as a highly reliable and sustainable energy reserve (Acar *et al.*, 2014). Although hydrogen is the most plentiful and lightest element, it is mostly found in compounds rather than free form. As a fuel, hydrogen combustion emits solely water vapour, thus eliminating harmful pollutants and significantly curbing greenhouse gases and air pollution (Hassan *et al.*, 2024; Nagar *et al.*, 2023).

Hydrogen's versatility extends to numerous applications, such as fuel cell vehicles, chemical processing, food manufacturing, aerospace technologies, material processing, portable energy solutions and backup power systems, as depicted in Figure 5.1. Within fuel cells, hydrogen efficiently produces electricity while emitting zero pollutants (Dincer *et al.*, 2011a).

5.1.1 Microbial fuel cells: their role in biohydrogen production

The concept of harnessing energy through microbial processes dates back to 1911, introduced by MC Potter, who demonstrated microorganisms' capability to convert organic material into usable energy. Given hydrogen's anticipated critical role in future energy strategies,

Figure 5.1 Hydrogen's role as an energy carrier

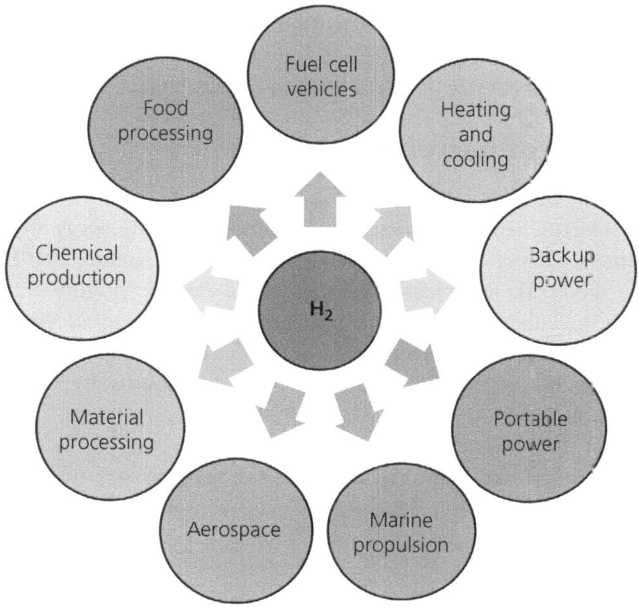

biohydrogen (hydrogen derived through biological means) is gaining importance as a sustainable energy resource. MFCs offer a promising pathway for biohydrogen production, utilising microbial membrane potentials to generate bioelectricity under controlled electron-accepting environments. An MFC is a bioelectrochemical device that uses microorganisms to convert chemical energy from organic compounds into electricity through anaerobic redox processes (Kumari and Das, 2017; Rahman *et al.*, 2016; Venkata Mohan *et al.*, 2010).

Within MFC systems, microbial respiration (reduction) and fermentation (oxidation) occur separately, divided by an ion-selective membrane, commonly a proton exchange membrane (PEM). Artificial electron acceptors or electrodes facilitate bioelectricity generation. The absence of oxygen at the cathode not only allows power production but also supports the synthesis of diverse valuable compounds (Rozendal *et al.*, 2006). Depending on electron acceptor conditions, MFCs can extend their application beyond wastewater treatment, aiding in the synthesis of various chemicals. Some reactions require minimal potentials, achievable through microbial-generated potentials, whereas others demand external electrical input to surpass higher energy thresholds (Rozendal *et al.*, 2008).

Hydrogen production through cathodic reduction has been extensively explored in microbial electrolysis cells (MECs), and specialised variants of MFCs have been designed explicitly for biohydrogen generation. MECs integrate microbial metabolic activity with electrochemical processes, using microbial-produced electricity complemented by minimal external voltage, thereby enhancing proton reduction at the cathode (Ahmed *et al.*, 2021).

This chapter underscores microbial fuel cells' potential as sustainable biohydrogen production systems. By converting organic substrates into electricity through microbial action, MFCs only require slight external voltage inputs to stimulate hydrogen evolution at the cathode. Such methods have considerable potential for decentralised, scalable biohydrogen production, significantly reducing dependency on fossil fuels and advancing a sustainable bioeconomy.

5.2. Principles of microbial fuel cells

5.2.1 Microbial fuel cell operation

An MFC is a bioelectrochemical device that converts chemical energy from organic substrates into electricity by way of microbial catalysis (Dutta and Kundu, 2014). MFCs oxidise organic matter at the anode and create hydrogen at the cathode (Figure 5.2). Electroactive microorganisms at the anode oxidise organic substrates and transmit electrons to electrodes by way of microbial metabolic pathways (Selim *et al.*, 2017). Biological catalysts near the cathode enhance reduction processes by mixing protons (transported by PEM) with electrons and oxygen to produce water.

Wang *et al.* (2015) found that microbial cells can release electrons from a variety of substrates, including glucose, acetate, ethanol, cysteine, complicated wastewater and bovine serum albumin compositions.

The quantity of hydrogen produced anaerobically is directly related to electrical power generation. At the cathode, electrons and protons unite to form hydrogen gas rather than

Figure 5.2 Microbial fuel cell

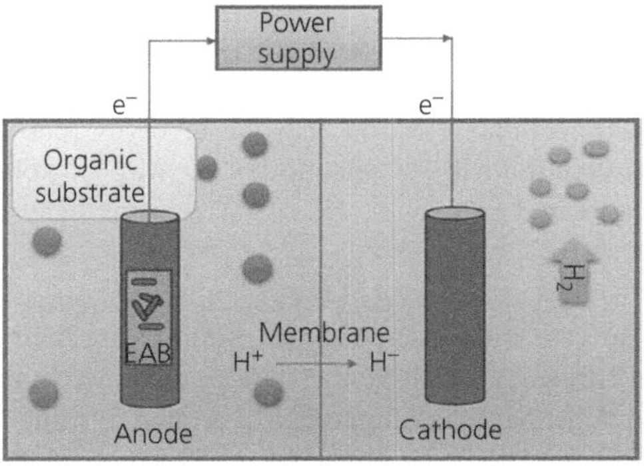

electricity, a process known as electro-hydrogenesis, characterised by bacterial electron emission (Hassan *et al.*, 2021). MFCs uniquely combine bacterial organic matter electrolysis and externally assisted electrochemical reactions within a single unit.

The corresponding electrochemical reactions at the anode and cathode are summarised below.

Anode reaction: $CH_3COO^- + 4H_2O \rightarrow 2HCO_3^- + 9H^+ + 8e^-$ (5.1)
Cathode reaction: $8H^+ + 8e^- \rightarrow 4H_2$ (5.2)
Overall reaction: $CH_3COO^- + 4H_2O \rightarrow 2HCO_3^- + H^+ + 4H_2$ (5.3)

At the anode, acetate ions react with water to form bicarbonate ions, protons and electrons. At the cathode, the protons and electrons combine to generate hydrogen gas. The overall reaction involves acetate ions and water converting into bicarbonate ions, hydrogen ions and hydrogen gas.

A typical MFC comprises an anode and cathode, separated by a PEM. Microorganisms engage in metabolic activities, both aerobic and anaerobic, utilising substrates and generating reducing equivalents (protons and electrons) carried by redox cofactors such as NAD^+, FAD^+ and FMN^+. These reactions generate adenosine triphosphate (ATP) essential for cellular respiration (Kim *et al.*, 2008; Nelson *et al.*, 2008). Glycolysis, occurring under aerobic and anaerobic conditions, breaks glucose into pyruvate. Aerobic metabolism continues by way of the tricarboxylic acid (TCA) cycle, releasing CO_2 and H_2O, whereas anaerobic metabolism yields CH_4, CO_2 and H_2O. Both processes generate energy-rich reducing equivalents directed toward terminal electron acceptors (TEA), creating a proton-driven force that powers the ATP synthesis.

Electron flow depends on TEA availability and hierarchy; in oxygen-rich conditions, O_2 possessing the highest biological reduction potential functions as the electron acceptor, driving oxidative phosphorylation (Kim *et al.*, 2008; Nelson *et al.*, 2008). Without oxygen, electrons are redirected through alternative available acceptors (Schröder, 2007). Crucially, microbial adhesion to electrode surfaces and subsequent electron transfer mechanisms underpin MFC technology. Electrons must traverse microbial membranes to electrodes either directly by way of reduced substances or through electron shuttling mediated by membrane-bound enzymes. This extracellular electron transfer involves redox-active intermediates, linking microorganisms electronically to electrodes by way of soluble redox mediators, metabolites or membrane-bound proteins.

5.2.1.1 Single-chamber microbial fuel cell

Single-chamber MFCs operate without membranes, simplifying their design but complicating hydrogen separation efficiency (Sharma *et al.*, 2023) as shown in Figure 5.3. As substrate concentration declines, anodic bacteria often reuse hydrogen forming internal electron loops that decrease hydrogen recovery and energy efficiency (Kadier *et al.*, 2014a). Single-chamber setups also risk enhancing hydrogenotrophic methanogenesis, causing unwanted methane production, reducing product quality and conversion rates (Rousseau *et al.*, 2020).

Various methods have been proposed to control methanogenesis, including the application of antibiotics, using pure microbial cultures, maintaining lower temperatures or acidic environments, employing positively charged anodes, restricting bicarbonate availability and applying periodic aeration (Yang *et al.*, 2021). Despite these limitations, single-chamber MFCs provide benefits like lower production costs, elimination of membrane fouling and reduced biodegradation rates (Ghasemi *et al.*, 2020). Guo *et al.* (2020) achieved hydrogen production of 3.16 ± 0.03 m^3 H_2/m^3 reactor/day using single-chamber, membrane-free MFCs equipped with graphite brush anodes, operated at an applied external voltage of 0.8 V. Enhanced efficiency was observed by minimising internal resistance and electrode spacing, particularly using *Geobacter sulfurreducens* at 1.1 V (Kadier *et al.*, 2018). Optimal hydrogen output in single-chamber MFCs involves narrow electrode spacing to minimise potential losses.

Figure 5.3 Single-chamber MFC

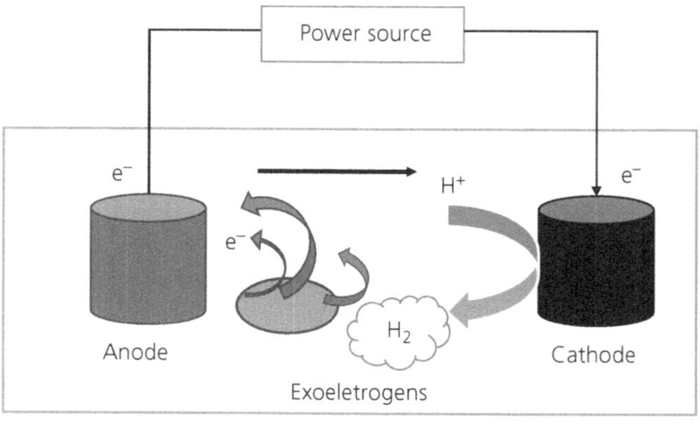

5.2.1.2 Double-chamber MFC

Double-chamber MFCs incorporate a membrane that partitions the anode and cathode chambers, minimising catholyte and anolyte interaction, reducing microbial contamination at the anode and enhancing hydrogen purity (see Figure 5.4) Diverse configurations of double-chamber MFCs, such as disc, cube, rectangular, bottle and concentric tube shapes, have been extensively researched (Sinbuathong *et al.*, 2021; Singh *et al.*, 2018). Temperature and pH significantly affect metabolic activity and hydrogen output in these systems. Typically, higher pH at the cathode and lower pH at the anode are optimal (Sun *et al.*, 2009). Although double-chamber MFCs have higher membrane costs and elevated resistance, they enable efficient hydrogen production at a suitable pH of around 6.5 for various substrates. However, complexity and practical challenges associated with scaling up remain important considerations.

5.2.2 Microbial processes producing hydrogen

Four principal biological methods for generating biohydrogen include (*a*) biophotolysis, which uses cyanobacteria and algae; (*b*) photofermentation, employing photosynthetic bacteria; (*c*) dark fermentation by anaerobic bacteria; and (*d*) integrated biophotolysis, which combines two sequential photosynthetic reactions, similar to the natural photosynthesis seen in plants, which harnesses sunlight to split water molecules directly, releasing hydrogen gas as a by product – that is, combining two photosystems from plant-type photosynthesis for water splitting and hydrogen generation by algae and cyanobacteria as seen in Figure 5.5.

Microbial metabolism significantly influences hydrogen production within MFCs, where diverse pathways convert organic substrates into hydrogen. Hydrogen generation in MFCs primarily involves fermentative and electroactive bacteria that oxidise substrates, transferring electrons to the anode and generating protons.

Key metabolic processes include dark fermentation, photofermentation and electrohydrogenesis. Dark fermentation involves facultative anaerobes like *Clostridium* and *Enterobacter*

Figure 5.4 Double-chamber MFC

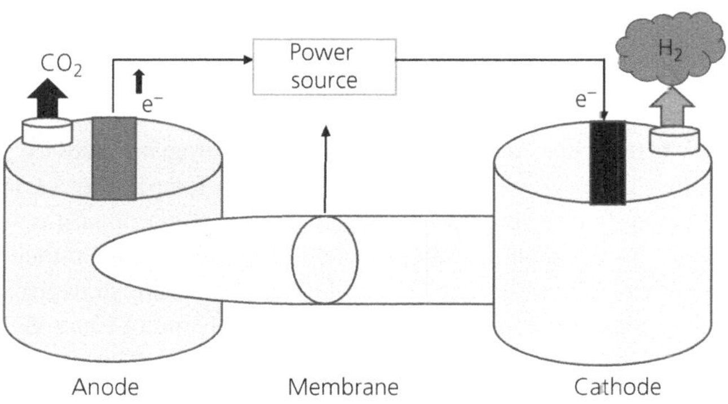

Figure 5.5 Microbial processes for biohydrogen generation

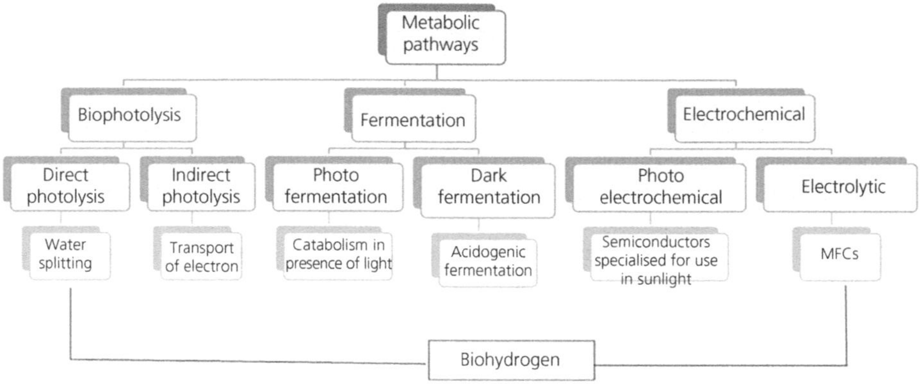

metabolising carbohydrates into hydrogen and volatile fatty acids. Photofermentative bacteria, such as *Rhodobacter*, enhance hydrogen yields by utilising organic acids under anaerobic, illuminated conditions. Electrohydrogenesis, driven by exoelectrogenic bacteria like *Geobacter* and *Shewanella*, facilitates direct electron transfer to the anode, enabling hydrogen production at the cathode.

Optimising substrate composition, redox potential and electrochemical conditions significantly enhances microbial hydrogen production efficiency in MFCs.

5.2.3 Microbial groups involved in hydrogen production

Significant advances in bioelectrochemical hydrogen production have been achieved through the use of specialised microorganisms, particularly extremophiles and thermophiles. Among the most well-studied electrogenic bacteria are *Shewanella oneidensis* and *Geobacter sulfurreducens*, both of which are known for their efficient electron transfer capabilities. Other bacteria that have demonstrated potential in MFC applications include *Acetobacterium woodie*, *Ochrobactrum* species, *Acetobacterium woodie*, *Rhodoferax ferrireducens*, *Rhodopseudomonas palustris* and *Sphingomonas* strain DJ.

Several studies indicate that MECs utilising *Alkalibacter* species as exoelectrogens have achieved hydrogen production rates of 2.6 L H_2/L reactor per day (Rago *et al.*, 2019). Research by Heidrich *et al.* (2014) also demonstrated that exoelectrogenic activity increases with temperature in MFCs using greywater as a substrate, with temperature variations between 1°C and 22°C affecting hydrogen output. A separate study using domestic wastewater as a substrate at 13°C–21°C recorded a hydrogen yield of 0.017 L/L/d. Further improvements in biohydrogen production have been achieved through co-culture strategies. Hasibar *et al.* (2020) developed an engineered microbial consortium consisting of *Enterobacter aerogenes* and *Clostridium acetobutylicum* to enhance hydrogen generation. This co-culture was incorporated into a microbial electrolysis cell, where the application of a low voltage (0.8 V) significantly boosted hydrogen production.

Various microbial groups contribute to hydrogen production in MFCs, primarily fermentative, exoelectrogenic and photoheterotrophic bacteria.

5.2.3.1 Fermentative bacteria

Fermentative bacteria, including *Clostridium*, *Enterobacter* and *Klebsiella*, metabolise carbohydrates into hydrogen and organic by products by way of pathways like Embden-Meyerhof-Parnas and pentose phosphate. These bacteria's efficiency is affected by factors like substrate type, pH and hydrogen partial pressure.

- *Clostridium* spp. – strict anaerobes that produce hydrogen through pyruvate fermentation, using protons as electron acceptors.
- *Enterobacter* spp. – facultative anaerobes that generate hydrogen by way of mixed-acid fermentation, often achieving higher yields in microaerobic conditions.
- *Klebsiella* spp. – capable of producing hydrogen while metabolising diverse carbon sources, including lignocellulosic hydrolysates.

pH, substrate availability and other factors all have an impact on hydrogen production by fermentative bacteria. Accumulation of volatile fatty acids (VFAs) can inhibit hydrogen synthesis, necessitating interactions with other microbial groups to utilise these byproducts effectively.

5.2.3.2 Exoelectrogenic bacteria

Exoelectrogenic bacteria enhance hydrogen production in MFCs by transport of electrons to an external electrode, thereby improving cathodic hydrogen evolution. These microorganisms break down organic compounds through oxidation, subsequently transferring electrons to external acceptors like anodes or metals.

- *Geobacter* sp. – gram-negative bacteria capable of transferring electrons through conductive pili (nanowires) or direct contact with electrode surfaces.
- *Shewanella* sp. – versatile bacteria that can reduce metal oxides and facilitate extracellular electron transfer using cytochrome proteins.

By forming biofilms on the anode, these bacteria enhance electron transfer efficiency, boosting current generation and hydrogen recovery. Pairing exoelectrogenic microorganisms with hydrogen-evolving catalysts in MFCs further improves biohydrogen yields.

5.2.3.3 Photoheterotrophic bacteria

Photoheterotrophic bacteria, including *Rhodobacter* and *Rhodopseudomonas*, generate hydrogen by way of photofermentation, a process in which microorganisms utilise organic acids as electron donors under anaerobic conditions and exposure to light, thereby facilitating hydrogen production. Under anaerobic conditions and in the presence of light, these microorganisms use organic acids as electron donors to drive hydrogen synthesis.

- *Rhodobacter sphaeroides* – a purple non-sulfur bacterium that efficiently converts volatile fatty acids such as butyric and acetic acid into hydrogen under light exposure.
- *Rhodopseudomonas palustris* – exhibits metabolic flexibility, producing hydrogen from various organic compounds.

Integrating these bacteria into MFCs allows the efficient utilisation of organic acids generated during fermentation, improving hydrogen recovery while reducing waste accumulation.

5.3. Substrates for biohydrogen production

Biohydrogen production depends on various renewable and waste-based feedstocks, including agricultural residues, organic waste streams, and industrial and domestic wastewater. Organic waste materials such as fruit and vegetable peels, food scraps and lignocellulosic biomass serve as abundant carbon sources for microbial fermentation (Dikshit *et al.*, 2023).

Agricultural byproducts like crop straw, husks, corn stover and sugarcane bagasse are particularly valuable due to their high carbohydrate content. Additionally, industrial wastewater from food, dairy and beverage processing industries contains biodegradable organic matter, which can be converted into hydrogen through anaerobic microbial activity (Akca *et al.*, 2021). Similarly, domestic wastewater, including sewage and municipal effluents, supports the growth of hydrogen-producing bacteria in anaerobic conditions.

Utilising these waste-derived substrates not only promotes sustainable hydrogen production but also contributes to efficient waste management and environmental sustainability.

5.4. Factors influencing biohydrogen production

5.4.1 Microorganisms used as an inoculum

Microbial communities play a vital role in MFCs by converting organic substrates into biohydrogen. However, factors in the environment, particularly variations in pH and temperature, can greatly impact microbial efficiency, as shown in Figure 5.6. Both single and

Figure 5.6 Parameters affecting biohydrogen generation in MFCs

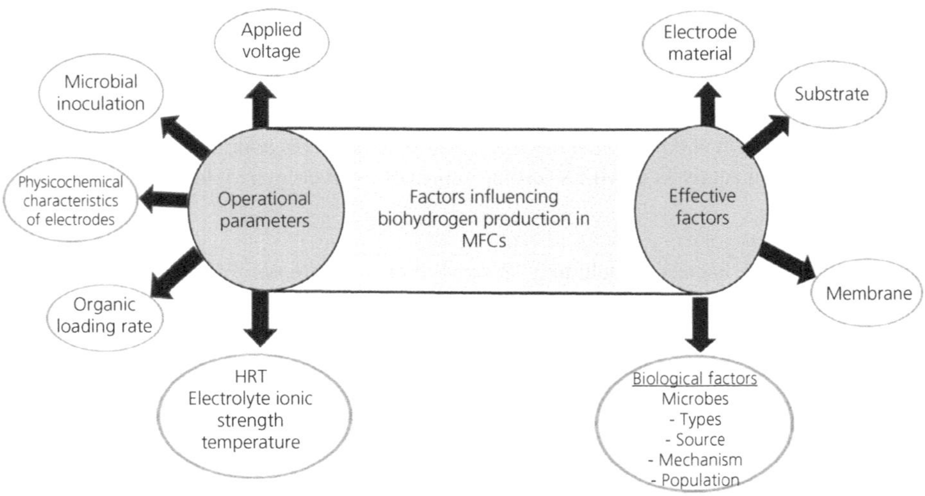

mixed microbial cultures are utilised to develop biofilms on electrodes in MFCs. Pure microbial cultures provide controlled conditions, minimising methane production, whereas mixed cultures demonstrate better adaptability and resilience under varying environmental conditions.

Employing a pure microbial strain like *Pseudomonas stutzeri* effectively reduces methanogenesis. According to research by Radhika *et al.* (2022), incorporating this microorganism into MFCs enhanced biohydrogen output by 128% and decreased methane formation by approximately 76.28%. Conversely, mixed microbial communities offer robust performance and adaptability, essential for practical field applications. Predominant microbial phyla identified in MFC systems include *Proteobacteria*, *Firmicutes* and *Bacteroidetes*. Electroactive genera such as *Acidaminococcus*, *Desulfocapsa*, *Desulfovibrio* and *Acetobacterium* significantly contribute to anodic biofilms (Kumar *et al.*, 2017), although substrate competition among microbial species may impact overall system performance.

Specific bacterial species have shown considerable promise in MFC-driven dark fermentation. Species like *Thermoanaerobacterium thermosaccharolyticum* and *Thermoanaerobacterium aciditolerans* have exhibited high biohydrogen productivity (Banu *et al.*, 2018). Over sustained operating periods, populations of *Thermoanerobacter brockii* decrease, while *Clostridium* species become dominant. Furthermore, *Geobacter*, *Desulfovibrio* and *Thermoanaerobacterium thermosaccharolyticum* are prevalent in secondary fermentation stages. Additional significant biohydrogen-producing bacteria include *Geobacillus* and *Shewanella* (Wang *et al.*, 2017). *Propionibacterium* sp., *Lactobacillus* sp. and *Clostridium* sp. also contribute positively to hydrogen yields, while *Acidaminococcus* enhances biohydrogen production rates (Almatouq *et al.*, 2020a).

5.4.2 Physical factors

5.4.2.1 Materials used as electrodes

An MFC setup typically comprises an anode and cathode arranged in close proximity within the reactor. An external voltage, typically ranging from 0.2 to 1.2 V, is applied to drive biohydrogen production. Electrode materials commonly employed include carbon cloth, graphite felt and platinum. Fujinawa *et al.* (2019) reported that the addition of 2% conductive nanoparticles effectively decreases hydrogen loss and limits methane generation by impairing methanogenic cells. Compared to conventional carbon nanoparticles, chemical inhibitors offer a more economical and efficient alternative for controlling methanogens making them promising candidates for electrode enhancements.

5.4.2.2 Materials for anode

Several critical factors influence the selection of anode materials, including electrical conductivity, microbial adhesion capability, corrosion resistance, surface area, affordability and environmental sustainability. Commonly used anode materials encompass carbon-based electrodes such as graphite felts, graphite rods, carbon brushes, carbon cloth and activated carbon, as seen in Table 5.1. Apostolopoulos *et al.* (2021) report that optimising anode materials can yield biohydrogen production rates around 0.31 ± 0.08 mmol/L/day.

Table 5.1 Comparative analysis of biohydrogen production from different electrode configurations in MFCs

Serial no.	Anode material	Cathode material	MFC type	Biohydrogen output (m^3/m^3/day)	Reference
1	Carbon-based fabric	Platinum-loaded carbon	Single-chambered	0.56	–
2	Graphite electrode	Biologically active cathode	Double-chambered	0.89	Huang *et al.*, 2021
3	Graphite brush electrode	Carbon fabric air cathode	Single-chambered	0.25–0.37	Apostolopoulos *et al.*, 2021
4	Graphite electrode	Nickel foam cathode	Double-chambered	0.63	Jwa *et al.*, 2019
5	Heat-treated graphite brush	Platinum-coated stainless-steel mesh	Double-chambered	2.2	Kadier *et al.*, 2014
6	Nitrogen-treated carbon fabric	Stainless-steel mesh	Single-chambered	1.31	Yu *et al.*, 2018
7	Moist carbon fabric	Nickel-molybdenum cathode	Single-chambered	2.1	Almatouq *et al.*, 2020
8	Plasma-modified carbon felt	Platinum-loaded carbon	Double-chambered	0.56	Kadier *et al.*, 2016

Carbon felt is widely recognised for biofilm formation in MFCs, but its high electrical resistance limits its effectiveness. Graphite, though more costly, exhibits superior stability and conductivity compared to conventional carbon materials. Graphite brushes, in particular, have become favourable due to their availability, extensive surface area and lower resistance (Aarthy *et al.*, 2020). Single-chamber MFCs utilising graphite brushes achieved hydrogen production rates (HPR) of 1.7 ± 0.1 m^3-H_2/m^3/d with Coulombic efficiencies (CE) around 78% at an applied voltage of 0.6V (Rozenfeld *et al.*, 2019). Another study involving pre-treated carbon felt as an anode reported a hydrogen yield of 0.44 mmol/L/d and a CE of approximately 75% (Kumar *et al.*, 2013).

Various alternative anode materials, such as steel mesh, nickel foam and titanium electrodes, have been investigated for their superior electrical conductivity, corrosion resistance and porous structure (Table 5.1). However, these materials typically require pretreatment at elevated temperatures using ammonia to boost hydrogen production and shorten activation periods (Zhou *et al.*, 2011). Among them, carbon cloth stands out as a favourable option due to its durability and flexibility.

5.4.2.3 Cathode material

In MFCs, the anode compartment breaks down organic substances, releasing protons and electrons. Electrons move through an external circuit to the cathode chamber, where they combine with protons to generate biohydrogen (Watanabe, 2008). The cathode material significantly impacts hydrogen production efficiency and formation of byproducts (Llobet *et al.*, 2013).

Materials such as titanium, silver mesh and nickel foam were assessed for cathodic roles. Carbon-based materials exhibit slower hydrogen evolution rates (HER) due to their higher overpotential. Platinum (Pt) catalysts help overcome this limitation. One study on double-chamber MFCs using carbon paper coated with Pt catalysts achieved an HPR of 14.64 ± 0.14 mL/L/d and a CE of 97% at 0.8V (Li *et al.*, 2013). Despite its effectiveness, platinum is expensive, less available and prone to contamination by sulfides and cyanides, which limits its practical application in MECs.

5.4.2.4 Membrane

Impact ion exchange membranes, particularly proton exchange membranes (PEMs), substantially influence MFC performance. Ideal membranes must have low resistance, efficient proton transport capabilities, and effectively prevent oxygen and substrate crossover. Nafion and Ultrex membranes are extensively utilised because of their exceptional cation conductivity, mechanical durability and selectivity for protons. Nafion membranes contain hydrophilic sulfonate groups (SO_3) attached to a hydrophobic fluorocarbon polymer backbone, which significantly enhances their performance (Jana and Muthukumar, 2023).

Various membranes, including cation exchange membranes, anion exchange membranes, reverse osmosis (RO), forward osmosis (FO), osmotic membranes, reverse electrodialysis and membrane bioreactors (MBRs), have been explored in MEC setups (Sharma *et al.*, 2024). Additional alternatives like AMI-7001, nanofibre-reinforced PEM composites, bipolar and charge-mosaic membranes have also been investigated (Karthikeyan *et al.*, 2017). In MFCs, external power is supplied to convert spent media's organic acids into hydrogen (Das and Veziroglu, 2008). Porous membranes are particularly effective at minimising substrate and oxygen crossover compared to membrane-less systems. Additionally, recent research by Zhen *et al.* (2024) explored carbon-based biocathodes for methane synthesis in double-chamber MFCs.

5.4.3 Operational factors affecting MFC performance

5.4.3.1 Substrate choices

Substrate selection critically influences biohydrogen yields in MFC systems. Typical substrates include wastewater effluent, carbon-rich compounds and nutrient-dense media. MFCs efficiently generate hydrogen from various organic sources (Passos *et al.*, 2019). Substrates such as glucose, butyric acid, acetic acid, lactic acid and fermentation residues have been tested, with acetate achieving the highest power densities.

Microalgae have become promising candidates due to their photosynthetic capabilities, harnessing sunlight and water to produce biohydrogen alongside ethanol and carbon dioxide. While there are some operational challenges, optimising microalgal substrates could significantly boost hydrogen yields and economic feasibility (Jorge *et al.*, 2016). Dark fermentation processes also effectively convert organic matter into biohydrogen.

Complex substrates like milk, glycol, starch and hybrid feedstocks have been extensively tested for hydrogen production in MEC and MFC configurations (Laikova *et al.*, 2024). Milk and hybrid substrates exhibited higher potential for hydrogen recovery compared to starch and glycol (Ginkel *et al.*, 2001). Food waste (FW) has also undergone evaluation within MFC systems employing continuous anaerobic digestion (AD). Comparative studies demonstrated higher hydrogen outputs in MFC systems than conventional AD alone, although organic matter breakdown occurred more rapidly in standard AD processes (Kyazze *et al.*, 2006) (Table 5.2).

5.4.3.2 Physicochemical properties of electrodes

Internal resistance within MFCs is influenced by multiple electrode parameters, including position, surface area, spacing, conductivity, morphology and activation resistance. Optimising these parameters is critical for minimising resistance and enhancing system performance (Arun *et al.*, 2024). Cheng *et al.* (2020) found that reducing electrode spacing significantly increased hydrogen production; specifically, hydrogen recovery improved from 64% at 3.5 cm spacing to 98% at 2 cm spacing. Further spacing reduction to 1 cm, however, showed

Table 5.2 Various studies on the effect of temperature, pH and substrate (wastewater) on hydrogen yield

Serial no.	Microbial inoculum	Type of substrate	pH	Operating temperature (°C)	Hydrogen yield	Reference
1	Anaerobic sludge	Waste from vegetables and fruits	–	35	191 mol H_2/ mol glucose	Chu *et al.*, 2013
2	Anaerobic granular sludge	Cheese whey	7.25	26.6	4554.5 ± 105 mL H_2/L	Vendruscolo, 2015
3	Mixed activated sludge	Corn cob waste	7.0 ± 0.2	37	114 mL H_2/L	Pant *et al.*, 2010
4	Raw agricultural digestate	Cotton waste	6.4	38	0.917 mL/g VSS	Chu *et al.*, 2013
5	Photosynthetic mixed consortium HAU-M1	Corn stover	6.5	35	345 mL H_2/L	Vendruscolo, 2015
6	*Enterobacter aerogenes* MTCC 2822	Rhizopus-treated rice bran waste	–	27 ± 1	0.0045 mol H_2/g biomass	Pandey *et al.*, 2016
7	*Rhodobacter sphaeroides*	Maize straw	7.1–7.6	30	72.69 mL H_2/L/h	Zheng and Nirmalakhandan, 2010

negligible improvement due to limited anode surface area available for bacterial activity. Similar research supports these findings, confirming that electrode spacing is crucial for optimised biohydrogen yields.

Bora *et al.* (2022) explored a novel electrocatalyst, NiFe-layered double hydroxide, highlighting its effective surface area of 1.03 m^2 for biohydrogen enhancement. Additionally, platinum-coated nickel foam electrodes were found to significantly improve electrochemical properties and hydrogen yields. Electrode arrangement also influences MFC efficiency; placing graphite anodes on either side of the cathode resulted in higher hydrogen production rates (5.56 m^3 H_2/m^3/day) compared to configurations where anodes were aligned on one side. Improved proton transfer efficiency with parallel electrode setups accounts for these observed performance improvements (Watanabe, 2008).

5.4.3.3 Mechanism for electron transfer

MFCs utilise electron transfer mechanisms involving microorganisms that oxidise organic substrates to produce electrons, protons, water and CO_2 under anaerobic conditions (Zhao *et al.*, 2018). Electron transfer takes place by way of direct or indirect routes. Direct transfer of electrons requires physical interaction between bacteria and electrodes, while indirect transfer employs soluble mediators.

Notable direct electron transfer methods include nanowires and membrane-bound proteins. Species such as *Shewanella* and *Geobacter* exhibit pili structures (nanowires), facilitating direct extracellular electron transfer. *Geobacter sulfurreducens* specifically employs redox-active proteins associated with the PilA gene, enabling iron reduction (Wu *et al.*, 2023). Recent insights reveal that nanopili comprise polymerised hexaheme cytochrome complexes. Conversely, indirect electron transfer uses electron carriers like flavins, quinones and phenazines to shuttle electrons from microbial cells to metabolic intermediates before transfer to cytochromes in periplasmic or outer membrane regions (Hua *et al.*, 2019). Ongoing research aims to further elucidate these extracellular electron transfer processes under varied conditions.

5.4.3.4 External resistance and its effect

External resistance substantially affects power output, chemical oxygen demand (COD) reduction and microbial activity in MFCs. Changes in external resistance alter cell voltage and current production through modulation of electrode potentials, where

- maximum voltage occurs at infinite external resistance (open circuit)
- reducing external resistance lowers applied voltage but enhances current flow by facilitating electron transport
- excessively low external resistance can result in current exceeding optimal electron transfer rates, thus restricting overall current production
- anode potentials are significantly affected by external resistance changes, impacting biofilm development and anode performance.

5.4.3.5 pH

Electrolyte pH and conductivity significantly influence MFC efficiency. Ionic strength directly impacts proton transfer and electrical conductivity; enhanced buffer concentrations, such as

NaCl, reduce internal resistance and improve performance (Kadier *et al.*, 2014, 2022). Sharma *et al.* (2023) reported optimal hydrogen recovery at pH 6.5, although Chu *et al.* (2013) found improved hydrogen production at catholyte pH 5.0 compared to pH 7.0. Microbial metabolism and electron transfer efficiency are highly sensitive to pH fluctuations, suggesting the necessity for further research on pH optimisation in MFCs.

5.4.3.6 Temperature

Temperature significantly influences microbial metabolism and MFC performance, with microbial activity heavily dependent on operating temperatures. Wu *et al.* (2023) demonstrated that moderate temperatures favour biohydrogen production compared to methanogenic MFCs, which prefer higher temperatures. Sharma *et al.* (2023) observed increased hydrogen yields when temperatures rose from 23°C to 28°C, emphasising the importance of optimal temperature management for MFC operation.

5.4.3.7 Hydraulic retention time and load rate

Hydraulic retention time (HRT) influences hydrodynamic conditions, organic loading and dissolved oxygen in MFCs, significantly affecting hydrogen yields. Catalyst loading on electrodes also boosts hydrogen output, as evidenced by maximum hydrogen production with a platinum catalyst loading of 0.2 g/cm^2 (Ndayisenga *et al.*, 2022). Optimal COD loading (60 g/L/day) with an HRT of 2 days was sufficient for enhanced hydrogen generation (Jain and He, 2020). Further research demonstrated improved yields by reducing HRT from 10 to 4 hours due to increased electrode exoelectrogen populations (Carrera *et al.*, 2023; Sangeetha *et al.*, 2017).

5.4.3.8 Applied voltage

Applied voltage critically affects MFC efficiency since biohydrogen generation is a non-spontaneous electrochemical reaction requiring external energy input. Using potentiostats or galvanostats helps maintain optimal anode potentials, controlling microbial biofilm formation and current generation. Optimal applied voltages generally range from 0.4 to 1.2 V (Zhang *et al.*, 2019). Excessive voltage (above 2.0 V) triggers oxygen production, damaging electrode materials and resembling conventional electrolysis (Ren *et al.*, 2013). Kumar *et al.* (2019) highlighted that applied voltage variations influence microbial metabolism and cell integrity. Escapa *et al.* (2016) emphasised that enzyme activities, notably dehydrogenases critical to biohydrogen production, are voltage-sensitive, with maximal outputs observed around 1.0 V.

5.4.3.9 Microbial inoculation

Microbial community composition directly impacts MFC performance. Using pure or mixed microbial cultures can enhance hydrogen yields. Effective strategies against competing methanogenic activity include chemical inhibitors like 2-bromoethanesulfonate and neomycin sulfate, increasing biohydrogen production efficiency (Das *et al.*, 2008). Common inoculation approaches involve transferring stable biofilms, using wastewater sludge, pure cultures or effluent from existing MEC/MFC setups, as wastewater contains rich microbial diversity advantageous for sustainable biohydrogen generation (Li *et al.*, 2013).

5.4.3.10 Ion strength or concentration of electrolyte

Electrolyte conductivity significantly affects MEC/MFC efficiency, influencing microbial activity and internal resistance. Enhanced proton transfer by way of increased catholyte concentration improves hydrogen production. Effective buffering solutions, notably phosphate buffers, optimise hydrogen yields due to their high buffering capacities (Zhang *et al.*, 2019). Adjusting ionic strength and electrolyte concentration through NaCl or concentrated buffers enhances system conductivity, reducing internal resistance and boosting hydrogen recovery (Almatouq *et al.*, 2020; Rani *et al.*, 2021).

5.5. Challenges and limitations

Biohydrogen is progressively regarded as an attractive alternative to fossil fuels, with MFCs and MECs recognised as being promising green technologies for biohydrogen production from diverse substrates. These systems facilitate hydrogen generation while concurrently treating wastewater. Despite their potential, numerous hurdles restrict practical deployment, including scalability issues, high upfront costs, dependency on energy sources, operational efficiencies and system reliability (Sun *et al.*, 2009). Advances have significantly improved MFC designs, microbial consortium insights and electrochemical performances; however, economic feasibility remains a pivotal determinant for the broader adoption of microbial electrochemical technologies. Optimising reactor designs, power output, electrode materials and microbial consortia selection are critical to boosting MFC efficiency for varied applications (Huang *et al.*, 2021).

Understanding microbial dynamics is essential to control non-electrogenic microbial growth, which diminishes power generation. Although mixed microbial cultures reduce costs, they often decrease power output over time due to methanogenic proliferation (Bora *et al.*, 2022). Low power generation poses a significant barrier in MFC implementation. Serially connecting MFC units may cause voltage reversal; hence, parallel configurations with effective power management strategies are necessary to achieve practical electricity outputs while reducing costs.

High initial capital investment, driven mainly by costly construction materials, hampers large-scale MFC adoption for biohydrogen production. Employing less expensive electrode alternatives, including affordable, durable and scalable membranes, cathodes and anodes, could notably enhance economic viability. Further research is necessary to optimise electron and proton transport within anode compartments. Additionally, cathode material expenses, their toxicity and environmental impacts pose considerable industrial application challenges. Biocathodes, as substitutes for conventional cathode materials, require additional research to enhance their efficiency and recovery processes.

MFCs face technical barriers such as low hydrogen recovery rates, membrane fouling and elevated material costs.

5.5.1 Low hydrogen recovery costs

A critical issue in MFCs is suboptimal hydrogen recovery, primarily due to competitive microbial activities. Methanogens, which flourish in anaerobic conditions, consume hydrogen, lowering overall yields. Additional constraints involve ineffective electron transfers and mass transport limitations. Potential solutions include

- application of methanogenic inhibitors like 2-bromoethanesulfonate (BES)
- improving electrode materials and expanding surface area for efficient electron transfer
- enhancing reactor designs to reduce hydrogen losses and maximise gas capture.

5.5.2 Membrane fouling and maintenance

MFC performance significantly suffers from membrane fouling which impedes ion-exchange efficiency and increases internal resistance. Microbial colonisation causing biofouling and inorganic scaling from salts can exacerbate system deterioration. Maintenance strategies include

- periodic chemical or physical membrane cleaning
- implementing anti-fouling coatings or specialised ion-selective membranes
- optimising operational conditions to reduce microbial growth and salt precipitation.

5.6. Environmental and economic considerations

Techno-economic analyses and lifecycle assessments (LCAs) have been undertaken to evaluate MFC commercialisation viability (Ganeshan *et al.*, 2023). Two major cost barriers identified include electrode materials and reactor design. Anode materials represent approximately 59% of capital expenses, while electricity accounts for up to 69% of operational costs (Cheng *et al.*, 2020). Although high cathode catalyst costs remain challenging, biocathodes show promise as cost-effective, stable and sustainable alternatives (Bora *et al.*, 2022). Nickel powder has been considered as an affordable cathode material. Membrane costs in double-chamber systems remain significant, necessitating further improvements in cost-effectiveness and performance.

LCAs have presented conflicting results regarding MFCs' environmental impacts. Some indicate minimal construction-related emissions, while others point to considerable greenhouse gas (GHG) emissions during installation phases (Santos *et al.*, 2021). Utilising locally produced electricity or transporting it by way of national grids can reduce electricity network congestion. MECs can exploit inexpensive, off-peak energy periods, but variability in renewable energy and wastewater composition necessitates robust control strategies for integrating MFCs into existing energy infrastructures (Tiwari *et al.*, 2024).

5.7. Future prospects

Feedstock choice and reactor design significantly impact MFC performance. Optimising critical factors, including electrode materials and configurations, is essential for enhanced efficiency. High-performance MFCs require anodes with extensive surface areas and stable microbial populations. Enhanced conductivity through metal current collectors and innovative two-dimensional (2D) anode structures could support scalability. Surface-treated anodes and stainless steel have been proposed as economically viable, durable electrode materials.

5.7.1 Integration with other bioenergy systems

Integrating MFCs with anaerobic digestion and algal biorefineries can elevate energy recovery and sustainability. Combining MFCs with biogas systems enables residual organic matter conversion into methane and hydrogen, optimising resource utilisation. Employing wastewater rich in organic substances reduces operational costs and boosts efficiency.

5.7.2 Hybrid technologies for hydrogen

Hybrid systems that merge MFCs with photocatalysis, solar-driven water splitting or thermochemical approaches can significantly enhance hydrogen generation. Photocatalytic integrations can boost electron transfer efficiency, offering a scalable and economically viable hydrogen production solution.

5.8. Conclusions

Microbial electrochemical systems, notably MFCs, have increasingly gained attention for biohydrogen production, presenting numerous advantages, such as high hydrogen recovery, versatile substrate compatibility, low energy demands and concurrent waste treatment.

This chapter identifies pivotal operational parameters influencing MFC efficacy. The performance of the system is heavily influenced by the activity of exoelectrogenic microorganisms, making it essential to better understand electron transfer processes to enhance substrate utilisation and hydrogen production.

Stainless steel emerges as a prevalent cathode material due to economic viability and durability. Investigations into diverse substrates, including waste-derived materials, highlight potential for energy recovery and waste mitigation. Ongoing research must further evaluate critical parameters, including HRT, electrode characteristics, organic loading rates, temperature and electrolyte compositions, to optimise system performance. Proper reactor designs and effective control of operational parameters can elevate electrochemical efficiency, with optimised electrode materials minimising ohmic losses.

Future research should clarify detailed performance mechanisms of MFC technologies. MFCs offer a sustainable approach to biohydrogen production by utilising renewable organic waste, supporting electricity generation simultaneously. Integrating microbial consortium optimisation, electrode innovations and refined process controls further augments hydrogen yields and overall system efficiency. Coupling MFCs with waste valorisation promotes circular bioeconomy objectives, reducing fossil fuel dependence and environmental footprints. Despite current challenges, continuous advancements in electrode designs, biocatalyst engineering and integrated hybrid technologies will considerably enhance biohydrogen production feasibility, establishing microbial electrochemical systems as a sustainable solution for clean energy generation.

REFERENCES

Aarthy M, Rajesh T and Thirunavoukkarasu M (2020) Critical review on microbial fuel cells for concomitant reduction of hexavalent chromium and bioelectricity generation. *Journal of Chemical Technology & Biotechnology* **95**(5): 1298–1307, 10.1002/jctb.6257.

Acar C and Dincer I (2014) Comparative assessment of hydrogen production methods from renewable and non-renewable sources. *International Journal of Hydrogen Energy* **39**(1): 1–12, 10.1016/J.IJHYDENE.2013.10.060.

Acar J, Özkan US and Demir P (2014) Electrochemical analysis of microbial fuel cells for hydrogen and electricity generation. *International Journal of Electrochemical Science*, **9**: 4503–4518.

Ahmed SF, Rafa N, Mofijur M *et al.* (2021). Biohydrogen production from biomass sources: metabolic pathways and economic analysis. *Frontiers in Energy Research* **9**, 10.3389/fenrg.2021.753878.

Akca MS, Bostancı O, Aydin AK, Koyuncu I and Altinbas M (2021) BioH2 production from food waste by anaerobic membrane bioreactor. *International Journal of Hydrogen Energy* **46**(55): 27941–27955, 10.1016/J.IJHYDENE.2021.06.042.

Almatouq A, Babatunde AO, Khajah M, Webster G and Alfodari, M (2020) Microbial community structure of anode electrodes in microbial fuel cells and microbial electrolysis cells. *Journal of Water Process Engineering* **34**: 101140, 10.1016/J.JWPE.2020.101140.

Almatouq MA, Dlugolecki P and ter Brake HJM (2020) Performance enhancement of microbial electrochemical systems for energy and hydrogen production. *Energy Conversion and Management* **205**: 112383.

Apostolopoulos I, Bampos G, Soto Beobide A *et al.* (2021) The effect of anode material on the performance of a hydrogen producing microbial electrolysis cell, operating with synthetic and real wastewaters. *Energies* **14**(24), 10.3390/en14248375.

Arun J, SundarRajan PS, Grace Pavithra K *et al.* (2024) New insights into microbial electrolysis cells (MEC) and microbial fuel cells (MFC) for simultaneous wastewater treatment and green fuel (hydrogen) generation. *Fuel* **355**: 129530, 10.1016/J.FUEL.2023.129530.

Banu JR, Yukesh Kannah R, Dinesh Kumar M *et al.* (2018) Recent advances on biogranules formation in dark hydrogen fermentation system: Mechanism of formation and microbial characteristics. *Bioresource Technology* **268**: 787–796, 10.1016/J.BIORTECH.2018.07.034.

Bora A, Mohanrasu K, Angelin Swetha T *et al.* (2022) Microbial electrolysis cell (MEC): reactor configurations, recent advances and strategies in biohydrogen production. *Fuel* **328**: 125269, 10.1016/J.FUEL.2022.125269.

Carrera, J, Hernández A and Martín L (2023) Microbial fuel cells for renewable bioenergy and hydrogen production: recent developments and challenges. *Renewable Energy Reviews* **158**, 112123.

Cheng J, Li, H, Ding L *et al.* (2020) Improving hydrogen and methane co-generation in cascading dark fermentation and anaerobic digestion: the effect of magnetite nanoparticles on microbial electron transfer and syntrophism. *Chemical Engineering Journal* **397**: 125394, 10.1016/J.CEJ.2020.125394.

Das D and Veziroglu TN (2008) Advances in biological hydrogen production processes. *International Journal of Hydrogen Energy* **33**(21): 6046–6057, 10.1016/J.IJHYDENE.2008.07.098.

Dikshit PK, Poddar MK and Chakma S (2023) Biohydrogen production from waste substrates and its techno-economic analysis. In *Hydrogen Economy: Processes, Supply Chain, Life Cycle Analysis and Energy Transition for Sustainability* (Scipion A, Manzardo A and Ren J (eds)), 2nd edn. Academic Press, Amsterdam, the Netherlands, pp. 399–429, 10.1016/B978-0-323-99514-6.00015-7.

Dincer I and Acar C (2011) Review and evaluation of sustainable hydrogen production methods and technologies. *International Journal of Hydrogen Energy* **36**(9): 5989–6005.

Dos Passos VF, Marcilio R, Aquino-Neto S *et al.* (2019) Hydrogen and electrical energy co-generation by a cooperative fermentation system comprising Clostridium and microbial fuel cell inoculated with port drainage sediment. *Bioresource Technology* **277**: 94–103, 10.1016/J.BIORTECH.2019.01.031.

Dutta K and Kundu PP (2014) A review on aromatic conducting polymers-based catalyst supporting matrices for application in microbial fuel cells. *Polymer Reviews* **54**(3): 401–435, 10.1080/15583724.2014.881372.

Escapa A, Mateos R, Martínez EJ and Blanes J (2016) Microbial electrolysis cells: an emerging technology for wastewater treatment and energy recovery. From laboratory to pilot plant and beyond. *Renewable and Sustainable Energy Reviews* **55**: 942–956, 10.1016/J.RSER.2015.11.029.

Fujinawa K, Nagoya M, Kouzuma A and Watanabe K (2019) Conductive carbon nanoparticles inhibit methanogens and stabilize hydrogen production in microbial electrolysis cells. *Applied Microbiology and Biotechnology* **103**(15): 6385–6392, 10.1007/s00253-019-09946-1.

Ganeshan S, Kumar P and Singh R (2023) Nanomaterials and microbial electrochemical technologies: a step toward sustainable biohydrogen. *International Journal of Hydrogen Energy* **48**(55): 33344–33369.

Ghasemi B, Yaghmaei S, Abdi K, Mardanpour MM and Haddadi SA (2020) Introducing an affordable catalyst for biohydrogen production in microbial electrolysis cells. *Journal of Bioscience and Bioengineering* **129**(1): 67–76, 10.1016/J.JBIOSC.2019.07.001.

Ginkel SV, Sung S and Lay J-J (2001) Biohydrogen production as a function of pH and substrate concentration. *Environmental Science & Technology* **35**(24): 4726–4730, 10.1021/es001979r.

Hasibar B, Ergal I, Abreu Dias S *et al.* (2020) Competing acetate consumption and production inside a microbial electrolysis cell. *Journal of Environmental Chemical Engineering* **8**(4): 103847, 10.1016/J.JECE.2020.103847.

Hassan RYA, Febbraio F and Andreescu S (2021) Microbial electrochemical systems: principles, construction and biosensing applications. *Sensors* **21**(4), 10.3390/s21041279.

Heidrich ES, Edwards SR, Dolfing J, Cotterill SE and Curtis TP (2014) Performance of a pilot scale microbial electrolysis cell fed on domestic wastewater at ambient temperatures for a 12 month period. *Bioresource Technology* **173**: 87–95, 10.1016/J.BIORTECH.2014.09.083.

Hua T, Li S, Li F, Zhou Q and Ondon BS (2019) Microbial electrolysis cell as an emerging versatile technology: a review on its potential application, advance and challenge. *Journal of Chemical Technology & Biotechnology* **94**(6): 1697–1711, 10.1002/jctb.5898.

Huang X, Duan C, Duan W *et al.* (2021) Role of electrode materials on performance and microbial characteristics in the constructed wetland coupled microbial fuel cell (CW-MFC): a review. *Journal of Cleaner Production* **301**: 126951, 10.1016/J.JCLEPRO.2021.126951.

Jain A and He Z (2020) Improving hydrogen production in microbial electrolysis cells through hydraulic connection with thermoelectric generators. *Process Biochemistry* **94**: 51–57, 10.1016/J.PROCBIO.2020.04.008.

Jana S and Muthukumar P (2023) Design, development and hydrogen storage performance testing of a tube bundle metal hydride reactor. *Journal of Energy Storage* **63**: 106936, 10.1016/J.EST.2023.106936.

Jiang T, Huang S and Yang J (2019) Structural carbon emissions from industry and energy systems in China: an input-output analysis. *Journal of Cleaner Production* **240**: 118116. https://doi.org/10.1016/j.jclepro.2019.118116

Jorge R and Hazael G (2016) Microalgae as a feedstock for biohydrogen production through microbial fuel cells. *Renewable and Sustainable Energy Reviews* **58**: 200–215.

Jwa E, Yun YM, Kim H *et al.* (2019) Domestic wastewater treatment in a tubular microbial electrolysis cell with a membrane electrode assembly. *International Journal of Hydrogen Energy* **44**(2): 652–660, 10.1016/J.IJHYDENE.2018.11.036.

Kadier A, Isa MH and Salmiati (2014) Optimization of microbial fuel cell performance: electrodes, membranes, and operational factors. *Journal of Environmental Chemical Engineering* **2**(3): 1824–1835.

Kadier A, Kalil MS, Abdeshahian P *et al.* (2016) Recent advances and emerging challenges in microbial electrolysis cells (MECs) for microbial production of hydrogen and value-added chemicals. *Renewable and Sustainable Energy Reviews* **61**: 501–525, 10.1016/J.RSER.2016.04.017.

Kadier A, Kalil MS, Chandrasekhar K *et al.* (2018) Surpassing the current limitations of high purity H2 production in microbial electrolysis cell (MECs): Strategies for inhibiting growth of methanogens. *Bioelectrochemistry* **119**: 211–219, 10.1016/J.BIOELECHEM.2017.09.014.

Kadier A, Simayi Y, Kalil MS, Abdeshahian P and Hamid AA (2014) A review of the substrates used in microbial electrolysis cells (MECs) for producing sustainable and clean hydrogen gas. *Renewable Energy* **71**: 466–472, 10.1016/J.RENENE.2014.05.052.

Kadier A, Wang J, Chandrasekhar K *et al.* (2022) Performance optimization of microbial electrolysis cell (MEC) for palm oil mill effluent (POME) wastewater treatment and sustainable Bio-H2 production using response surface methodology (RSM). *International Journal of Hydrogen Energy* **47**(34): 15464–15479, 10.1016/J.IJHYDENE.2021.09.259.

Karthikeyan R, Cheng KY, Selvam A, Bose A and Wong JWC (2017) Bioelectrohydrogenesis and inhibition of methanogenic activity in microbial electrolysis cells – a review. *Biotechnology Advances* **35**(6): 758–771, 10.1016/J.BIOTECHADV.2017.07.004.

Kim H-J, Chang IS and Moon H (2008) A mediator-less microbial fuel cell using a metal-reducing bacterium, Shewanella putrefaciens. *Enzyme and Microbial Technology* **42**(2): 176–181.

Kumar G, Bakonyi P, Zhen G *et al.* (2017) Microbial electrochemical systems for sustainable biohydrogen production: Surveying the experiences from a start-up viewpoint. *Renewable and Sustainable Energy Reviews* **70**: 589–597, 10.1016/J.RSER.2016.11.107.

Kumar SS, Kumar V, Malyan SK *et al.* (2019) Microbial fuel cells (MFCs) for bioelectrochemical treatment of different wastewater streams. *Fuel* **254**: 115526, 10.1016/J.FUEL.2019.05.109.

Kumari S and Das D (2017) Improvement of biohydrogen production using acidogenic culture. *International Journal of Hydrogen Energy* **42**(7): 4083–4094, 10.1016/J.IJHYDENE.2016.09.021.

Kyazze G, Martinez-Perez N, Dinsdale R *et al.* (2006) Influence of substrate concentration on the stability and yield of continuous biohydrogen production. *Biotechnology and Bioengineering* **93**(5): 971–979, 10.1002/bit.20802.

Laikova AA and Zhuravleva EA (2024) Substrate composition and effects on biohydrogen production. *Biohydrogen – Advances and Processes* (pp. 181–214). Springer Nature, Cham, Switzerland, 10.1007/978-3-031-49818-3_8.

Laikova ON, Ivanov VV, Petrov AB and Sidorova EL (2024) Advances in microbial electrochemical systems for sustainable hydrogen production. *Journal of Renewable Bioenergy*, **12**(3): 150–169.

Li X, Wang Q, Wu S, Hu Z and Bai J (2024) Critical upstream technologies for hydrogen energy industry: research progress on ammonia decomposition catalysts. *Sustainable Chemistry and Pharmacy* **38**: 101492, 10.1016/J.SCP.2024.101492.

Ndayisenga F, Yu Z, Wang B *et al.* (2022) Thermophilic-operating environment promotes hydrogen-producing microbial growth in a lignocellulose-fed DF-MEC system for enhanced biohydrogen evolution. *Process Safety and Environmental Protection* **167**: 213–224, 10.1016/J.PSEP.2022.09.026.

Nelson K, Marsili E and Bond DR (2008) *Electron transfer mechanisms in microbial fuel cells: implications for hydrogen production. Electroanalysis* **20**(3), 297–304.

Pandey P, Shinde VN, Deopurkar RL *et al.* (2016) Recent advances in the use of different substrates in microbial fuel cells toward wastewater treatment and simultaneous energy recovery. *Applied Energy* **168**: 706–723, 10.1016/J.APENERGY.2016.01.056.

Pant D, Van Bogaert G, Diels L and Vanbroekhoven K (2010) A review of the substrates used in microbial fuel cells (MFCs) for sustainable energy production. *Bioresource Technology* **101**(6): 1533–1543, 10.1016/J.BIORTECH.2009.10.017.

Radhika D, Shivakumar A, Kasai DR, Koutavarapu R and Peera SG (2022) Microbial electrolysis cell as a diverse technology: overview of prospective applications, advancements, and challenges. *Energies* **15**(7), 10.3390/en15072611.

Rago L, Pant D and Schievano A (2019) *Electro-Fermentation—Microbial Electrochemistry as New Frontier in Biomass Refineries and Industrial Fermentations* (pp. 265–288), 10.1016/B978-0-12-817941-3.00014-0.

Rahman SNA, Masdar MS, Rosli MI *et al.* (2016) Overview biohydrogen technologies and application in fuel cell technology. *Renewable and Sustainable Energy Reviews* **66**: 137–162, 10.1016/J.RSER.2016.07.047.

Rani G, Krishna K and Yogalakshmi KN (2021) Enhancing the electrochemical performance of Fe_3O_4 nanoparticles layered carbon electrodes in microbial electrolysis cell. *Journal of Environmental Chemical Engineering* **9**(6): 106326, 10.1016/J.JECE.2021.106326.

Ren L, Siegert M, Ivanov I, Pisciotta JML and Logan BE (2013) Treatability studies on different refinery wastewater samples using high-throughput microbial electrolysis cells (MECs). *Bioresource Technology* **136**: 322–328, 10.1016/J.BIORTECH.2013.03.060.

Rousseau R, Etcheverry L, Roubaud E *et al.* (2020) Microbial electrolysis cell (MEC): strengths, weaknesses and research needs from electrochemical engineering standpoint. *Applied Energy* **257**: 113938, 10.1016/J.APENERGY.2019.113938.

Rozendal RA, Hamelers HVM, Euverink GJW, Metz SJ and Buisman CJN (2006) Principle and perspectives of hydrogen production through biocatalyzed electrolysis. *International Journal of Hydrogen Energy* **31**(12): 1632–1640, 10.1016/J.IJHYDENE.2005.12.006.

Rozendal RA, Hamelers HVM, Rabaey K, Keller J and Buisman CJN (2008) Towards practical implementation of bioelectrochemical wastewater treatment. *Trends in Biotechnology* **26**(8): 450–459, 10.1016/J.TIBTECH.2008.04.008.

Rozenfeld S, Ouaknin Hirsch L, Gandu B *et al.* (2019) Improvement of microbial electrolysis cell activity by using anode based on combined plasma-pretreated carbon cloth and stainless steel. *Energies* **12**(10), 10.3390/en12101968.

Sangeetha T, Guo Z, Liu W *et al.* (2017) Energy recovery evaluation in an up flow microbial electrolysis coupled anaerobic digestion (ME-AD) reactor: Role of electrode positions and hydraulic retention times. *Applied Energy* **206**: 1214–1224, 10.1016/J.APENERGY.2017.10.026.

Santos FJA, de Andrade MA and Cavalcanti EHL (2021) Biohydrogen and bioelectricity co-generation in bioelectrochemical systems: trends and challenges. *Journal of Cleaner Production* **289**: 125684.

Schröder U (2007) Anodic electron transfer mechanisms in microbial fuel cells and their energy efficiency. *Physical Chemistry Chemical Physics* **9**(21): 2619–2629, 10.1039/B703627M.

Selim HMM, Kamal AM, Ali DMM and Hassan RY (2017) Bioelectrochemical systems for measuring microbial cellular functions. *Electroanalysis* **29**(6): 1498–1505, 10.1002/elan.201700110.

Sharma A, Hussain Mehdi SE, Pandit S, Eun-Oh S and Natarajan V (2024) Factors affecting hydrogen production in microbial electrolysis cell (MEC): a review. *International Journal of Hydrogen Energy* **61**: 1473–1484, 10.1016/J.IJHYDENE.2024.02.193.

Sharma GD, Verma M, Taheri B, Chopra R and Parihar JS (2023) Socio-economic aspects of hydrogen energy: An integrative review. *Technological Forecasting and Social Change* **192**: 122574, 10.1016/J.TECHFORE.2023.122574.

Sharma P, Jain A, Bora BJ *et al.* (2023) Application of modern approaches to the synthesis of biohydrogen from organic waste. *International Journal of Hydrogen Energy* **48**(55): 21189–21213, 10.1016/J.IJHYDENE.2023.03.029.

Sinbuathong N and Sillapacharoenkul (2021). Dark fermentation of starch factory wastewater with acid- and base-treated mixed microorganisms for biohydrogen production. *International Journal of Hydrogen Energy* **46**(31): 16622–16630, 10.1016/J.IJHYDENE.2020.06.109.

Singh R, White D, Demirel Y *et al.* (2018) Uncoupling fermentative synthesis of molecular hydrogen from biomass formation in Thermotoga maritima. *Applied and Environmental Microbiology* **84**(17): e00998-18, 10.1128/AEM.00998-18.

Sun M, Sheng GP, Mu ZX *et al.* (2009) Manipulating the hydrogen production from acetate in a microbial electrolysis cell–microbial fuel cell-coupled system. *Journal of Power Sources* **191**(2): 338–343, 10.1016/J.JPOWSOUR.2009.01.087.

Tiwari A, Yadav N, Jadhav DA *et al.* (2024) A critical review on the advancement of the development of low-cost membranes to be utilized in microbial fuel cells. *Water* **16**(11): 1597. https://doi.org/10.3390/w16111597

Vendruscolo F (2015) Starch: a potential substrate for biohydrogen production. *International Journal of Energy Research* **39**(3): 293–302, 10.1002/er.3224.

Venkata Mohan S, Mohanakrishna G, Velvizhi G, Babu VL and Sarma PN (2010) Bio-catalyzed electrochemical treatment of real field dairy wastewater with simultaneous power generation. *Biochemical Engineering Journal* **51**(12): 32–39, 10.1016/J.BEJ.2010.04.012.

Wang H, Park J-D and Ren ZJ (2015) Practical energy harvesting for microbial fuel cells: a review. *Environmental Science & Technology* **49**(6): 3267–3277, 10.1021/es5047765.

Wang J, Song X, Wang Y *et al.* (2017) Bioenergy generation and rhizodegradation as affected by microbial community distribution in a coupled constructed wetland-microbial fuel cell system associated with three macrophytes. *Science of The Total Environment* 607–608: 53–62, 10.1016/J.SCITOTENV.2017.06.243.

Watanabe K (2008) Recent developments in microbial fuel cell technologies for sustainable bioenergy. *Journal of Bioscience and Bioengineering* **106**(6): 528–536, 10.1263/JBB.106.528.

Wu Y, Zhu X, Wang X *et al.* (2023) A new electron shuttling pathway mediated by lipophilic phenoxazine via the interaction with periplasmic and inner membrane proteins of Shewanella oneidensis MR-1. *Environmental Science & Technology* **57**(6): 2636–2646, 10.1021/acs.est.2c07862.

Yang E, Omar Mohamed H, Park SG *et al.* (2021) A review on self-sustainable microbial electrolysis cells for electro-biohydrogen production via coupling with carbon-neutral renewable energy technologies. *Bioresource Technology* **320**: 124363, 10.1016/J.BIORTECH.2020.124363.

Yu Z, Leng X, Zhao S *et al.* (2018) A review on the applications of microbial electrolysis cells in anaerobic digestion. *Bioresource Technology* **255**: 340–348, 10.1016/J.BIORTECH.2018.02.003.

Zhang G, Zhou Y and Yang F (2019) Hydrogen production from microbial fuel cells-ammonia electrolysis cell coupled system fed with landfill leachate using Mo2C/N-doped graphene nanocomposite as HER catalyst. *Electrochimica Acta* **299**: 672–681, 10.1016/J.ELECTACTA.2019.01.055.

Zhang Y and Angelidaki I (2014) Microbial electrolysis cells (MECs) for efficient biohydrogen production from organic waste: a review. *Bioresource Technology*, **150**: 411–422.

Zhao J, Zhang C, Sun C *et al.* (2018) Electron transfer mechanism of biocathode in a bioelectrochemical system coupled with chemical absorption for NO removal. *Bioresource Technology* **254**: 16–22, 10.1016/J.BIORTECH.2018.01.066.

Zhen Z, Ou X, Wang Y and Zhou S (2024) Assessing transition pathways of hydrogen production in china with a probabilistic framework. *Environmental Science & Technology* **58**(30): 13263–13272, 10.1021/acs.est.4c01304.

Zheng X and Nirmalakhandan N (2010) Cattle wastes as substrates for bicelectricity production via microbial fuel cells. *Biotechnology Letters* **32**(12): 1809–1814, 10.1007/s10529-010-0360-3.

Zhou M, Chi M, Luo J, He H and Jin T (2011) An overview of electrode materials in microbial fuel cells. *Journal of Power Sources* **196**(10): 4427–4435, 10.1016/J.JPOWSOUR.2011.01.012.

Microbial Processes in Environmental Engineering: From Theory to Practice

Rajneesh Kumar, Ram Sharan Singh and Maulin P Shah
ISBN 978-1-83708-327-5
https://doi.org/10.1108/978-1-83708-326-820261006

Chapter 6
Nanobiomaterials for microbial fuel cells and their application in bioenergy production

Shelly Singh, Shilpa Sharma and Ashok K Dubey
Department of Biological Sciences and Engineering, Netaji Subhas University of Technology, Dwarka, New Delhi, India.

Abstract

Microbial fuel cells (MFCs), potential bioelectrochemical systems that utilise microorganisms as catalysts to transform organic matter into electricity, are being explored as sustainable energy alternatives. For this, nanobiomaterials such as biogenic nanoparticles, nanozymes, carbon nanoparticles and so on can be explored to enhance the applicability, efficiency and scalability of MFCs for bioenergy production. These nanobiomaterials provide special physicochemical properties including enhanced conductivity, surface area and biocompatibility which enables them to enhance the MFC's overall performance. This chapter provides a comprehensive review on recent advancements in the manufacturing, functionalisation, economic viability and integration of nanobiomaterials for modification of MFCs, increasing their efficiency by lowering internal resistance, improving microbial adhesion and promoting electron transfer. Special consideration is given to the hybrid development of biological and synthetic components creating synergistic effects to upgrade overall performance. To assess the sustainability and economic viability of upgrading nanobiomaterial-enhanced MFCs for industrial applications, the chapter further explores the usability of nanobiomaterials in customised MFC designs for the generation of bioenergy, and the challenges associated with material stability and their environmental impact. By integrating advances in biotechnology with nanotechnology, this discipline has the opportunity to address the world's energy challenges and make a substantial contribution to the shift to a green bioeconomy.

Keywords: Microbial fuel cells; bioelectrochemical system; nanobiomaterial; bioenergy.

6.1. Introduction

The escalating global crisis, driven by the vast and mismanaged use of conventional energy resources, has spurred the need for the development of alternative energy generation technologies. Although renewable energy resources such as solar, wind and water hold potential, intrinsic limitations exist. In the case of solar technology, constraints of material sustainability, limited lifespan and high production costs exist. Wind energy requires robust energy storage solutions, whereas water energy systems, despite having huge potential, possess significant costs during extraction, treatment and transport, impacting their sustainability and efficiency (Elkhrachy *et al.*, 2023; Ferguson, 2008; Larsen *et al.*, 2019).

Addressing these limitations necessitates the exploration of unique, innovative and sustainable bioenergy solutions converting organic matter sources into energy. Microbial fuel cells (MFCs) offer a fascinating avenue for bioenergy generation, leveraging the biocatalytic activity of microorganisms to produce electricity. MFC can utilise diverse organic substrates as fuel sources, converting the chemical bond energy directly into an electrical current through biological processes, thus avoiding the need for traditional metal-based catalysts and presenting a cost-effective and environmentally friendly alternative (Slate *et al.*, 2019).

At present, the advancement of MFCs is a slow-growing process due to expensive electrode materials, low power generation and restrictions in scale-up. Overcoming these hindrances is essential for the widespread adoption of MFCs as a viable energy source.

The pursuit of sustainable energy technologies has placed a heightened emphasis on the development of high-performance MFCs. However, the performance efficiency of MFCs is governed by various critical factors, including the selection of electroactive microorganisms, the choice of biocompatible and highly conductive electrode materials, the optimisation of MFC design and the regulation of key operational parameters. Refining these aspects will enhance the electron transfer efficiency of MFCs leading to the maximisation of their power output and increase in the system's stability, thereby enhancing their potential commercialisation and large-scale implementation.

An innovative approach to integrating nanotechnology and biomaterials has emerged as an optimal strategy. Specifically, the development and application of nanobiomaterials, combining the unique properties of nanomaterials with biological components, offers unparalleled opportunities for modifying and optimising MFC technology. These materials, leveraging principles of increased surface area-to-volume ratio, improved conductivity and biocompatibility, are capable of significantly enhancing the electron transfer mechanisms, enhancing microbial adhesion and optimising the electrode materials. By precisely engineering materials at the nanoscale, researchers aim to overcome the inherent limitations of conventional MFCs and achieve higher power output and efficiency.

This chapter investigates the diverse applications of nanobiomaterials in MFC technology upgrade, detailing their impact on the key performance parameters and overall system advancement, emphasising their potential for environmentally sustainable bioenergy production.

6.2. Fundamentals of microbial fuel cells

6.2.1 Principle of operation

MFC technology utilises the ability of electroactive microorganisms to produce bioelectricity by harnessing their metabolic activities. The basic principle of MFCs depends on the oxidation of the organic substrates produced by microorganisms, where microorganisms play the role of biocatalysts to alter chemical energy into electrical energy (Kunwar *et al.*, 2023). The electrons get transferred to an external electrode during the cellular respiration of the microorganisms using soluble electron mediators or direct electron transfer mechanisms. An electrochemical potential is established due to the separation of electron and proton transfer pathways leading to the generation of an electric current (Prathiba *et al.*, 2024).

6.2.2 The working mechanism

The MFC contains two chambers, the anode and the cathode, which are separated by a proton exchange membrane (PEM). In the anodic chamber, release of electrons and protons takes place by way of oxidation of the organic matter by microorganisms.

Redox-active compounds, such as NADH (nicotinamide adenine dinucleotide + hydrogen) and FAD (flavine adenine dinucleotide), are involved in the microbial electron transport chains that transfer the electrons. These electrons later migrate towards the anode instead of getting utilised in the metabolic pathways by way of extracellular or membrane-bound electron transfer mechanisms; correspondingly the PEM-generated protons transfer to the cathodic chambers. The electrons then start traversing towards the cathode through an external circuit creating an electric current flow. Once these electrons reach the cathode, they are combined with the protons and the dissolved oxygen to form water as shown in Figure 6.1.

Figure 6.1 Schematic diagram of MFCs

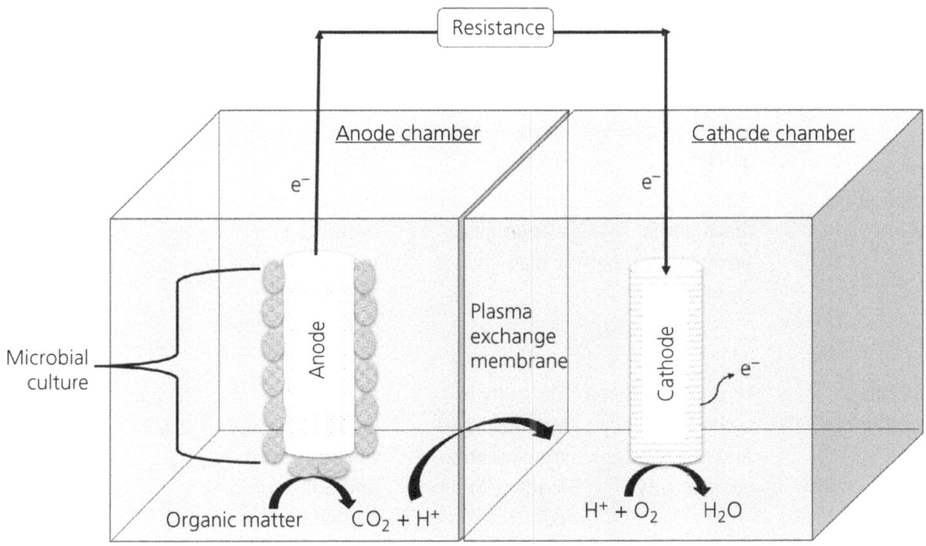

The electron transfer efficiency depends on the electrochemical gradient between the anode and cathode thus enabling the production of bioelectricity. Therefore, the performance of the MFC hinges on the electrode interactions, availability of the redox mediators and the electron transfer kinetics. These mechanisms further depend on several factors, such as the material of the electrode, type of organic substrate and environmental conditions influencing the electrical output of the MFC (Pandya *et al.*, 2024).

6.2.3 Key challenges in MFC technology

Despite wide research on the exploration of MFCs as a sustainable energy generator, challenges such as limited electron transfer efficiency, low power density and material degradation still exist, necessitating a requirement for advanced materials, typically nanobiomaterials, to enhance their performance. For this, carbon-based electrodes are commonly used as they possess good conductivity, high surface area and are biocompatible. However, further modification is required to meet societal energy demands in a sustainable fashion. Table 6.1 lists the advantages and disadvantages of the latest electrode materials used in MFCs.

6.2.4 Nanobiomaterials in MFCs

Over the past decade, improvements in biomaterials have significantly impacted the advancement of nanotechnology. This development has led to innovations in various sectors, including the production of bioenergy and biofuels (Khoo *et al.*, 2020).

Table 6.1 Currently used materials for the development of MFCs

Electrode material	Advantages	Disadvantages	Impact on MFC performance	Reference
Graphene	Increased surface area and flexibility	Heightened cost and potential toxicity	Increased current density leads to great power output	Aiswaria *et al.*, 2022
Graphitic carbon nitrides	High number of N-atoms leads to increased catalytic active sites	Limited electrocatalytic oxygen reduction activity	Improved electron transfer efficiency	Zhang *et al.*, 2017
Conducting polymers	High surface areas, large flexibility and high conductivity	High cost and challenging scalability	High energy density, electrochemical stability and environmental friendliness	Zhao *et al.*, 2017
Carbon nanotubes (CNTs)	High surface area, biocompatibility and electrical conductivity	Difficulty in synthesis and manipulation leading to high cost	Enhanced current density, power output and adhesion of bacteria	Murjani *et al.*, 2022

Currently used electrode materials, such as carbon-based and metal-based, exhibit limitations such as high cost, low biocompatibility and limited surface area for microbial adhesion. To address these, nanotechnology and novel biomaterials are being explored as an alternative source to enhance the power output and performance of MFCs as shown in Figure 6.2.

Modification of electrodes using nanotechnology and nanomaterials has garnered significant consideration for enhancing MFC performance. The surface area-to-volume ratio increases at the nanoscale level leading to enhanced electron transfer and power output. Additionally, nanoparticle-based electrodes decrease microbial adhesion, improve corrosion resistance and enhance long-term stability (Kižys *et al.*, 2023). They possess unique physicochemical properties resulting in superior electrical conductivity and biocompatibility making them ideal for MFC optimisation.

Recently, bacterial cellulose has been explored owing to its high porosity, mechanical strength and biocompatibility. However, pure bacterial cellulose lacks intrinsic electrical conductivity which resists its direct application as an electrode material. To address this, a nanotechnology approach was taken to incorporate conductive fillers in the form of nanoparticles, nanowires, nanocrystals and nanofibrils, which significantly improved the electron transfer efficiency, electrical conductivity and overall MFC performance (Hu *et al.*, 2014; Vilela *et al.*, 2019). A few examples of bio-based nanomaterials used for upgrading MFC performance are listed in Table 6.2.

6.3. Modifications in MFC components

6.3.1 Nanostructured anodes

In MFCs, an anode is considered as the most critical component, as microbial biofilm formation and transfer of electrons take place here. To enhance the rate of transfer of electrons, nanomaterials such as carbon nanotubes, nanocellulose and nanocomposites have been explored as they provide increased surface area for the attachment of the biofilm.

Figure 6.2 Nanobiomaterials used for MFCs

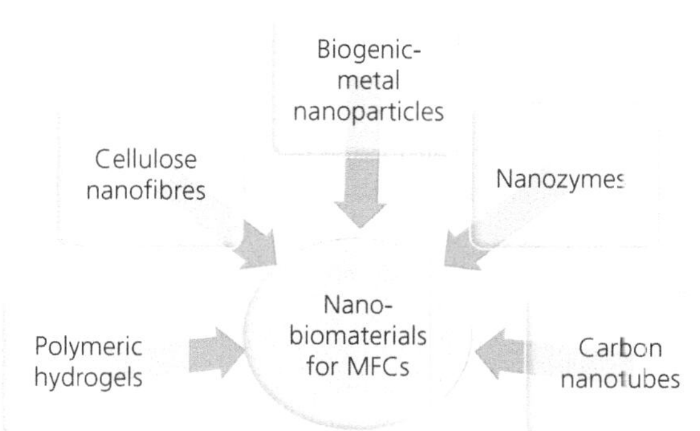

Table 6.2 Bio-based nanomaterials for increased performance of MFCs

Electrode material	Advantages	Disadvantages	Impact on MFC performance	Reference
Biogenic-gold nanoparticles	High power density and short start-up time	High cost	Enhanced electric generation	Wu *et al.*, 2018
Nanoenzyme	High operational stability and high catalytic activity	Limited types of nanozymes available	Enhanced power generation	Le and Kim, 2021
Biogenic-nano-palladium	High catalytic activity, easy synthesis, environmental sustainability	Sensitivity to environmental conditions	Enhanced power density and reduced electron resistance	Matsena, *et al.*, 2020; Quan *et al.*, 2015
Biogenic silver nanoparticles	High power density, low electrical resistance, greater chemical oxygen demand (COD) and removal efficacy	May develop stability and aggregation issues	Improved alkalinity of the MFC cathodes and improved conductivity	Elkhrachy *et al.*, 2023
Biogenic TiO_2 nanoparticles	High specific surface area, chemical stability and biocompatibility	Might show limited performance under dark conditions	Long-term durability and high-power production	Kumar *et al.*, 2023

In a very recent study, iron-catalysed, nitrogen-doped carbon nanotubes were electrochemically activated on carbon cloth (A-FeNCNTs@CC) to construct a hydrophilic and capacitive anode. This achieved a power density of 3.07 W/m^2 and COD removal of 84.7 mg/L/d with greater biocompatibility, facilitating the adhesion of *Geobacter*, thereby displaying a start-up time of 1.8 days and enhancing the extracellular electron transfer (Wang *et al.*, 2025).

In another recent study, corn stalk biochar loaded with four-layer multi-walled carbon nanotubes (MWCNTs), employed as an anodic material in the MFC, exhibited enhanced power production (2.68 W/m^2) performance, charge transfer resistance of 3.56 Ω and lowest solution resistance of 9.86 Ω, due to a well-structured conductive network and efficient electron transport (Yan *et al.*, 2024).

Similarly, enhanced power output and power density of MFCs, by the modification of anodes using nanochitin aerogel and carbonised nanocellulose paper, was reported (Yeam-suksawat *et al.*, 2024). Cellulose nanofibres, nanocrystals and aerogels have been known to provide greater electrode flexibility, high strength, aspect ratio and biocompatibility (Rana *et al.*, 2024).

Functionalisation of conventional materials such as graphene, metal oxides and so on with biomolecules may further enhance microbial adhesion, thereby increasing electron transfer. Recently, a study investigated the modification of conventional graphite anodes (GA) using a composite of chia seed nanopowder (CSNP) and acid-activated MWCNTs. This functionalised nanocomposite was reported to enhance electrical conductivity and electron transfer ability, while CSNP contributed to the surface porosity leading to high microbial adhesion promoting efficient charge transport and improving overall MFC performance (Abbas and Ismail, 2024).

In another study, a novel anode modification by immobilising riboflavin onto the carbon cloth using bio inspired and layer-by-layer self-assembled peptide nanotubes led to greater power generation, accelerated electron transfer with decreased internal resistance and enhanced stability of the MFC (Xu and Quan, 2016). A macroalgae, *Laminaria saccharina*, was used as a biocatalyst in the anodic chamber with glucose as the electron donor to produce bioenergy in the form of electricity (Gadhamshetty *et al.*, 2013).

It can be concluded that anode modification using biological or biologically derived materials presents a vast opportunity to sustainably elevate MFC efficiency.

6.3.2 Cathode catalysts

For high-performing MFCs, efficient oxygen reduction reactions (ORRs) at the cathode are desirable. Biogenic nanomaterials, the nanoparticles that are synthesised by way of utilisation of biological entities such as plant extracts, algae, fungi or bacteria, offer a sustainable alternative for the modification of cathode efficiency.

Recently, cobalt phthalocyanines and iron-supported MWCNTs having ORR activity comparable to the conventional platinum catalysts were developed, signifying their potential as sustainable alternatives in MFCs (Bhowmick *et al.*, 2019). In another study, a mixture of algae and aloe vera at 5 mg/cm^2 used to develop antimicrobial biogenic silver nanoparticles was employed onto the cathodes, inhibiting unwanted microbial growth on the cathode's surface. This improved the ORRs, increasing power density (66.5 mW/m^2) and the stability of the MFC (Das *et al.*, 2024).

Similarly, in a very recent study, biogenic gold nanoparticles were used to modify conventional graphene-oxide (GO) electrodes. The biogenic gold nanoparticles graphene oxide-based MFC displayed superior performance – that is, 13.19 times higher power density and 1.03 times higher Cr(VI) removal rate than those of chemical gold nanoparticles-based graphene oxide-based electrodes (Tang *et al.*, 2024).

In a study, *Saccharina japonica*, a brown macroalgae, was metabolised for the production of bioelectricity with a maximum power density of 1820 mW/m^2 (Gebreslassie *et al.*, 2021). Another study reported enhanced MFC performance with a high power density of 0.044 mW/m^3 and low volume specific resistance, using carbon cloth electrodes modified with biogenic palladium nanoparticles [Pd(0)] formed by the sulfate-reducing bacteria isolated from a wastewater treatment plant (Malunga-Makatu Khanyisile *et al.*, 2023).

Innovative dual-electrode modification – that is, targeting both anode and cathode for enhancing MFC performance – has also been explored recently by incorporating gold nanoparticles and polyaniline, resulting in significant improvement in hydrophilicity and conductivity and, hence, increased power density. The cathode was engineered using a combination of chitosan and DNA origami resulting in efficient electron transfer and reducing the need for oxidiser replacements. Both these modifications resulted in a substantial increase of 52.42% in the overall power density of the MFC, presenting an efficient strategy for enhancing large-scale MFC systems due to their ease of preparation and scalability (Thapa *et al.*, 2024).

6.3.3 Electrolytes and membranes

Electrolytes and protein exchange membranes are essential for maintaining charge balance and separating cathodic and anodic environments. However, they have certain limitations such as biofouling, oxygen crossover, may suffer from dehydration and are significantly expensive. Therefore, the integration of bioinspired materials can reduce the cost and enhance the overall performance of the MFCs.

In this context, researchers have focused on improving MFC technology by increasing the electron transfer mechanisms and enhancing properties of the constituent materials – for example, crosslinked hydrogels have been utilised for superior ionic conductivity and water retention. Recently, mussel-mimetic polydopamine (PDA) chemistry has been utilised for bio inspired modifications for membrane surface functionalisation as mussels display excellent adhesion properties. This mussel-mimetic PDA coating attaches to several surfaces providing a natural adhesion mechanism (Abid *et al.*, 2017; Ding *et al.*, 2016).

In one study, PDA (0.03%) was combined with TiO_2:ZnO (1:1 ratio) nanoparticles to modify PEM for MFCs. PDA acted as a bioadhesive and the resultant membrane enhanced the power density by 2.5 times, current generation by 5.7 times, decreased the MFC operation cycle from five to three days and doubled the lifespan of the PEM of the MFCs, thereby enhancing the overall performance of the MFCs (Bahamonde Soria *et al.*, 2022). In another very recent study, chondroitin sulfate, a biomaterial, was used to develop a low-cost proton-exchange membrane with high performance, demonstrating its potential to be used in MFCs (Erenler *et al.*, 2024).

6.3.4 Role of enzymatic nanobiomaterials

The natural ability of enzymes as catalysts in combination with the latest nanotechnology techniques has also been explored to create hybrid materials for MFCs. Nanomaterials that possess intrinsic enzyme-like catalytic properties mimicking the functioning of natural enzymes are defined as 'nanozymes' or 'enzymatic nanobiomaterials'. Nanozymes can be synthesised using various methods such as co-precipitation, hydro-solvothermal, sol-gel, microwave-assisted and thermal decomposition processes (Sen *et al.*, 2024).

The catalytic activity of the nanozymes can be tailored by controlling their size, hence nanozymes with unique physico chemical characteristics have been developed for the revitalisation of traditional enzyme engineering.

The incorporation of nanozymes onto both anode and cathode electrodes has emerged as a promising strategy. At the anode, nanozymes are employed to increase the surface area, thereby providing a large number of sites for microbial adhesion leading to robust biofilm formation, crucial for extracellular electron transfer. Whereas, at the cathode, replacement of conventional and expensive platinum catalysts with enzymatic biocathodes are being explored for enhanced ORR (Le Goff *et al.*, 2015). Recently, a biomimetic nanozyme with Fe-N-S-C active sites, inspired by the hemes of c-type cytochromes, was developed as an anode material for MFCs. It was found to exhibit high power density (start-up time, 3.5 days; power density, 2366 ± 34 mW/m^2) and extracellular electron transfer comparing to that of traditional carbon cloth (start-up time, 5.6 days; power density, 1009 ± 26 mW/m^2) (Xiang *et al.*, 2023).

Immbolisation of enzymes on nanomaterials within the electrode structure is another vital aspect of MFC upgradation (Nemiwal *et al.*, 2022). This technique creates a microenvironment and retains the enzyme on the electrode, which helps in preserving enzymes' stability and activity over long periods of time. Several techniques, like covalent attachment, encapsulation, adsorption and crosslinking, have been used for the immobilisation of the enzymes (Maghraby *et al.*, 2023). In a recent report, a bioanode (a hybrid of three-dimensional graphene with carbon nanotubes) was fabricated for the immbolisation of glucose oxidase on a carbon electrode for the development of efficient biofuel cells, displaying a power density of 253.36 $\mu W/cm^2$ at 0.65 V (Babadi *et al.*, 2024). While nanozymes offer substantial enhancements to MFC performance, their implementation is hindered by some practical challenges enlisted below.

- *Enzyme stability in the presence of nanomaterials* – interaction between nanozymes and the nanomaterial may compromise the structural integrity and catalytic activity of the enzyme leading to decreased operational stability.
- *Biocompatibility* – nanozyme-modified MFC system compatibility with the microbial community and the surrounding environment is essential to avoid adverse effects on cell viability and overall MFC performance.
- *Scalability* – synthesis and integration of nanozymes into large-scale MFC systems pose significant engineering challenges questioning their feasibility in industrial applications.
- *Cost-effectiveness* – integration of nanozymes with MFCs has to compete with the cost-effectiveness of the existing technologies.

Enzymatic nanobiomaterials hold significant potential for transforming MFC technology. By enhancing the catalytic activity, electron transfer efficiency and stability, these materials offer a promising avenue to overcome the limitations of conventional MFCs, and attain higher power efficiency and bioenergy generation. While challenges related to scalability, cost and long-term stability remain, ongoing research efforts are steadily addressing these hurdles. The advantages offered by nanozymes, including their tunability, potential for multi-functionality and, in some cases, cost-effectiveness, position them as valuable tools for the development of next-generation MFCs.

As research continues to advance in this interdisciplinary field, nanozymes will function to play a crucial role in realising the full potential of MFC technology for sustainable energy production and environmental remediation.

6.4. Environmental and economic implications

6.4.1 Sustainability and green chemistry

The incorporation of nanobiomaterials in the upgrade of MFC technology aligns with the principles of sustainability by leveraging natural processes for material synthesis and green chemistry, reducing energy inputs, minimising waste and utilising renewable resources.

Efficient enhancement of MFCs using nanobiomaterials facilitates a more significant conversion of organic waste into electrical energy, thereby reducing the volume of waste that would otherwise be directed to landfills and decreasing the release of greenhouse gases released during traditional waste treatment practices (Sonawane *et al.*, 2022). Furthermore, the application of nanobiomaterials permits the recovery of both bioelectricity and valuable materials from waste streams, representing a potentially appealing alternative to existing waste treatment procedures.

The sustainability of nanobiomaterials in MFCs is also linked to reduced energy consumption during the production of bioelectricity in comparison to conventional materials. Also, the usage of cost-effective and naturally abundant materials like chitosan, biochar and so on for the modification of electrodes for MFCs is capable of reducing the energy footprint associated with the process (Chakraborty *et al.*, 2020; Khan *et al.*, 2020). Beyond the reduction in waste volume and energy consumption, nanobiomaterials have significant potential in bioremediation applications such as wastewater treatment plants (Pérez *et al.*, 2023). They are capable of oxidising organic pollutants and heavy metals present in wastewater.

A study reported that biomass derived from *Microcystis aeruginosa* acted as an electron donor in the MFC and was utilised for waste treatment and bioenergy generation with maximum current density of 83 mW/m^2 (Ali *et al.*, 2020). Another novel approach to enhance the function of MFCs for sustainable energy production was the use of quorum sensing, in which cell-to-cell communication takes place by way of auto-inducers between the microbial species, leading to enrichment of the bio-anode and increasing MFC efficiency which can also be applied for wastewater treatment (Sivasankar *et al.*, 2019). This simultaneous ability of MFCs to remove pollutants and produce bioenergy is promising for sustainable technology in the area of bioremediation.

6.4.2 Economic feasibility

The initial expense for the usage of nanobiomaterials in MFC modification can be a concern due to their production costs and ongoing progress in large-scale synthesis procedures. However, they are considered economically sustainable solutions for widespread industrial adoption due to the significant gains in energy efficiency leading to increased power output and greater energy density.

As previously discussed, studies have shown remarkable enhancement in power density, electron transfer rates and reduced internal resistance, leading to an overall increase in power output. The enhanced electron transfer rates boost the catalytic activity of the modified electrodes resulting in improved stability and performance of MFCs. Another important aspect is the extended operational lifespan and reduced maintenance requirements which improve the economic viability of nanobiomaterial-based MFCs due to the lowering of operational costs. Also, the conversion of biomass into bioenergy is a viable option to reduce carbon emissions and contribute to a low-carbon outcome which promotes bioeconomy by enhancing waste management sustainably (Ye *et al.*, 2024).

Nanobiomaterial-enhanced MFCs offer economic competitiveness, making them economically viable for large-scale productions and providing long-term benefits.

6.5. Challenges and future directions

The future challenges for MFCs are with manufacturing issues, economic viability and environmental impacts.

6.5.1 Manufacturing

Despite the great potential for nanobiomaterials in the sustainable modification of MFCs, there exist certain challenges that need to be addressed. Firstly, the correct operation of the manufacturing processes of the nanobiomaterials needs attention. It is pertinent to ensure that proper mixing, coating and drying are taking place for the uniform dispersion and stability of the nanobiomaterials, reducing the chances of agglomeration which may negatively affect the performance of the MFC (Nyabadza *et al.*, 2023). Also, meticulous control of the synthesis parameters needs to be maintained to develop defect-free products to prevent low yields and throughput which may cause difficulty in scaling up. Therefore, the traditional production process of MFCs requires significant re-engineering.

6.5.2 Economic viability

Advanced nanomaterials influence the economic viability significantly due to the high cost of production. Therefore, researchers are exploring and identifying low-cost nanobiomaterials that have competitive energy generation compared to traditional materials – for example, chitosan could be used as a cathode design and green-synthesised nanozyme-modified anodes could be considered as a value-added approach for achieving cost reduction while improving performance (Hammi *et al.*, 2020). Additionally, the integration of readily available resources, like bentonite clay and red soil, in membrane construction may show effectiveness in MFC applications, making them more cost-effective and accessible (Jenani *et al.*, 2024).

6.5.3 Environmental impact

The probable environmental and health consequences of using nanobiomaterials in MFC technology raise critical considerations that need to be systematically addressed for their sustainable development and extensive application. Concerns have been raised by several studies regarding the potential toxicity of some nanomaterials, including their dangerous effects on the human body on inhalation (Xuan *et al.*, 2023). Therefore, exposure risks during handling, manufacturing and disposal necessitate the implementation of strict protocols.

Furthermore, ongoing research is trying to speculate their impact on air, water and soil quality. At present, no specific limits for nanomaterial emissions have been established by the respective authorities, necessitating a comprehensive assessment of the long-term effects and understanding of the lifecycle of all biomaterials in MFC modification, ensuring their sustainable, safe and environmentally responsible application.

6.6. Conclusions

In conclusion, this chapter has assessed the major challenges associated with conventional MFC technology and the potential role of nanobiomaterials in addressing these limitations, particularly for bioenergy production enhancement.

The integral advantages of nanobiomaterials, including their high surface area-to-volume ratio, enhanced conductivity, stability and biocompatibility, have demonstrated significant improvement in MFC performance by way of modifications of anodes, cathodes, electrolytes and membranes, resulting in enhanced bioenergy output. Explicitly, the integration of carbon nanotubes, nanozymes, graphene and biogenic nanoparticles has shown substantial improvements in power density, electron transfer efficiency and operational stability of MFCs. The identification of nanobiomaterials such as bacterial cellulose and chitin aerogels highlights the shift of MFC development towards a sustainable environment.

Although nanobiomaterial-modified MFCs may involve high manufacturing costs initially, their elevated energy output and lower demand for maintenance supports their economic viability, thereby offering a cost-effective solution for bioenergy generation in a sustainable manner. Nevertheless, several challenges remain such as the scalability of nanobiomaterial synthesis for large-scale bioenergy systems, economic viability of large-scale production, long-term environmental impact in bioenergy applications and potential toxicity risks associated with nanomaterial handling and disposal.

Upcoming research should focus on the optimisation of manufacturing processes to reduce costs, development of standardised protocols for environmental impact assessments and exploring novel low-cost nanobiomaterials tailored for efficient bioenergy conversion. For this, interdisciplinary efforts involving microbiology, materials science and environmental engineering are essential to fully comprehend the potential of nanobiomaterials in MFC technology modification to maximise bioenergy sustainability and yield. Addressing these challenges will pave the way for the development of highly efficient, sustainable and economically viable MFCs for diverse applications in bioenergy production and environmental bioremediation.

REFERENCES

Abbas WMA and Ismail ZZ (2024) Enhanced bioenergy recovery by innovative application of chia seeds nanopowder for anode modification in microbial fuel cell treating hospital wastewater. *Environmental Technology:* 1–12, 10.1080/09593330.2024.243248

Abid HS, Johnson DJ, Hashaikeh R *et al.* (2017) A review of efforts to reduce membrane fouling by control of feed spacer characteristics. *Desalination* **420**: 384–402, 10.1016/j.desal.2017.07.019

Aiswaria PA, Naina Mohamed S, Singaravelu DL *et al.* (2022) A review on graphene/graphene oxide supported electrodes for microbial fuel cell applications: Challenges and prospects. *Chemosphere* **296**: 133983, 10.1016/j.chemosphere.2022.133983

Ali J, Wang L., Waseem H *et al.* (2020) Turning harmful algal biomass to electricity by microbial fuel cell: a sustainable approach for waste management. *Environmental Pollution* **266**: 115373, 10.1016/j.envpol.2020.115373

Babadi AA, Fakhlaei R., Rahmati S *et al.* (2024) A high-power hybrid carbon nanotube/three-dimensional reduced graphene oxide glucose/O_2 enzymatic biofuel cell. *Electrochimica Acta* **506**: 145054, 10.1016/j.electacta.2024.145054

Bahamonde Soria R, Chinchin BD, Arboleda D *et al.* (2022) Effect of the bio-inspired modification of low-cost membranes with TiO_2:ZnO as microbial fuel cell membranes. *Chemosphere* **291**: 132840, 10.1016/j.chemosphere.2021.132840

Bhowmick GD, Kibena-Põldsepp E, Matisen L *et al.* (2019) Multi-walled carbon nanotube and carbide-derived carbon supported metal phthalocyanines as cathode catalysts for microbial fuel cell applications. *Sustainable Energy & Fuels* **3**(12): 3525–3537, 10.1039/C9SE00574A

Chakraborty I, Sathe SM, Dubey BK *et al.* (2020) Waste-derived biochar: applications and future perspective in microbial fuel cells. *Bioresource Technology* **312**: 123587, 10.1016/j.biortech.2020.123587

Das S, Tripathi A and Ghangrekar MM (2024) Application of biomimetically synthesized silver nanoparticles as cathode catalyst, quorum-quencher, and anti-biofouling agent for the performance boosting of microbial fuel cell. *Chemosphere* **352**: 141392, 10.1016/j.chemosphere.2024.141392

Ding YH, Floren M and Tan W (2016) Mussel-inspired polydopamine for bio-surface functionalization. *Biosurface and Biotribology* **2**(4): 121–136, 10.1016/j.bsbt.2016.11.001

Elkhrachy I, Singh V, Kumar A *et al.* (2023) Use of biogenic silver nanoparticles on the cathode to improve bioelectricity production in microbial fuel cells. *Frontiers in Chemistry* **11**: 1273161, 10.3389/fchem.2023.1273161

Ferguson ARB (2008) Wind power: benefits and limitations. In *Biofuels, Solar and Wind as Renewable Energy Systems* (Pimentel D (ed.)). Springer, Dordrecht, the Netherlands, pp. 133–151, 10.1007/978-1-4020-8654-0_6

Gadhamshetty V, Belanger D, Gardiner C.-J *et al.* (2013) Evaluation of Laminaria-based microbial fuel cells (LbMs) for electricity production. *Bioresource Technology* **127**: 378–385, 10.1016/j.biortech.2012.09.079

Gebreslassie TR, Nguyen PKT, Yoon HH *et al.* (2021) Co-production of hydrogen and electricity from macroalgae by simultaneous dark fermentation and microbial fuel cell. *Bioresource Technology* **336**: 125269, 10.1016/j.biortech.2021 125269

Hammi N, Chen S, Dumeignil F *et al.* (2020) Chitosan as a sustainable precursor for nitrogen-containing carbon nanomaterials: synthesis and uses. *Materials Today Sustainability* **10**: 100053, 10.1016/j.mtsust.2020.100053

Hu W, Chen S, Yang J *et al.* (2014) Functionalized bacterial cellulose derivatives and nanocomposites. *Carbohydrate Polymers* **101**: 1043–1060, 10.1016/j.carbpol.2013.09.102

Jenani R, Karishmaa S, Babu Ponnusami A *et al.* (2024) A recent development of low-cost membranes for microbial fuel cell applications. *Desalination and Water Treatment* **320**: 100698, 10.1016/j.dwt.2024.100698

Khan A, Goepel M, Colmenares JC *et al.* (2020) Chitosan-based N-doped carbon materials for electrocatalytic and photocatalytic applications. *ACS Sustainable Chemistry & Engineering* **8**(12): 4708–4727, 10.1021/acssuschemeng.9b07522

Khoo KS *et al.* (2020) Nanomaterials utilization in biomass for biofuel and bioenergy production. *Energies* **13**(4): 892, 10.3390/en13040892

Kižys K, Zinovičius A, Jakštys B *et al.* (2023) Microbial biofuel cells: fundamental principles development and recent obstacles. *Biosensors* **13**(2): 221, 10.3390/bios13020221

Kumar A, Siddiqui T, Pandit S *et al.* (2023) Application of biogenic TiO_2 nanoparticles as ORR catalysts on cathode for enhanced performance of microbial fuel cell. *Catalysts* **13**(6): 937, 10.3390/catal13060937

Kunwar S, Pandey N, Bhatnagar P *et al.* (2023) A concise review on wastewater treatment through microbial fuel cell: sustainable and holistic approach. *Environmental Science and Pollution Research* **31**(5): 6723–6737, 10.1007/s11356-023-31696-x

Larsen MAD, Petrovic S, Engström RE *et al.* (2019) Challenges of data availability: analysing the water-energy nexus in electricity generation. *Energy Strategy Reviews* **26**: 100426, 10.1016/j.esr.2019.100426

Le G and Kim MI (2021) Research progress and prospects of nanozyme-based glucose biofuel cells. *Nanomaterials* **11**(8): 2116, 10.3390/nano11082116

Malunga-Makatu Khanyisile B, Chirwa Evans N and Tichapondwa Shepherd M (2023) The application of Bio-pd(0) as a biogenic cathode in a microbial fuel cell. *Chemical Engineering Transactions* **99**: 283–288, 10.3303/CET2399048

Matsena MT, Tichapondwa SM and Chirwa EMN (2020) Synthesis of biogenic palladium nanoparticles using Citrobacter sp. for application as anode electrocatalyst in a microbial fuel cell. *Catalysts* **10**(8): 838, 10.3390/catal10080838

Murjani BO, Kadu PS, Bansod M *et al.* (2022) Carbon nanotubes in biomedical applications: current status, promises, and challenges. *Carbon Letters* **32**(5): 1207–1226, 10.1007/s42823-022-00364-4

Nyabadza A, McCarthy É, Makhesana M *et al.* (2023) A review of physical, chemical and biological synthesis methods of bimetallic nanoparticles and applications in sensing, water treatment, biomedicine, catalysis and hydrogen storage. *Advances in Colloid and Interface Science* **321**: 103010, 10.1016/j.cis.2023.103010

Pandya RS, Kaur T, Bhattacharya R *et al.* (2024) Harnessing microorganisms for bioenergy with microbial fuel cells: Powering the future. *Water-Energy Nexus* **7**: 1–12, 10.1016/j.wen.2023.11.004

Pérez H, Quintero García OJ, Amezcua-Allieri MA *et al.* (2023) Nanotechnology as an efficient and effective alternative for wastewater treatment: an overview. *Water Science & Technology* **87**(12): 2971–3001, 10.2166/wst.2023.179

Prathiba S, Kumar S and Rangasamy G (2024) Recent advancements on the development of microbial fuel cells: Anode modification and scale-up challenges in upgrading anode electrode. *International Journal of Hydrogen Energy:* **S0360319924005597**, 10.1016/j.ijhydene.2024.02.144

Quan X, Sun B and Xu H (2015) Anode decoration with biogenic Pd nanoparticles improved power generation in microbial fuel cells. *Electrochimica Acta* **182**: 815–820, 10.1016/j.electacta.2015.09.157

Rana AK, Hart P and Thakur VK (2024) Nanocellulose/zero, one- and two-dimensional inorganic additive based electrodes for advanced supercapacitors. *Materials Today Electronics* **8**: 100103, 10.1016/j.mtelec.2024.100103

Sivasankar P, Poongodi S, Seedevi P *et al.* (2019) Bioremediation of wastewater through a quorum sensing triggered MFC: a sustainable measure for waste to energy concept. *Journal of Environmental Management* **237**: 84–93, 10.1016/j.jenvman.2019.01.075

Slate AJ, Whitehead KA, Brownson DAC *et al.* (2019) Microbial fuel cells: An overview of current technology. *Renewable and Sustainable Energy Reviews* **101**: 60–81, 10.1016/j.rser.2018.09.044

Sonawane JM, Mahadevan R, Pandey A *et al.* (2022) Recent progress in microbial fuel cells using substrates from diverse sources. *Heliyon* **8**(12): e12353, 10.1016/j.heliyon.2022.e12353

Tang S, Zhuang X, Zhao W *et al.* (2024) Catalytic role of biogenic gold nanoparticles in improving Cr (VI) removal efficiency of biocathode microbial fuel cells. *Journal of Chemical Technology & Biotechnology* **99**(6): 1364–1376, 10.1002/jctb.7635

Thapa K, Liu W, Zhang Y *et al.* (2024) Boosting the power performance of microbial fuel cells by using dual nanomaterial-modified carbon felt electrodes. *Energy & Fuels* **38**(21): 21412–21422, 10.1021/acs.energyfuels.4c03755

Vilela C, Silvestre AJD, Figueiredo FML *et al.* (2019) Nanocellulose-based materials as components of polymer electrolyte fuel cells. *Journal of Materials Chemistry A* **7**(35): 20045–20074, 10.1039/C9TA07466J

Wang Y, Yu M, Wang Y *et al.* (2025) Electrochemically activated carbon nanotube anodes for enhanced microbial fuel cell performance. *Next Energy* **8**: 100255, 10.1016/j.nxener.2025.100255

Wu Q, Zhang G, Wang B *et al.* (2018) Production and identification of Inthomycin B produced by a deep-sea sediment-derived Streptomyces sp. YB104 based on cultivation-dependent approach. *Current Microbiology* **75**(7): 942–951, 10.1007/s00284-018-1469-1

Xu H and Quan X (2016) Anode modification with peptide nanotubes encapsulating riboflavin enhanced power generation in microbial fuel cells. *International Journal of Hydrogen Energy* **41**(3): 1966–1973, 10.1016/j.ijhydene.2015.11.124

Xuan L, Ju Z, Skonieczna M and Zhou P (2023) Nanoparticles-induced potential toxicity on human health: applications, toxicity mechanisms, and evaluation models. *MedComm* **4**(4): e327, 10.1002/mco2.327

Yan J, Zhang M, Chen X *et al.* (2024) Multi-walled carbon nanotubes modified corn straw biochar as high-performance anode in microbial fuel cells. *Journal of Environmental Chemical Engineering* **12**(5): 113316, 10.1016/j.jece.2024.113316

Ye Y, Guo W, Ngo HH *et al.* (2024) Biofuel production for circular bioeconomy: present scenario and future scope. *Science of The Total Environment* **935**: 172863, 10.1016/j.scitotenv.2024.172863

Yeamsuksawat T, Kim HJ and Eom Y (2024) Shape-tunable and sustainable carbon materials derived from nanocellulose and nanochitin: carbonization, structures, and applications. *Materials Today Energy* **43**: 101604, 10.1016/j.mtener.2024.101604

Zhang Y, Liu L, van der Bruggen B *et al.* (2017) Nanocarbon based composite electrodes and their application in microbial fuel cells. *Journal of Materials Chemistry A* **5**(25): 12673–12698, 10.1039/C7TA01511A

Zhao C, Gai P, Song R *et al.* (2017) Nanostructured material-based biofuel cells: recent advances and future prospects. *Chemical Society Reviews* **46**(5): 1545–1564, 10.1039/C6CS00044D

Microbial Processes in Environmental Engineering: From Theory to Practice

Rajneesh Kumar, Ram Sharan Singh and Maulin P Shah
ISBN 978-1-83708-327-5
https://doi.org/10.1108/978-1-83708-326-820261007

Chapter 7
Advanced microbial technologies for bioremediation of emerging contaminants

Anjali Mishra, Aditi Singh, Sudha Suman, Rajneesh Kumar and Ram Sharan Singh

Department of Chemical Engineering and Technology, Indian Institute of Technology (BHU) Varanasi, Uttar Pradesh, India

Abstract

In recent years, growing global pollution has introduced emerging contaminants (ECs) into the aquatic environment, which pose a threat to the environment. Conventional methods are inadequate to remediate these toxic substances and have limited EC removal efficiency and cost-effectiveness. On the other hand, classical microbial technologies using bacteria, algae and fungi were used for the remediation of ECs, but they are also inefficient in degrading recalcitrant pollutants, which necessities the development of advanced technologies to remediate and sustainably reduce aquatic environmental toxicity. Various advanced methods combined with microbial technology show higher efficiency in the degradation of harmful products into less harmful products, whether it is a single pollutant or a multiple pollutant system, and also prove useful in wastewater treatment. These advanced microbial techniques have emerged as a transformative field that acknowledges environmental challenges through the innovative application of microorganisms. Therefore, this chapter discusses emerging contaminants, the microbial role in bioremediation, and advanced microbial technologies such as the enzymatic method, genetic engineering, microbiome engineering, omics technology, nanobioremediation, bioelectrochemical systems and the role of synthetic biology in enhancing microbial efficiency. Additionally, microbial technology enabling the utilisation of microbial biomass for the production of value-added products, contributing to sustainable and innovative solutions, is also discussed.

Keywords: Bioremediation; emerging contaminants; advanced microbial techniques; microbiome engineering; synthetic biology.

7.1. Introduction

In recent decades, the rapid growth in industrialisation, urbanisation, modern agricultural practices and healthcare sector advancements, has introduced a range of emerging pollutants into aquatic ecosystems contributing to environmental degradation (Kumar *et al.*, 2022). These contaminants pose a serious threat to the natural environment, threatening aquatic biodiversity and the health of living organisms due to their persistence, chemical nature and toxicity at extremely low concentrations (ng/L to μg/L) (Liu *et al.*, 2025; Pande *et al.*, 2020). Emerging contaminants (ECs) are defined as the variety of natural and anthropogenic compounds recently identified and increasingly found in aquatic environments. Examples of ECs include pharmaceuticals, personal care products (PCPs), microplastics, pesticides, endocrine-disrupting chemicals (EDCs) and per/polyfluoroalkyl substances (PFASs).

These chemicals often lack proper regulation and sufficient data regarding their presence, behaviour and environmentally harmful effects (Mishra *et al.*, 2025; Visvanathan *et al.*, 2024). Due to the increasing concern about ECs, there is increasing attention on the remediation of these contaminants from the environment. Thus, various conventional remediation methods are applied, such as physicochemical and biological processes, among which a biological process, namely bioremediation, is the most suitable due to its eco-friendly nature, sustainability, cost-effectiveness, lower energy requirement, low carbon dioxide (CO_2) emissions and its social acceptability.

In contrast, physicochemical processes produce more carbon dioxide, have a high energy requirement and are costly (Bala *et al.*, 2022). Bioremediation is an engineered process utilising microorganisms (bacteria, algae and fungi) to metabolise organic pollutants into intermediary metabolites or mineralise them into less stable compounds like carbon dioxide and water. Bioremediation processes require human intervention, optimising conditions like nutrient availability, aeration and temperature to enhance pollutant removal. During the process, contaminants are broken down, immobilised and detoxified using microbial enzymes produced by microorganisms. Under optimal conditions, these microorganisms can independently degrade pollutants like pesticides, antibiotics, PFAS, parabens and others (Visvanathan *et al.*, 2024).

Recent studies have shown that synergistic associations of these microorganisms can enhance the biodegradation process by combining their action and overcoming their individual limitations. This is achieved by increased enzyme formation, reduced reactive oxygen species (ROS) production, their adaption to extreme environments and counteracting feedback inhibition caused by intermediate metabolites. These mixed cultures of microorganisms can be formulated to degrade recalcitrant pollutants and utilised for multi-pollutant treatment. They are applicable in real-world applications, which signifies potential for efficient degrading of ECs. Mixed cultures can comprise bacterial consortia, bacterial-microalgal consortia and bacterial-fungal consortia (Li *et al.*, 2022).

Researchers have been focusing recently on advanced microbial technologies that are emerging techniques to address the challenges caused by classical methods for effective removal. These advanced methods aim to achieve faster and more efficient removal of ECs, leveraging the unique metabolic capabilities of microorganisms. By enhancing bioremediation processes, these technologies hold promise for mitigating the impacts of these contaminants while ensuring environmental sustainability (Ayilara and Babalola, 2023). Enzyme-mediated degradation performs biocatalytic transformation, which removes more undesired and toxic bioproducts than chemical remediation (Gan *et al.*, 2022). Advancements in molecular biology have developed a way to utilise genetically engineered microorganisms (GEMs) in bioremediation. The microbial capabilities have been enhanced through modifications in catabolic pathways, metabolic routes, insertion or overexpression of catabolic genes, accumulation of toxic intermediates prevention and degradation efficiency (Singh *et al.*, 2025).

On the other hand, synthetic biology designs gene circuits, programmable cells and biomolecular components engineered for targeted functions, such as enhanced degradation pathways or pollutant-sensing mechanisms. These tools enable efficient bioremediation of specific emerging pollutants (such as microplastics and pharmaceuticals) or a diverse range of contaminants in complex environments (Nguyen *et al.*, 2023).

Microbiome engineering is a unique and crucial technique that incorporates the power of microbial communication through quorum sensing. Microbiome engineering involves engineering a particular set of microorganisms either inter- or intra species that can perform the desired function of degrading specific pollutants (Singha and Shukla, 2023). In addition, metagenomics, transcriptomics, proteomics and metabolomics (known as omics techniques) are techniques that analyse microbial functions through deoxyribonucleic acid (DNA), ribose nucleic acid (RNA), proteins and metabolites in environmental samples.

Tools like fluxomics and next-generation sequencing provide insights into microbial activity, diversity and gene expression, advancing research in microbial roles and their applications in bioremediation (Sharma *et al.*, 2022). Similarly, nanobioremediation is a synergistic approach which combines nanotechnology and microbial remediation, helping to improve catalytic activity, reactive and adsorptive properties, which significantly degrade ECs (Ambaye *et al.*, 2025). These advancements of microbial technology provide a sustainable and eco-friendly approach, together with better removal efficiency.

However, different sectors require more nuanced insights into the particular advancements. So, this chapter delves into types of emerging contaminants and their toxic effects, microbial methods to remediate emerging contaminants, and more profound insights into advanced microbial technologies, such as enzyme-mediated degradation, genetic engineering, synthetic biology, microbiome engineering, microbial fuel cells and multi-omics approaches for enhanced EC degradation. It also discusses the use of microbial biomass to produce value-added bioproducts such as biofuels, biodiesel, biohydrogen, bioplastics, biocatalysts and so on, along with challenges of and future perspectives on the advancements of microbial technologies.

Figure 7.1 illustrates the pathway from sources of emerging contaminants, their types and their harmful effects to mitigating strategies, microbial bioremediation and advanced remediation techniques. It also highlights the utilisation of microbial biomass for the production of value-added products, promoting sustainability.

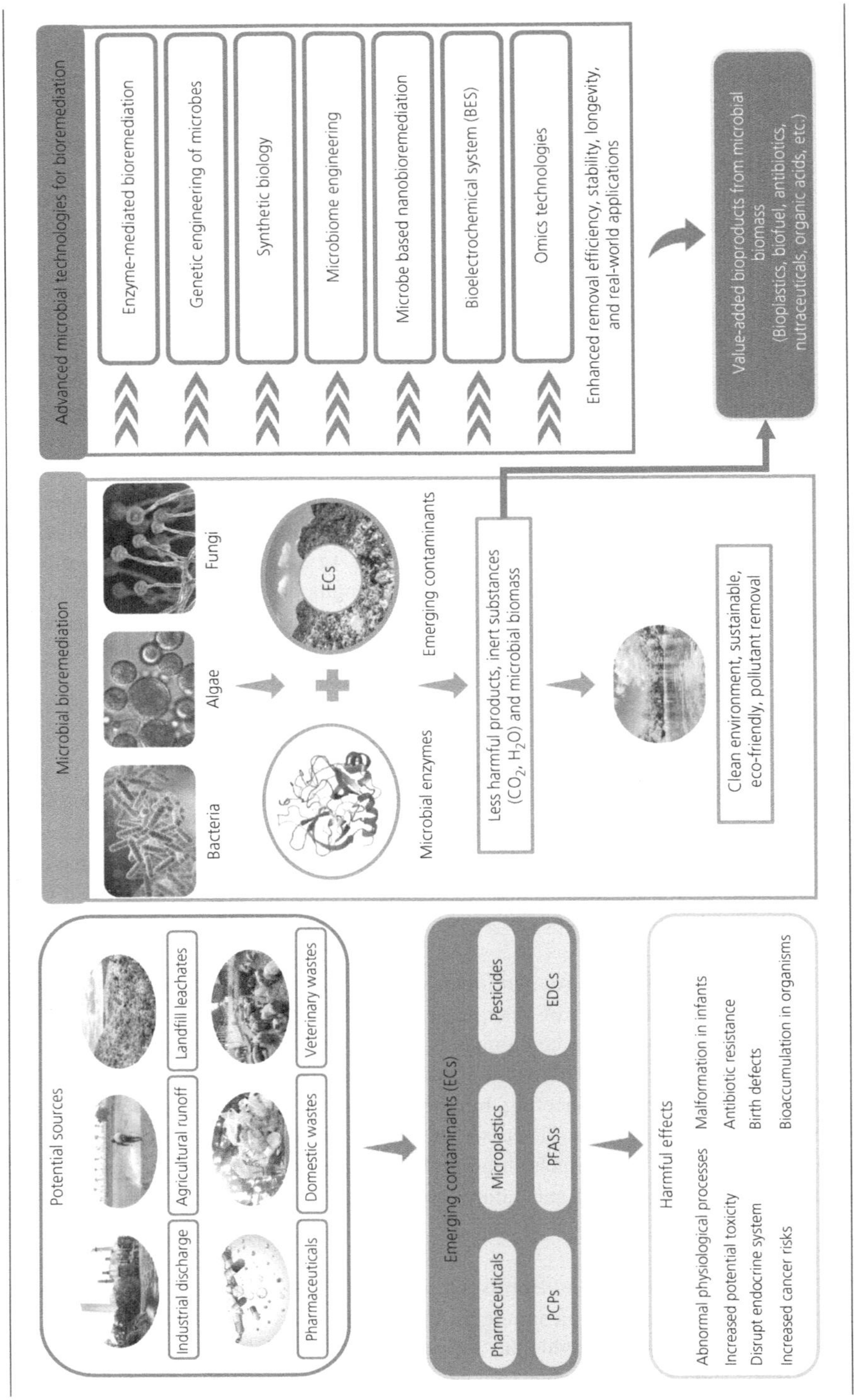

Figure 7.1 Graphical abstract: emerging contaminants to sustainable solutions

7.2. Occurrence of ECs in aquatic environments

ECs are new pollutants that include pharmaceuticals, microplastics, EDCs, PCPs, pesticides and PFASs, and their concentrations in water range from ng/L to µg/L (Liu *et al.*, 2025). These ECs include substances that negatively impact the biological systems of living organisms, including aquatic and terrestrial animals, including humans, and tend to bioaccumulate in living organisms (Rodriguez-Narvaez *et al.*, 2017). Different types of ECs, their sources, concentration in the aquatic environment and toxic effects on living organisms are summarised in Table 7.1.

Table 7.1 Emerging contaminants: sources, concentrations in aquatic environments, toxicity (continued on next page)

ECs		Source	Concentration	Toxic effect	Reference
Microplastics	Fragments	Surface water	0–320 000 items/km^2	*Carassius auratus* reduction in weight, severe breakage of the dermal layer with haemorrhages, histopathological changes	Elizalde-Velázquez and Gómez-Oliván, 2021; Swain *et al.*, 2025
	Microplastic (PE)	Lagoons	453.0–335.2 items/m^3	The plant growth and nitrogen content are significantly decreased, impairing water absorption and nutrients	Ge *et al.*, 2021; Mishra *et al.*, 2022
	Microplastic (microfibre)	Coastal areas	176–2842 mm	Gut blockage in mussels	Mishra *et al.*, 2022
Pharmaceuticals	Ibuprofen	WWTP influent	114.00 µg/L	Decrease in zebrafish *Danio rerio* hatch rate at 55 hpf (63%, 500 µg/l)	Munzhelele *et al.*, 2025
	Diclofenac	WWTP influent	2.21–25.00 µg/L	Cytopathological liver damage of rainbow trout *Oncorhynchus mykiss*	
	Metformin	WWTP influent	1421 to 325 000 ng/L	MET exhibited cytotoxic properties and caused a mild genotoxic effect	Gewurtz *et al.*, 2025; Yuzbasioglu *et al.*, 2022
	Ofloxacin	WWTP influent	0.1 to 0.5 µg/L	Antioxidant system inhibition. impacts on cell growth photosynthetic rate and chlorophyll content of *Microcystis aeruginosa*	Camara *et al.*, 2025

Table 7.1 Continued

ECs		Source	Concentration	Toxic effect	Reference
PCPs	Triclosan	Surface water	<0.1–2300 ng/L	Altered cell shape and reduced survival rate of Euglena *Closterium ehrenbergii*	Srain *et al.*, 2021
	Musk xylene	Surface water	1.1–180 ng/L	Affected heart rate and growth of zebrafish *Danio rerio*	Srain *et al.*, 2021
	DEET	Surface water	33.4 μg/L	Slightly toxic to *S. vittatum*, it may also reduce or modify an insect's oviposition patterns	Wang *et al.*, 2021; Weeks *et al.*, 2012
PFASs	PFOA	Surface water	6.36 ng/L	Tumorigenesis in liver cells	Li *et al.*, 2017; Sims *et al.*, 2022
	PFOS	Surface water	3.04 ng/L	Impacts on chlorophyll accumulation and soluble protein synthesis	Qu *et al.*, 2010; Sims *et al.*, 2022
	PFHxA	Drinking water	22 ng/L	Decreased survival rate in *Danio rerio* by 7×106 – 1×107 μg/l	Domingo and Nadal, 2019; Kreychman *et al.*, 2024
	PFHpA	Drinking water	23 ng/L	Reduced locomotor activity in fish at 0.1 μM	Domingo and Nadal, 2019; Huang *et al.*, 2023
EDCs	Parabens	Surface water	15–400 ng/L	Acute toxicity in *Daphnia magna*	Metcalfe *et al.*, 2022
	EHMC	Surface water	2.7–224 ng/L	Adult and larval zebrafish *Danio rerio* showed thyroid disruption (T3 and T4 downregulated), nephrotoxicity and neurotoxicity	Chu *et al.*, 2021
	Benzophenone 3	Surface water	2.3–545 ng/L	Modification in endocrine signalling gene, ultraspiracle, in the aquatic invertebrate *Chironomus riparius*	Kim and Choi, 2014; Metcalfe *et al.*, 2022
	Bisphenol A	Ground and surface water	Up to 20 μg/L	Enhances breast cancer risk in humans	Kabir *et al.*, 2015

[PE = polyethylene; EHMC = 2-ethyl-hexyl-4-trimethoxycinnamate; DEET = N, N diethyl-meta-toluamide; PFHxA = perfluorohexanoic acid; PFHpA = perfluoroheptanoic acid; PFOA = perfluorooctanoic acid; PFOS = perfluorooctane sulfonate]

7.3. Microbial role in bioremediation

In bioremediation processes, microorganisms degrade toxic pollutants by using them as a source of carbon and convert them into less toxic and stable forms. The process also produces several other metabolites/byproducts further utilised by other microorganisms through mineralisation. Bioremediation depends on microbial metabolism, which catalyses the transformation of pollutants through enzymatic reactions (Saeed *et al.*, 2022), occurring by way of aerobic (oxygen-dependent) or anaerobic (oxygen-independent) pathways. Microorganisms grow and complete their lifecycle in harsh, contaminated environments, increasing their efficiency in degrading pollutants in contaminated sites. Bioremediation can occur in situ (directly at site) or ex situ (after removing contaminants to a controlled environment). Microorganisms remediate pollutants through an immobilisation process (complexation, bioaccumulation, biosorption and precipitation (solidification)) and a mobilisation process (bioleaching, bioaugmentation, biostimulation, enzymatic oxidation and reduction) (Ayilara and Babalola, 2023).

Most studies have utilised bacteria in bioremediation due to their versatile nature of degradation, rapid growth rates, genetic adaptability and diverse metabolic pathways. Various microbial genera used in bioremediation include *Alcaligenes*, *Achromobacter*, *Arthrobacter*, *Bacillus*, *Pseudomonas*, *Xanthobacter*, *Corynebacterium*, *Mycobacterium*, *Nitrosomonas*, *Flavobacterium* and so on. Major factors such as oxygen, temperature, moisture content, nutrients, pH, concentration of contaminants, physicochemical properties of contaminants and biological factors influence the processes (Bala *et al.*, 2022).

Microalgae, which are eukaryotic, unicellular and photosynthetic organisms that can grow in harsh conditions like wastewater or contaminated sites, are effective in bioremediation. They support bacterial degradation by producing oxygen and have potential for biomass recovery for biofuel production and other products. Their remediation efficiency depends on species and cell wall structure. This biodegradation process can occur intra- or extracellularly by microalgal enzymes (Dubey *et al.*, 2023).

Bioremediation using microalgae generally involves species of *Chlorella*, *Scenedesmus*, *Microcystis*, *Spirulina*, *Nannochloris*, *Cyanobacteria* and so on. Various factors which influence micro algal bioremediation include pH, temperature, light intensity, redox, aerobic condition, adsorbent size and concentration of ECs (Gondi *et al.*, 2022). Bioadsorption, bioaccumulation, biouptake, photodegradation and biodegradation/biotransformation are the potential mechanisms of EC removal using microalgae (Kurade *et al.*, 2016). Furthermore, fungi are omnipresent, chemoheterotrophic organisms; they generally thrive in aerobic conditions and are common throughout aquatic and terrestrial environments. Extracellular and intracellular enzymatic machinery of fungi perform oxidative reactions and also excrete acids that allow a wide range of ECs to degrade and metabolise. Fungal enzymes show low substrate specificity; this feature makes them suitable to degrade ECs as their energy and carbon source. *Aspergillus*, *Fusarium*, *Cladosporium*, *Trametes*, *P. chrysosporium*, *Penicillium*, *Alternaria*, *Trichoderma* and so on are common genera in bioremediation (Vaksmaa *et al.*, 2023).

More efficient degradation of ECs has been achieved recently using microbial consortia that comprise bacteria, microalgae and fungi or in combinations thereof, each contributing its specific metabolic functions. These microorganisms form diverse communities in nature which are more stable and resilient to stresses and various environmental changes. These microbial consortia support each other by way of sharing nutrients, enhanced enzymatic functions, and enhanced support from core partners to utilise and degrade emerging pollutants (Li *et al.*, 2021). These consortia interact with each other through quorum sensing, which helps microbial communities to communicate with each other by producing signalling molecules that accumulate in their surroundings and assist in cell communication (González-González *et al.*, 2022; Santo *et al.*, 2022).

Researchers design synthetic microbial consortia for the bioremediation of ECs to utilise their synergistic interaction within microbial consortia and their multi-component system, enhancing the efficiency, stability, scalability, broader substrate ranges and resilience for EC degradation (Cao *et al.*, 2022).

7.4. Advanced techniques in microbial remediation

The uncertainty of microbial behaviour and the slow process necessitates an advancement with new technologies and changes which efficiently degrade ECs with certain, fast, predictive and optimised degradation. The advanced techniques for microbial remediation are illustrated in Figure 7.2 and a summary of all techniques is presented in Table 7.2.

7.4.1 Enzyme-mediated biodegradation

Enzyme-mediated bioremediation enhances pollutant degradation and offers a sustainable alternative to microbial remediation, which provides synthetic pathways, is economical, generates less waste and is more energy efficient. Enzymatic bioremediation, using purified free microbial enzymes, is more efficient than conventional methods, as it overcomes limitations associated with microbial-mediated degradation, such as the slow degradation process and microbial growth requiring optimal conditions such as moisture, temperature, pH, redox potential and pollutant chemical composition.

In contrast, enzyme-based approaches accelerate pollutant breakdown without these limitations. Free enzyme has the capability to function at different temperatures and pH ranges. However, free enzymes have less operational stability due to their faster decay in a cell-free environment and have poor recovery after the bioremediation process completes (Somu *et al.*, 2022). Researchers suggested immobilisation of free enzymes to overcome these limitations, which involves attaching purified enzymes to a matrix that contains biodegradable or non-biodegradable carriers which can be granular, fibrous or a membrane, to improve stability, productivity, lifetime and reusability. Such conditions ensure an increase in enzymatic activity and enzyme immobilisation increases the enzyme's resistance to fluctuations in temperature, pH and substrate concentration, thereby extending its operational lifetime and enhancing productivity (Netto *et al.*, 2013).

Immobilisation of enzymes is technically done by physical and chemical methods. Physical techniques include entrapment or encapsulation, binding to the carrier, cross-linking and attachment, which can be ionic, covalent or physical adsorption to the carrier. Chemical

Figure 7.2 An overview of advanced microbial technologies: (*a*) enzyme-mediated bioremediation, (*b*) genetic engineering of microorganisms, (*c*) synthetic biology-mediated, (*d*) microbiome engineering, (*e*) bioelectrochemical system-mediated, (*f*) nanobioremediation and (*g*) omics technique-mediated bioremediation

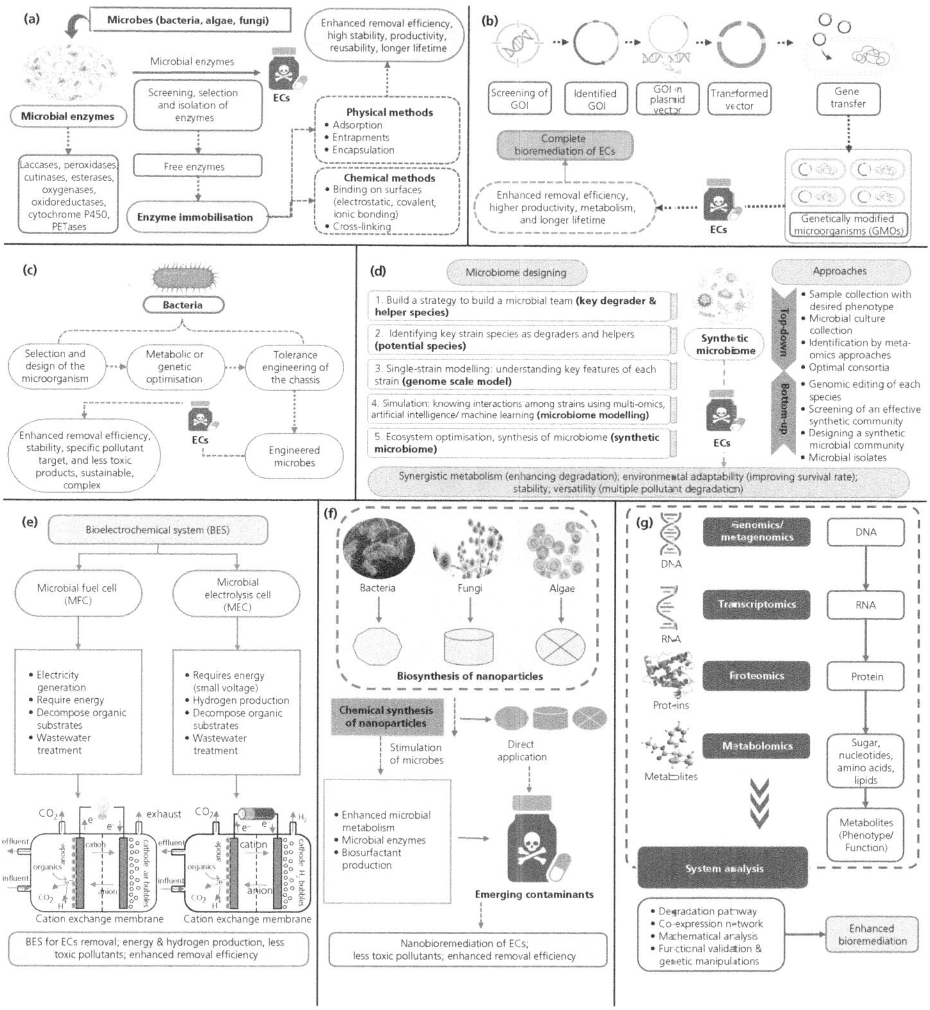

binding occurs by cross-linking enzymes to proteins or another enzymes, working as a bifunctional reagent with enhanced activity, stability and function (Chia *et al.*, 2024).

Generally, enzyme-mediated remediation utilises microbial enzymes from bacteria, fungi and algae to enhance efficient EC degradation by lowering activation energy. Enzymes or biocatalysts are globular proteins that enhance the rate of reaction and convert substrate into end

Table 7.2 Advanced microbial technologies used for bioremediation of ECs (continued on next page)

Technique	Microorganism	Method	Pollutant	Degradation efficiency	Reference
Enzyme-mediated degradation	*Tramete versicolor*	Laccase enzyme isolated and immobilised by covalent binding on SiO_2	Bisphenol A	>85% removal	Chang *et al.*, 2019
	Aspergillus sp.	Immobilisation of laccase onto r-GO (reduced graphene oxide) bio-functionalised sponges by covalent binding showed significant removal	EDCs (Triclosan and Bisphenol A)	TCS 84%; BPA – 87% after 24 h	Moayedi and Yousefi, 2025
	Escherichia coli	Erythromycin-degrading enzyme EreB esterase is immobilised on Cu-BTC synthetic metal-organic framework (MOF) to degrade the pollutant	Erythromycin (antibiotic)	Enhanced degradation efficiency	Ni *et al.*, 2023
Genetically modified organisms	*Escherichia coli*	Tetracycline-degrading enzyme (TetX) from *Bacteroides fragilis* was recombinantly expressed in *E.coli* and enhances expression level of TetX to degrade	Tetracycline	Degradation of tetracycline with high efficiency	Mu *et al.*, 2018
Synthetic biology-mediated	*Pseudomonas putida* KT2440	Synthetic biology chassis, genetically engineered, metabolic pathway engineering, genomic barcoding, controlled cell adhesion, redox metabolism optimisation, stress management	May be useful for emerging contaminants (pharmaceuticals, EDCs, microplastics)	–	Martínez-García and de Lorenzo, 2024
	Escherichia coli strain DH5α (for plasmid DNA extraction) and *E.coli* strain BL21 (DE3) (for protein expression)	Engineer the FerTiG (modular enzyme assembly), consists of three modules: 1. TC degradation module (Tet(X4)), 2. glucose dehydrogenase (GDH) (cofactor recycling module) and 3. ferritin (protection module)	Tetracycline	Enhanced removal efficiency	Ren *et al.*, 2025

Table 7.2 Continued

Technique	Microorganism	Method	Pollutant	Degradation efficiency	Reference
Microbiome engineering	*Desmodesmus* sp. algal-bacterial symbiotic system	Microbiome synthesis, deformation of polymer and pangenomic analysis resulted in identification of biodegradation pathway and genes related to chemotaxis	Micro PVC (polyvinyl chloride), PE (polyethene), PS (polystyrene), PET (polyethylene terephthalate)	Enhanced removal efficiency	Fan *et al.*, 2024
	Constructed natural microbiome of 18 key species (*Comamonas*, *Pigmentiphaga*, *Pseudoxanthomonas*, *Pseudar throbacter*, *Bacillus, Sphingomonas*, *Sphingobacterium*, *Streptomyces*, *Lysinibacillus*, *Sinomonas*, *Arthrobacter*, *Bradyrhizobium*, *Nocardioides*, *Aliihoeflea*, *Acinetobacter*, *Achromobacter* and *Pseudomonas*)	Developed metabolic modelling pipeline superCC, helped in documenting metabolic interactions within microbiomes and simulates performance, microbiome design and constructed herbicide-degrading microbiome	Herbicide – bromoxynil octanoate (BO)	–	Ruan *et al.*, 2024

Table 7.2 Continued

Technique	Microorganism	Method	Pollutant	Degradation efficiency	Reference
Microbial nanobioremediation	*Chlorella vulgaris* and *Scenedesmus* spp.	Microalgae and ME-nFe (carbon-encapsulated iron nanoparticles). electrostatic interaction, degradation	Ofloxacin, Irbesartan, Diclofenac, Clarithromycin, lamotrigine, metoprolol	98%; 65-80%	Mantovani *et al.*, 2024
	White-rot fungi *Pleurotus ostreatus*	Fungi with γ-Fe_2O_3 nanoparticles, γ-Fe_2O_3 was bioreduced to Fe_3O_4, produced hydroxyl radical	Bisphenol A	32% in 3 days	Li and Zhang, 2016
	Pseudomonas aeruginosa	With NP Zr-O_2; cell adsorption	Tetracycline	52.6%	Debnath *et al.*, 2020
Bioelectrochemical system	*Proteobacteria* and *Synergistetes*	Microbial electrolysis cell (MEC) – anaerobic process, integrated degradation process involves breaking the S–N bond in SMX for metabolism and cleaving the N–O bond in its isoxazole ring under sulfate-reducing conditions.	Sulfamethoxazole	77.60%, 87.55%, and 92.53% in a 3-days;	Xue *et al.*, 2019
	Electrogenic bacteria, including *Geobacter sulfurreducens* and *Beta Proteobacteria*	Microbial fuel cell (MFC) – anaerobic process, reductive dechlorination pathway removes chlorine atoms from compounds through electron transfer, aiding in detoxification. Bacteria function in the presence of sodium acetate as a co-substrate	Hexachlorobenzene (toxic, refractory organic pesticide)	71.15%	Cao *et al.*, 2015
	Microbial consortium contains *Serratia*, *Petrimonas*, *Pseudochrobactrum* and *Sphingobacterium*	MEC – anaerobic process, catalase activity, molecular docking analysis	Tetracycline	91.98 % within 4 days	Wang *et al.*, 2025

Table 7.2 Continued

Technique	Microorganism	Method	Pollutant	Degradation efficiency	Reference
Omics technologies	Anaerobic sludge	MFC – anaerobic process	Nonylphenol	56.33 ± 0.97% in 24 days	De la Peña *et al.*, 2024
	Proteobacteria, *Actinobacteria*, *Acidobacteria*, *Firmicutes*, *Chloroflexi* and *Gemmatimonadetes*	16S rRNA/Illumina Miseq and Illumina Hiseq perfluorinated	Perfluorinated compounds (PFCs)		Cai *et al.*, 2020
	Microalgae (*Scenedesmus quadricauda*)	Transcriptomic analysis, removal through biosorption and bioaccumulation, functional key genes revealed involved in defence, degradation and energy response	PFOA (perfluorooctanoic acid)	58.2%	Ha *et al.*, 2025
	Bacterial consortia consisting of three bacteria: *Exiguobacterium* sp., *Ochrobactrum* sp. and *Halomonas* sp	Transcriptomic analysis for PET to investigate potential degradation methods using it on PE for significant degradation results	PET (polyethylene terephthalate), PE (polyethylene)	–	Gao and Sun, 2021

Table 7.2 Continued

Technique	Microorganism	Method	Pollutant	Degradation efficiency	Reference
	Bradyrhizobium, *Rhodopseudomonas*, *Mycobacterium*, *Cupriavidus*, *Pseudomonas*, and *Streptomyces*, *Starkeya*, *Rhodococcus*, *Rhizobium*, *Ochrobactrum*, *Sphingomonas*, *Methylobacillus*, *Stenotrophomonas* and *Alicycliphilus*	Shotgun metagenome sequencing and Illumina HiSeq, identified pollutant-degrading communities	Carbamazepine (CBZ), triclosan (TCS) and triclocarban (TCC)	80%	Thelusmond *et al.*, 2019

products. Enzymes have active sites (one or more) through covalent or non-covalent interactions. They modify pollutants structurally and reduce toxicity, converting them into harmless end products (Mousavi *et al.*, 2021).

Enzymes are produced in bacteria, fungi and algae both extracellularly and intracellularly, and these enzymes are mainly classified into six divisions: hydrolases, transferases, oxidoreductases, isomerases, lyases and ligases (synthetases). Oxidoreductases catalyse oxidation-reduction reactions, while transferases shift functional groups between molecules, hydrolases utilise water to break chemical bonds, lyases cleave bonds by elimination, forming double bonds, isomerases rearrange molecular structures geometrically or structurally, and ligases join two molecules (Karigar and Rao, 2011).

Two major classes of enzymes are used in bioremediation which contain potent biocatalytic capabilities to degrade organic compounds. These enzymes are oxidoreductases, which include enzymes such as peroxidases, laccases, oxygenases and tyrosinases, and hydrolases, which include enzymes such as cutinases, lipases, esterases, PETases and dehalogenases (Khalifa *et al.*, 2024). These enzymes are used after extraction from microbial species by immobilisation, and they show significantly better degradation efficiency than traditional microbial technology, which is slow and uncertain; the enzymatic method is fast, certain and sustainable.

7.4.2 Genetic engineering of microorganisms

Genetic engineering of microorganisms refers to genetically modified microorganisms (GMOs) with specific functions in which genes of interest (GOI) are fragmented, spliced and incorporated using molecular tools. Recombinant genes are inserted into the host cell (bacteria) with specific new phenotypes, with the desired function (Liu *et al.*, 2019; Mahdizade Ari *et al.*, 2024). The process comprises screening and cloning of GOI which are highly effective in degradation, alongside enhancing the expression of microbial enzymes with degradation functions, followed by protoplast fusion and other techniques that can be utilised to blend the advantages of bacteria for pollutant degradation. The final step leads to expressing degrading genes for different pollutants by way of engineered bacteria. GMOs show high degradation rate, maximal removal efficiency and high bacterial survival (Liu *et al.*, 2019; Zhao *et al.*, 2017).

Various technologies are used to genetically engineer microorganisms, such as clustered regularly interspaced short palindromic repeats (CRISPR-Cas9), transcription activators like effector nucleases (TALEN), bacterial artificial chromosomes (BACs) and protoplast fusion and transposition recombination (Rafeeq *et al.*, 2023). Small DNA fragments (<10 kb) are modified with polymerase chain reaction (PCR) and restriction enzymes, while larger segments (>50 kb) are edited using recombination methods like CRISPR-Cas9, which add, delete or substitute gene segments in plasmids or genomes. CRISPR-Cas systems provide RNA-directed genome editing, offering microorganism adaptive immunity.

Cas9 proteins introduce double-strand breaks (DSBs) at target sites, which are repaired through nonhomologous end joining (NHEJ) or homology-directed repair (HDR) pathways. NHEJ is error-prone, while HDR ensures precise genetic modifications, making it useful for bioremediation purposes. TALEN binds to a specific DNA sequence in the host genome, creating DSBs with sticky ends for stability.

Zinc finger nucleases (ZFNs) use a 30-amino-acid DNA-binding domain, and their Fok1 cleavage domain introduces DSBs at targeted locations (Sarma *et al.*, 2021). Foreign DNA integration relies on homologous recombination and transposition, using transposases which integrate small or large DNA fragments into the genomes of bacteria. However, efficiency decreases as fragment size increases. Different methods to transfer genetic material depend on the size of the DNA and the bacterial host properties. Common plasmid transfer methods used are electroporation and heat shock transfection in the host bacterium. Modified plasmids are transferred from donor bacteria to recipient bacteria.

Conjugative transfer and protoplast fusion allow plasmids to move between donor and recipient bacteria by way of a specialised structure, such as the Type IV secretion system, which helps in targeted transfer (Liu *et al.*, 2022). Researchers have used different systems like BAC and transformation-associated recombination, which enable heterologous expression of gene clusters for large fragments (>100 kb) (Mahdizade Ari *et al.*, 2024).

These biotechnological tools allow the precise modification of microbial genomes, improving their functionality for bioremediation.

7.4.3 Synthetic biology for enhanced biodegradation

Synthetic biology (synbio) modifies microorganisms using molecular biology tools to enhance their abilities. The National Human Genome Research Institute defines synbio as '*a field of science that involves redesigning organisms for useful purposes by engineering them to have new abilities.*' (Rylott and Bruce, 2020, p.1). Synbio helps to design and synthesise components such as gene promoters, transcription factors, enzymes and so on.

In the context of bioremediation, synthetic biology helps in designing biosensors for pollutant detection, creating resilient microorganisms that can grow in extreme environments, and improving pollutant degradation. Synthetic biology provides an understanding of fundamental biological processes and develops microorganisms for specific purposes, such as enhanced activity for bioremediation.

By assembling metabolic pathways and modifying enzymes, synthetic biology accelerates bioremediation. Advances like the CRISPR genome editing tool provide more advancements in the development of specific bioremediation strategies of ECs (Chunyan *et al.*, 2023). Engineered microbial biosensors and metabolic modifications enable the detection and breakdown of contaminant residues, helping in the remediation of ECs. Advancements in synthetic biology have been driven by improved DNA sequencing, synthesis and genome-scale engineering, enabling precise genetic modifications.

For bioremediation, extremophilic microorganisms are valuable as they produce robust enzymes capable of functioning in extreme conditions where conventional proteins would degrade. These adapted enzymes enhance pollutant breakdown, making synthetic biology a powerful tool for developing efficient environmental remediation strategies (Rylott and Bruce, 2020).

Metabolic reconstruction uses computational platforms to analyse and optimise microbial metabolism for bioremediation. Metabolic pathway analysis (MPA) is a mathematical model that helps predict metabolic flux and enzyme interactions during cellular growth. Tools like COBRA Toolbox, Matlab toolbox and OptKnock facilitate genome-scale engineering, knock-out genes responsible for desired compounds and identify missing metabolic reactions and optimise microbial functions.

Recent advances integrate omics data with computational algorithms to construct synthetic microbiomes for pollutant degradation. Bioinformatics tools, such as MetaRouter and OxDBase, support microbial bioremediation studies by mapping metabolic pathways and enzyme functions for pollutant degradation (Jaiswal and Shukla, 2020). Synthetic microbial communities are engineered to enhance bioremediation by selecting desired function-specific species. Understanding natural microbial interactions helps design optimised artificial communities. Engineered cooperation between strains improves pollutant breakdown, while quorum sensing regulates microbial communication. These communities offer sustainable solutions for ECs. Maintaining stability through synergistic interactions is key to their effectiveness (Cao *et al.*, 2022; Großkopf and Soyer, 2014; Kachienga *et al.*, 2018).

Genetic engineering enhances bioremediation by modifying microbial genomes, such as local/global genome, which facilitates precise addition, deletion or substitution of GOI for efficient pollutant degradation. Techniques like CRISPR-Cas, TALEN and ZFNs enable precise genome editing, improving microbial resilience and enzymatic activity. Among them, CRISPR-Cas is mainly used (Kanchiswamy *et al.*, 2016).

Metabolic engineering modifies existing metabolic pathways. This technique is used for the study of microbial enzymes, such as oxidases, oxidoreductases and esterases. The genetic approach leads to recombinant enzymes; altogether, these changes accelerate pollutant breakdown, while microbial consortia enhance degradation pathways (Dangi *et al.*, 2019). Synthetic genetic circuits and microbial biosensors enhance environmental bioremediation by utilising engineered microorganisms. Whole-cell biosensors identify ECs, like pesticides and pharmaceutical residues, through signal transduction mechanisms. These engineered systems optimise pollutant detection and degradation, advancing bioremediation strategies (Jaiswal and Shukla, 2020).

7.4.4 Microbiome engineering for bioremediation

Microbiome engineering is a robust technology to deal with recalcitrant ECs. The single-strain-based bioremediation approach faces several limitations, including low survival rates, reduced metabolism due to secondary metabolites or incomplete EC degradation. Synthetic microbiomes provide a significant advancement in bioremediation over traditional methods. Microbiomes are genetically diverse with versatile metabolism that synergistically interact to degrade complex pollutants.

Synergy prevents the accumulation of intermediates and reduces feedback inhibition, enhancing degradation activity, improving stability and survival rate of microbial inocula, and it is versatile in nature and can be utilised to degrade multiple pollutants simultaneously. Additionally, cross-feeding within the microbiome supports degrader growth, further improving overall efficiency (Xu and Jiang, 2024). The microbiome consists of all microbial community members inhabiting various environmental zones, including extreme conditions. The microorganisms' interaction with both biotic and abiotic factors forms relationships with their surroundings.

Microbiome engineering uses systematic approaches like top-down and bottom-up strategies to enhance microbial applications. Bottom-up approaches create artificially designed synthetic microbiomes by combining strains based on a deep understanding of the individual microbial strains and intra-/inter species. On the other hand, top-down approaches refine natural microbiomes by optimising their complexity for specific functional outcomes. It is the simplest and indirect method to modulate the microbiome.

Genome-scale metabolic models and simulation algorithms help in silico analysis of strain-specific metabolic activities and interspecies interactions within microbiomes (Ruan *et al.*, 2024). The core microbiome, identified from metagenomics data, includes both abundant and consistent low-abundance taxa. Keystone microorganisms help design synthetic microbial communities (SynCom) for targeted environmental functions. Microbial interactions through quorum sensing between microbial inter- and intra species, bacteria-fungi, bacteria-algae and plant-microorganism interactions influence microbiome-host engineering. This occurs due to the production of different kinds of extracellular signalling molecules that are accepted by other species, and various functions are performed. To synthesise microbiomes, omic technology plays a major role in identifying potential consortia to degrade complex recalcitrant pollutants, identifies metabolic pathways and helps in stress reduction even after changes in climatic or environmental conditions (Fan *et al.*, 2024; Singha and Shukla, 2023; Tariq *et al.*, 2025).

In recent advancements, microalgae and their microbiomes play an essential role in bioremediation. Naturally, microalgae with bacterial combinations enhance nutrient recycling, growth, pathogen defence and environmental stability. Artificial intelligence (AI) and machine learning (ML) leverage the optimisation of existing data to predict microbiome potential, its interaction and community stability (Koneru *et al.*, 2022). Therefore, microbiome engineering is a potential technology for the remediation process of ECs which has recently emerged. However, more research is required for the establishment of efficient removal methods.

7.4.5 Microorganism-based nanoremediation

Nanotechnology is an advanced and environmentally friendly technology that uses a high surface area and nanoscale size (1–100 nm) materials to remediate pollution. High surface area, stability, surface modification techniques, small size and biocompatibility play essential roles in bioremediation. Nanotechnology has applications in sectors such as healthcare, biomedical, electronics, fuel cells, nanosensors and so on.

Nanotechnology is also significantly being used in the field of bioremediation and treatment of ECs (Alsaba *et al.*, 2020). Nanoparticles (NPs) and bioremediation together provide a better approach for EC removal with higher efficiency than if deployed singly. Nanoparticles can be synthesised using physical processes that break larger particles into smaller ones, while chemical processes include laser pyrolysis. Biological synthesis relies on biomolecules to reduce metal ions and stabilise particles. Bio mediated synthesis involves peptides, polyphenols and amino acids, preventing nanoparticle agglomeration.

Microorganisms can also produce nanoparticles of various metals, although microbial synthesis requires optimal conditions. Intracellular synthesis is possible by disrupting microbial cells, while algae-derived phytochemicals aid in nanoparticle stabilisation (Yamini and Devi Rajeswari, 2023). NPs applied in the biodegradation process directly or through adsorption facilitate the growth of microorganisms or immobilise contaminants by producing enzymes (alcohol dehydrogenase, Mn peroxidase, laccase, catalase, lipase, glucose oxidase and pectate lyase). NPs enhance biosurfactant synthesis in microorganisms through hydrocarbon solubility modification, which promotes the removal of polycyclic aromatic hydrocarbons (PAHs) (Kumari and Singh, 2016).

Nanoparticle behaviour depends on its size, surface coating, chemical nature and shape, and also other factors like the nature of pollutants, temperature, environmental pH and type of media, all of which can affect its potential for bioremediation (Ayilara and Babalola, 2023). Nanotechnological applications in bioremediation have made a revolutionary advancement in environmental quality. However, their disadvantages include human health effects, bioaccumulation, loss of biodiversity and their transport across trophic chains due to their higher exposure during production/application after their use. Diverse international bodies like the European Observatory for Nanomaterials (EON), United States Environmental Protection Agency (USEPA), the ISO Technical Committee (TC) 229 ‘Nanotechnologies’ and other organisations have formed international cooperations to enhance the application and enforcement of existing regulations.

The advancement of new technologies and the incorporation of nanomaterials in the field of bioremediation maximises new opportunities for effective treatment of sewage, lakes, wastewater treatment, rivers and ponds (Vázquez-Núñez *et al.*, 2020).

7.4.6 Bioelectrochemical systems and microbial fuel cells

Bioelectrochemical systems (BES), also known as microbial electrochemical systems (MES), are an advanced innovative bioengineering technology for EC remediation. They utilise electroactive microorganisms or enzymes as biocatalysts within an electrochemical system to transform chemical energy into electrical energy or vice versa by way of catalysing organic compounds.

MES are broadly classified into two categories based on whether electricity is generated or provided – that is, microbial fuel cell (MFC) and microbial electrolysis cell (MEC), respectively. MFC and its variations, such as microbial desalination cell (MDC), plant microbial fuel cell (PMFC) and enzymatic fuel cell (EFC), require no external energy supply and produce electricity using microorganisms at the anode through decomposition of organics. MEC and microbial electrosynthesis cells (MESC) need external power to perform non-spontaneous chemical reactions that produce biohydrogen and organic chemicals (Choudhary *et al.*, 2024).

Among these technologies, MFCs and MECs are widely used in the treatment of wastewater and the remediation of ECs. Microbial enzymes act as biocatalysts in an oxygen-limited environment, attaching to the anode surface, which leads to the oxidation of organic compounds, thereby releasing electrons from substrates to the anode, travelling to the cathode (abiotic) compartment through the extracellular electron transport (EET). A proton exchange membrane separates the chambers, allowing hydrogen ions to move toward the cathode. The released electrons are reduced substrates like carbohydrates, hydrogen and methane, which travel through the electrode to an electron sink, which is oxygen. Ion transport maintains electroneutrality, enabling continuous electricity generation.

MFCs rely on electroactive bacterial biofilms to drive bioelectricity generation. Their efficiency depends on exoelectrogens (microorganisms generating electricity) which facilitate electron transfer from organic substrates (Gupta *et al.*, 2022; Kiely *et al.*, 2011; Srikanth *et al.*, 2016). Bacteria like *Clostridium cellulolyticum*, *Geobacter sulfurreducens*, *Clostridium butyricum* and *Shewanella putrefaciens*, and fungi like *Phanerochaete chrysosporium* and *Aspergillus awamori* are examples of electrogenic microorganisms (Choudhary *et al.*, 2024).

MECs are a modified version of MFCs. They facilitate the anaerobic breakdown of organic compounds, employing a solid electrode to accept electrons. Bacteria capable of electron transfer utilise the anode as an electron sink, generating electrons and protons that drive redox reactions to produce valuable compounds (Wilberforce *et al.*, 2021). MEC cathodes are anaerobic in nature and provide favourable conditions for the formation of hydrogen. Electrogenic biofilms act as biocatalysts, accelerating anodic reactions in MECs.

The anode transfers electrons to the cathode through an external wire, facilitating the production of hydroxide (OH^-) and hydrogen (H_2). Due to unfavourable thermodynamics, an external voltage is needed, but MECs require less energy than traditional water electrolysis since part of their energy comes from organic matter decomposition. Additionally, other compounds like methane can be produced if different substances undergo reduction at the cathode (Bajracharya *et al.*, 2016).

Thus, the potential of bioremediation of ECs using BES technology shows greater efficiency with the arrangement of different methods, and it is also being industrialised and used to treat wastewater.

7.4.7 Omics technologies in bioremediation

The study and identification of all biomolecules associated with microbial metabolism, including their physiology, growth, structure and dynamics, is known as 'omics' technology. Omics is an advanced technology, including genomics, metagenomics, transcriptomics, proteomics, metabolomics and fluxomics, that can help to develop mitigation strategies to remediate ECs. It is a type of molecular technology which helps in the detection of biomolecules like protein, DNA, RNA and metabolites from a single species or an entire microbial community (Sharma *et al.*, 2022).

In the process of bioremediation of ECs, it helps in identifying functional and key genes, interaction details and toxicity mechanisms. These omics approaches help in understanding the metabolic, physiological and genetic regulations of certain microorganisms which thrive in a polluted environment. It helps to know whether the interaction of microorganisms with their native microbiome is mutualistic, competitive or succession processes, which impact colonisation success. Omics study helps in formulating mitigating strategies to introduce inocula doses into an existing community. Investigation of the biotic and abiotic stressors present in contaminated sites is necessary, as these are the limiting conditions or proliferation conditions that impact the development of stress-tolerant strains (Alidoosti *et al.*, 2024).

Genomics and metagenomics are the study of genetic material, specifically DNA, collected from environmental samples using sequencing technology, which helps in the identification of specific genes involved in the degradation process (Bycroft *et al.*, 2017). To determine biodiversity and functionality of the microbiome, the phylotyping sequence approach is used. Functional metagenomics helps in the study of gene functions by analysing DNA from environmental samples (Prakash and Taylor, 2012; Xiang *et al.*, 2020). It identifies new genes, and gives insights into microbial diversity and metabolic processes which are essential for bioremediation.

Metagenomic study can also employ environmental DNA, amplified using specific primers to assess the specific gene of biodegradation which helps in finding potential biodegrading bacteria (Ghai *et al.*, 2024).

Transcriptomics and metatranscriptomics technologies analyse mRNA sequences to study microbial gene expression, operon structures and regulatory networks. Metatranscriptomics provides functional insights into microbial communities, aiding environmental studies, stress responses and pollutant degradation, which significantly helps to determine the genetic profile of certain microorganisms used in biodegradation. This can be used for the modification of microorganisms (Kaster and Sobol, 2020).

Transcriptomic analysis requires cDNA synthesis, by way of microarray cDNA hybridisation (Sharma *et al.*, 2022). Proteomics studies analyse protein profiles of microbial species to assess their responses and expressed phenotypes to contaminants by detecting changes in protein composition (Chakdar *et al.*, 2022). Techniques like polyacrylamide gel electrophoresis (PAGE), 2-DE and mass spectrometry (MS) are traditionally used. Recent advancements in technology have introduced new methods of protein identification such as matrix-assisted laser desorption ionisation time-of-flight mass spectrometry (MALDI-ToF-MS) and mudPIT (multidimensional-protein identification), which helps identifying proteins involved in pollutant degradation.

Metaproteomics examines microbial communities, providing insights into their functional roles in environmental remediation. Advances in proteomics have enhanced bioremediation research, improving understanding of microbial metabolism in contaminated ecosystems (Ghai *et al.*, 2024).

Metabolomics identifies metabolites in microbial samples during biodegradation, providing insights into metabolic pathways and biochemical activity. Its study provides insights into biodegradation pathways by identifying key enzymes and intermediates involved. Analysis of metabolites helps determine enzymatic reactions, guiding the formation of products and by products in EC biodegradation. It also assesses environmental stress impacts, aiding in understanding microbial responses and remediation strategies. It integrates with other omics techniques to assess toxicant mechanisms and microbial biodegradation, aiding risk assessment and pollution management (Pandey *et al.*, 2025).

The prediction of microbial activity in bioremediation through modelling has been advanced by omics techniques, improving the efficiency of EC remediation.

7.5. Bioproducts using microbial biomass

Microbial biomass is used to manufacture biofuels, bioplastics, proteins, organic acids and bioactive chemicals for industries such as energy, food and pharmaceuticals. Its manufacturing is sustainable and minimises reliance on fossil fuels while reusing agricultural and industrial waste.

Future advances in microbial biotechnology will increase strain efficiency and broaden applications. This innovation contributes to a circular economy by providing environmentally responsible and cost-effective solutions for numerous industrial sectors.

Microorganisms produce biomaterials under different environmental conditions, including bioplastics. Bioplastics are microbial polyesters that are an alternative to standard petrochemical plastics/polymers. Using microbial biomass, various types of bioplastics can be made, such as polyhydroxyalkanoates (PHAs), polyhydroxybutyrate (PHB) and polylactic acid (PLA), a starch-based polymer. They are produced by bacteria through fermentation and metabolic engineering by using substrates such as lignocellulosic biomass, fatty acids and agricultural waste (de Souza and Gupta, 2024).

Mycelium-based composites can be obtained from fungi, whereas algae produce polysaccharides and bio composite polymers (Rajpoot *et al.*, 2022; Verma *et al.*, 2023). These microorganisms produce secondary metabolites essential for antibiotics, anticancer drugs, antifungal, antiparasitic and cholesterol-lowering medications used to treat infections and other disorders (Ramírez-Rendon *et al.*, 2022). *Streptomyces*, *Penicillium* and other fungi metabolites contribute to infection control and also function as immune-suppressants (Kapoor *et al.*, 2020).

Algae-derived compounds show potential in treating drug-resistant bacteria, inflammation and neurological disorders, advancing pharmaceutical innovation (Barbalace *et al.*, 2019). Biofertilisers and biopesticides provide a sustainable alternative to chemical dependency on synthetic fertilisers, pesticides, herbicides and so on in agriculture. Microorganism-derived biopesticides naturally fight against diseases and pests, whereas some bacteria, such as nitrogen-fixing bacteria, are directly used as biofertilisers (Jehani *et al.*, 2023).

Biofuels generated from microbial biomass offer a novel approach to sustainable energy production. Bacteria, yeasts, fungi and microalgae can synthesise biofuels (biodiesel, bioethanol, biohydrogen and biogas) using a variety of metabolic pathways. Bio alcohols, primarily ethanol, are created by way of direct and indirect fermentation processes (Robak and Balcerek, 2018). These microorganisms store a lot of fatty acids, which can be trans-esterified to make biodiesel (Chintagunta *et al.*, 2021).

Anaerobic digestion of organic waste by bacteria produces biogas, which is mostly constituted of methane (Atelge *et al.*, 2018). Similarly, phototrophic microorganisms like cyanobacteria and green algae produce biohydrogen through photobiological processes (Bolatkhan *et al.*, 2019). Organic acids, enzymes, biocatalysts, pigments and natural dyes are the main classes of bioproducts. Organic acids (lactic acids, citric acids and acetic acids) are produced through a fermentation process (Zalán *et al.*, 2010).

Microorganism-derived enzymes, such as cellulases, amylases and proteases, are essential biocatalysts in sectors like food processing, detergents and textiles (Saxena, 2015). Natural pigments, such as monascus pigments (including red, orange and yellow chemicals) from *Monascus purpureus* and prodigiosin (a bright red pigment) from *Serratia marcescens*, are obtained from novel microbial strains (Gong *et al.*, 2023; Srimathi, 2017). Microorganisms are also a major source of nutraceuticals with the potential to create value-added compounds such as phytochemicals, prebiotics, polysaccharides and polyamino acids (Wang *et al.*, 2016).

Microbial biomass is also an excellent source of single cell protein (SCP), which contains necessary amino acids, protein-rich compounds and is easily digested, making it a useful nutritional supplement (Ugbogu, 2016). Therefore, utilising microorganisms in bioremediation significantly provides microbial biomass and can be used to make value-added bioproducts, which are sustainable and eco-friendly. Value-added bioproducts formed using biomass are shown in Figure 7.3.

Figure 7.3 Bioproducts made using microbial biomass

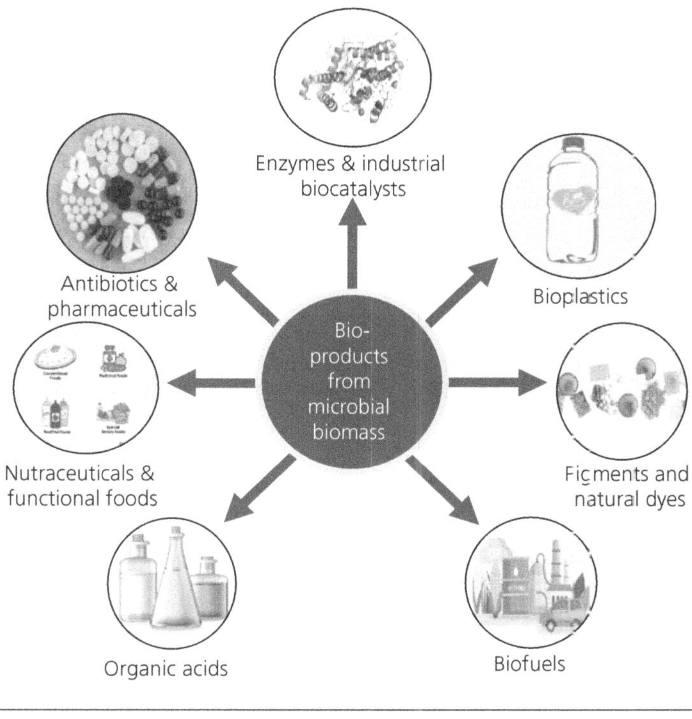

7.6. Challenges and future perspectives

Despite the promising potential of advanced microbial technologies for bioremediation, it has certain limits and a few challenges which must be considered for further advancements in real-world applications.

Enzyme-mediated remediation struggles with environmental instability and high production costs, while the genetically engineered microorganisms and synthetic biology approaches face significant biosafety, regulatory and ecological concerns over their release and persistence of modified organisms in natural ecosystems.

Microbiome engineering is limited by the complex mechanisms of microbial interactions and the uncertainty of community dynamics in situ. Furthermore, microorganism-based nano-remediation raises concerns about nanotoxicity and the long-term fate of engineered particles.

BESs and MFCs are limited by scalability issues, operational costs and the need for technological optimisation. Meanwhile, omics-based approaches, though rich in data, demand high-end computational resources and bioinformatics expertise to translate molecular insights into field-level strategies. These overlapping challenges highlight the need for integrated frameworks that address technical, ecological and regulatory barriers simultaneously to ensure safe, scalable and efficient application of microbial biotechnologies for emerging contaminant degradation (Alsaba *et al.*, 2020; Choudhary *et al.*, 2024; Chunyan *et al.*, 2023; Rafeeq *et al.*, 2023; Sharma *et al.*, 2022; Somu *et al.*, 2022; Xu and Jiang, 2024).

Researchers continue to investigate advanced technologies to improve the remediation of emerging contaminants. Establishing clear regulatory standards for harmful ECs, such as single-use plastics and EDCs, is essential. Future studies could focus on optimising bioreactor systems, incorporating algal reactors, and developing hybrid technologies like BES combined with algal-bacterial bioreactors for pilot-scale pollutant degradation.

The integration of multi-omics, AI, ML and systems biology will enable predictive microbiome engineering, refining microbial systems for more effective biodegradation. Also, development of a better understanding of bioinformatics data for practical application is needed.

Scalable bioreactor technologies and circular bioeconomy strategies have the potential to transform waste into valuable bioproducts, ensuring more sustainable environmental solutions.

7.7. Conclusions

Emerging contaminants like microplastics, EDCs, PCPs, PFASs and pharmaceuticals are increasing in the aquatic environment, creating severe harmful health impacts on aquatic and other living organisms, including humans. Microbial bioremediation offers an effective and sustainable approach to address the concern of emerging contaminants in aquatic environments.

Microorganisms like bacteria, algae, fungi and their combined consortia have a significant potential for breaking down complex pollutants, including microplastics, pharmaceuticals,

EDCs and PCPs. Advancements in microbial technologies, such as enzymatic degradation, genetic engineering, omics, synthetic biology, BES and nanotechnology, have further improved the efficiency and effectiveness of these processes.

While promising, challenges such as scalability, environmental safety, regulations and costs remain key hurdles to overcome. Furthermore, it is crucial to focus on real-world applications, combine these technologies with wastewater treatment and create value-added bioproducts from microbial biomass, ensuring the protection of environmental and public health. With consistent research and development, microbial remediation has the potential to restore ecological balance and improve the impacts on human life and aquatic systems.

REFERENCES

Alidoosti F, Giyahchi M, Moien S and Moghimi H (2024) Unlocking the potential of soil microbial communities for bioremediation of emerging organic contaminants: omics-based approaches. *Microbial Cell Factories* **23(1)**: 1–21, 10.1186/s12934-024-02485-z.

Alsaba MT, Al Dushaishi MF and Abbas AK (2020) A comprehensive review of nanoparticles applications in the oil and gas industry. *Journal of Petroleum Exploration and Production Technology* **10(4)**: 1389–1399, 10.1007/S13202-019-00825-Z/TABLES/6.

Ambaye TG, Hassani A, Vaccari M *et al.* (2025) Nano-bioremediation for the removal of inorganic and organic pollutants from the soil. *Environmental Science and Pollution Research* **32(14)**: 8698–8716, 10.1007/s11356-025-36188-8.

Atelge MR, Krisa D, Kumar G *et al.* (2018) Biogas production from organic waste: recent progress and perspectives. *Waste and Biomass Valorization* **11(3)**: 1019–1040, 10.1007/S12649-018-00546-0.

Ayilara MS and Babalola OO (2023) Bioremediation of environmental wastes: the role of microorganisms. *Frontiers in Agronomy* **5**(May): 1–15, 10.3389/fagro.2023.1183691.

Bajracharya S, Sharma M, Mohanakrishna G *et al.* (2016) An overview on emerging bioelectrochemical systems (BESs): Technology for sustainable electricity, waste remediation, resource recovery, chemical production and beyond. *Renewable Energy* **98**: 153–170, 10.1016/J.RENENE.2016.03.002.

Bala S, Garg D, Thirumalesh BV *et al.* (2022) Recent strategies for bioremediation of emerging pollutants: a review for a green and sustainable environment. *Toxics* **10(8)**: 484, 10.3390/TOXICS10080484.

Barbalace MC, Malaguti M, Giusti L *et al.* (2019) Anti-Inflammatory activities of marine algae in neurodegenerative diseases. *International Journal of Molecular Sciences* **20(12)**: 3061, 10.3390/IJMS20123061.

Bolatkhan K, Kossalbayev BD, Zayadan BK *et al.* (2019) Hydrogen production from phototrophic microorganisms: reality and perspectives. *International Journal of Hydrogen Energy* **44(12)**: 5799–5811, 10.1016/J.IJHYDENE.2019.01.092.

Bycroft C, Freeman C, Petkova D *et al.* (2017) Genome-wide genetic data on ~500,000 UK Biobank participants. *BioRxiv*, Cold Spring Harbor Laboratory, p. 166298, 10.1101/166298.

Cai Y, Chen H, Yuan R *et al.* (2020) Metagenomic analysis of soil microbial community under PFOA and PFOS stress. *Environmental Research* **188**: 109838, 10.1016/J.ENVRES.2020.109838.

Camara FA, Ramézani H, Mathieu N, Delpeux-Ouldriane S and Bhatia SK (2025) Unraveling ofloxacin behavior in aqueous environments: molecular dynamics of colloidal formation and surface adsorption mechanisms. *Langmuir*, 10.1021/acs.langmuir.4c04265.

Cao X, Song H liang, Yu C yan and Li X ning (2015) Simultaneous degradation of toxic refractory organic pesticide and bioelectricity generation using a soil microbial fuel cell. *Bioresource Technology*, **189**: 87–93 doi: 10.1016/J.BIORTECH.2015.03.148.

Cao Z, Yan W, Ding M and Yuan Y (2022) Construction of microbial consortia for microbial degradation of complex compounds. *Frontiers in Bioengineering and Biotechnology* **10**: 1051233, 10.3389/FBIOE.2022.1051233.

Chakdar H, Thapa S, Srivastava A and Shukla P (2022) Genomic and proteomic insights into the heavy metal bioremediation by cyanobacteria. *Journal of Hazardous Materials* **424**: 127609, 10.1016/J.JHAZMAT.2021.127609.

Chang KL, Teng TC, Fu, CK and Liu CH (2019) Improving biodegradation of bisphenol A by immobilization and inducer. *Process Safety and Environmental Protection* **128**: 128–134, 10.1016/J.PSEP.2019.05.038.

Chia XK, Hadibarata T, Kristanti RA *et al.* (2024) The function of microbial enzymes in breaking down soil contaminated with pesticides: a review. *Bioprocess and Biosystems Engineering* **47**(5): 597–620, 10.1007/S00449-024-02978-6/FIGURES/3.

Chintagunta AD, Zuccaro G, Kumar M *et al.* (2021) Biodiesel production from lignocellulosic biomass using oleaginous microorganisms: prospects for integrated biofuel production. *Frontiers in Microbiology* **12**: 658284, 10.3389/FMICB.2021.658284/XML/NLM.

Choudhary M, Verma P and Ray S (2024) A comprehensive review on bio-electrochemical systems for wastewater treatment: process, electricity generation and future aspect. *Environment, Development and Sustainability*, 10.1007/s10668-024-05866-x.

Chu S, Kwon BR, Lee YM, Zoh KD and Choi K (2021) Effects of 2-ethylhexyl-4-methoxycinnamate (EHMC) on thyroid hormones and genes associated with thyroid, neurotoxic, and nephrotoxic responses in adult and larval zebrafish (Danio rerio). *Chemosphere*, **263**: 128176, doi: 10.1016/j.chemosphere.2020.128176.

Chunyan X, Qaria MA, Qi X and Daochen Z (2023) The role of microorganisms in petroleum degradation: Current development and prospects. *Science of the Total Environment* **865**, October 2022, doi: 10.1016/j.scitotenv.2022.161112.

Dangi AK, Sharma B, Hill RT and Shukla P (2019) Bioremediation through microorganisms: systems biology and metabolic engineering approach. *Critical Reviews in Biotechnology* **39(1)**: 79–98, 10.1080/07388551.2018.1500997.

Debnath B, Majumdar M, Bhowmik M *et al.* (2020) The effective adsorption of tetracycline onto zirconia nanoparticles synthesized by novel microbial green technology. *Journal of Environmental Management* **261**: 110235, 10.1016/J.JENVMAN.2020.110235.

Domingo JL and Nadal M (2019) Human exposure to per- and polyfluoroalkyl substances (PFAS) through drinking water: a review of the recent scientific literature. *Environmental Research* **177**(August), 10.1016/j.envres.2019.108648.

Dubey S, Chen CW, Haldar D *et al.* (2023) Advancement in algal bioremediation for organic, inorganic, and emerging pollutants. *Environmental Pollution* **317**: 120840, 10.1016/J.ENVPOL.2022.120840.

Elizalde-Velázquez GA and Gómez-Oliván LM (2021) Microplastics in aquatic environments: a review on occurrence, distribution, toxic effects, and implications for human health. *Science of the Total Environment* **780**: 146551, 10.1016/j.scitotenv.2021.146551.

Fan X, Kong L, Wang J *et al.* (2024) Surface-programmed microbiome assembly in phycosphere to microplastics contamination. *Water Research* **262**: 122064, 10.1016/J.WATRES.2024.122064.

Gan JS, Bilal M, Li, XB *et al.* (2022) Peroxidases-based enticing biotechnological platforms for biodegradation and biotransformation of emerging contaminants. *Chemosphere* **307**: 136035, 10.1016/J.CHEMOSPHERE.2022.136035.

Gao R and Sun C (2021) A marine bacterial community capable of degrading poly(ethylene terephthalate) and polyethylene. *Journal of Hazardous Materials* **416**: 125928, 10.1016/J.JHAZMAT.2021.125928.

Ge J, Li H, Liu P *et al.* (2021) Review of the toxic effect of microplastics on terrestrial and aquatic plants. *Science of the Total Environment* **791**: 148333, 10.1016/j.scitotenv.2021.148333.

Gewurtz SB, Auyeung AS, Teslic S and Smyth SA (2025) Pharmaceuticals and personal care products in canadian municipal wastewater and biosolids: occurrence, fate, and time trends 2010–2013 to 2022. *Environmental Science and Pollution Research* **32**(9): 5022–5039, 10.1007/s11356-025-36007-0.

Ghai MK, Khatri A, Kumar K and Thakur IS (2024) Multi-omics and advance technologies in biodegradation of emerging contaminants and eco-estrogens in environmental waste. *Total Environment Advances* **11**(July): 200113, 10.1016/j.teadva.2024.200113.

Gondi R, Kavitha S, Yukesh Kannah R *et al.* (2022) Algal-based system for removal of emerging pollutants from wastewater: a review. *Bioresource Technology* **344**(PB): 126245, 10.1016/j.biortech.2021.126245.

Gong P, Shi R, Liu Y *et al.* (2023) Recent advances in monascus pigments produced by monascus purpureus: biosynthesis, fermentation, function, and application. *LWT* **185**: 115162, 10.1016/J.LWT.2023.115162.

González-González RB, Flores-Contreras EA, Parra-Saldívar R and Iqbal HMN (2022) Bio-removal of emerging pollutants by advanced bioremediation techniques. *Environmental Research* **214**(April), 10.1016/j.envres.2022.113936.

Großkopf T and Soyer OS (2014) Synthetic microbial communities. *Current Opinion in Microbiology* **18**(1): 72–77, 10.1016/J.MIB.2014.02.002.

Gupta SK, Rachna, Singh B *et al.* (2022) Bioelectrochemical technologies for removal of xenobiotics from wastewater. *Sustainable Energy Technologies and Assessments* **49**: 101652, 10.1016/J.SETA.2021.101652.

Ha, GS, Sim MG, Jeon BH and Baek G (2025) Bioremediation of perfluorooctanoic acid using microalgae with a transcriptomic approach. *Journal of Hazardous Materials* **488**: 137508, 10.1016/J.JHAZMAT.2025.137508.

Huang M, Ivantsova E, Konig I *et al.* (2023) Developmental and mitochondrial toxicity assessment of perfluoroheptanoic acid (PFHpA) in zebrafish (Danio rerio). *Environmental Toxicology and Pharmacology* **97**(July 2022): 104037, 10.1016/j.etap.2022.104037.

Jaiswal S and Shukla P (2020) Alternative strategies for microbial remediation of pollutants via synthetic biology. *Frontiers in Microbiology* **11**: 522718, 10.3389/FMICB.2020.00808/XML/NLM.

Jehani MD, Singh S, Archana TS, Kumar D and Kumar G (2023) Azospirillum—a free-living nitrogen-fixing bacterium. *Rhizobiome: Ecology, Management and Application* 285–308, 10.1016/B978-0-443-16030-1.00001-8.

Kabir ER, Rahman MS and Rahman I (2015) A review on endocrine disruptors and their possible impacts on human health.. *Environmental Toxicology and Pharmacology* **40(1)**, 241–258, 10.1016/j.etap.2015.06.009.

Kachienga L, Jitendra K and Momba M (2018) Metagenomic profiling for assessing microbial diversity and microbial adaptation to degradation of hydrocarbons in two South African petroleum-contaminated water aquifers. *Scientific Reports* **8(1)**: 1–6, 10.1038/s41598-018-25961-0.

Kanchiswamy CN, Maffei M, Malnoy M, Velasco R and Kim JS (2016) Fine-tuning next-generation genome editing tools. *Trends in Biotechnology* **34(7)**: 562–574, 10.1016/J.TIBTECH.2016.03.007/ASSET/FCFE1A79-4836-4599-B0BD-E49BC179A55B/MAIN.ASSETS/GR3.JPG.

Kapoor D, Sharma P, Sharma MMM, Kumari A and Kumar R (2020) Microorganisms in pharmaceutical industry. *Microbial Diversity, Interventions and Scope*, Springer, Singapore, pp. 259–299, 10.1007/978-981-15-4099-8_16.

Karigar CS and Rao SS (2011) Role of microbial enzymes in the bioremediation of pollutants: a review. *Enzyme Research* **2011(1)**: 805187, doi: 10.4061/2011/805187.

Kaster AK and Sobol MS (2020) Microbial single-cell omics: the crux of the matter. *Applied Microbiology and Biotechnology* **104(19)**: 8209–8220, 10.1007/S00253-020-10844-0.

Khalifa HAAO, Yoon HJ, Ki M-R *et al.* (2024) Microbial immobilized enzyme biocatalysts for multipollutant mitigation: harnessing nature's toolkit for environmental sustainability. *International Journal of Molecular Sciences* **25(16)**: 8616, 10.3390/IJMS25168616.

Kiely PD, Regan JM and Logan BE (2011) The electric picnic: synergistic requirements for exoelectrogenic microbial communities. *Current Opinion in Biotechnology* **22(3)**: 378–385, 10.1016/J.COPBIO.2011.03.003.

Kim S and Choi K (2014) Occurrences, toxicities, and ecological risks of benzophenone-3, a common component of organic sunscreen products: a mini-review. *Environment International* **70**: 143–157, 10.1016/j.envint.2014.05.015.

Koneru H, Bamba S, Bell A, Estrada-graf AA and Johnson ZI (2022) Integrating microbial communities into algal biotechnology: a pathway to enhanced commercialization. *Frontiers in Microbiology* **16**:1555579.

Kreychman M, Ivantsova E, Lu, A, Bisesi JH and Martyniuk CJ (2024) A comparative review of the toxicity mechanisms of perfluorohexanoic acid (PFHxA) and perfluorohexanesulphonic acid (PFHxS) in fish. *Comparative Biochemistry and Physiology Part C: Toxicology and Pharmacology* **279**(January) p. 109874, 10.1016/j.cbpc.2024.109874.

Kumar R, Saini GK and Jawed M (2022) Impact of emerging contaminants on biological wastewater treatment process. In *Relationship Between Microbes and the Environment for Sustainable Ecosystem Services, Volume 2* (Samuel J, Kumar A and Singh J (eds.)), Elsevier, Amsterdam, the Netherlands, pp. 17–40, 10.1016/B978-0-323-89937-6.00014-0.

Kumari B and Singh DP (2016) A review on multifaceted application of nanoparticles in the field of bioremediation of petroleum hydrocarbons. *Ecological Engineering* **97**: 98–105, 10.1016/J.ECOLENG.2016.08.006.

Kurade MB, Kim JR, Govindwar SP and Jeon BH (2016) Insights into microalgae mediated biodegradation of diazinon by chlorella vulgaris: microalgal tolerance to xenobiotic pollutants and metabolism. *Algal Research* **20**: 126–134, 10.1016/J.ALGAL.2016.10.003.

De la Peña E, Castillo-Zacarías C, Araújo RG *et al.* (2024) Exploring the potential of microbial fuel cell-assisted biodegradation of nonylphenol. *International Journal of Environmental Science and Technology*, Institute for Ionics **21(3)**: 2465–2474, 10.1007/S13762-023-05112-3/TABLES/4.

Li K, Gao P, Xiang P *et al.* (2017) Molecular mechanisms of PFOA-induced toxicity in animals and humans: implications for health risks. *Environment International* **99**: 43–54, 10.1016/j.envint.2016.11.014.

Li M and Zhang C (2016) γ-Fe_2O_3 nanoparticle-facilitated bisphenol A degradation by white rot fungus. *Science Bulletin* **61(6)**: 468–472, 10.1007/S11434-016-1021-2.

Li X, Feng C, Lei M *et al.* (2022) Bioremediation of organic/heavy metal contaminants by mixed cultures of microorganisms: a review. *Open Chemistry* **20(1)**: 793–807, 10.1515/CHEM-2022-0198/ASSET/GRAPHIC/J_CHEM-2022-0198_FIG_004.JPG.

Li X, Wu S, Dong Y, Fan H, Bai Z and Zhuang X (2021) Engineering microbial consortia towards bioremediation. *Water* **13**(20, p. 2928, 10.3390/W13202928.

Liu L, Bilal M, Duan X and Iqbal HMN (2019) Mitigation of environmental pollution by genetically engineered bacteria — current challenges and future perspectives. *Science of The Total Environment* **667**: 444–454, 10.1016/J.SCITOTENV.2019 02.390.

Liu Y, Feng J, Pan H, Zhang X and Zhang Y (2022) Genetically engineered bacterium: principles, practices, and prospects. *Frontiers in Microbiology* **13**, p. 997587, 10.3389/FMICB.2022.997587/XML/NLM.

Liu Z, Gan Y, Luo J *et al.* (2025) Current status of emerging contaminant models and their applications concerning the aquatic environment: a review. *Water* **17(1)**: 85, 10.3390/W17010085.

Mahdizade Ari M, Dadgar L, Elahi Z, Ghanavati R and Taheri B (2024) Genetically engineered microorganisms and their impact on human health. *International Journal of Clinical Practice*, p. 6638269, 10.1155/2024/6638269.

Mantovani M, Collina E, Passalacqua E, Lasagni M and Mezzanotte V (2024) Microalgal-based carbon encapsulated iron nanoparticles for the removal of pharmaceutical compounds from wastewater. *Journal of Environmental Management* **368**, p. 122171, 10.1016/J.JENVMAN.2024.122171.

Martínez-García E and de Lorenzo V (2024) Pseudomonas putida as a synthetic biology chassis and a metabolic engineering platform. *Current Opinion in Biotechnology* **85**: 103025, 10.1016/j.copbio.2023.103025.

Metcalfe CD, Bayen S, Desrosiers M *et al.* (2022) An introduction to the sources, fate, occurrence and effects of endocrine disrupting chemicals released into the environment. *Environmental Research* **207**, 10.1016/j.envres.2021.112658.

Mishra A, Tripathi P and Singh RS (2025) Potential environmental contamination by microplastics. *Environmental Science and Engineering* **Part F158**: 17–44, 10.1007/978-3-031-76949-8_2.

Mishra S, Swain S, Sahoo M, Mishra S and Das AP (2022) Microbial colonization and degradation of microplastics in aquatic ecosystem: a review. *Geomicrobiology Journal* **39**(3–5): 259–269, 10.1080/01490451.2021.1983670.

Moayedi M and Yousefi N (2025) Laccase immobilized on reduced graphene oxide sponges for simultaneous adsorption and enzymatic degradation of endocrine disrupting chemicals. *International Journal of Biological Macromolecules* **302**, p. 139984, 10.1016/J.IJBIOMAC.2025.139984.

Mousavi SM, Hashemi SA, Iman Moezzi SM *et al.* (2021) Recent advances in enzymes for the bioremediation of pollutants. *Biochemistry Research International*, 10.1155/2021/5599204.

Mu, Z, Zou Z, Yang Y, Wang W *et al.* (2018) A genetically engineered escherichia coli that senses and degrades tetracycline antibiotic residue. *Synthetic and Systems Biotechnology* **3(3)**: 196–203, 10.1016/J.SYNBIO.2018.05.001.

Munzhelele EP, Ayinde WB, Gitari WM, Pindihama GK and Mudzielwana R (2025) Occurrence of efavirenz, levonorgestrel, ibuprofen, and diclofenac in wastewaters of limpopo province, South Africa. *Heliyon* **11(2)**: e41524, 10.1016/j.heliyon.2024.e41524.

Netto CGCM, Toma HE and Andrade LH (2013) Superparamagnetic nanoparticles as versatile carriers and supporting materials for enzymes. *Journal of Molecular Catalysis B: Enzymatic* 85–86: 71–92, 10.1016/J.MOLCATB.2012.08.010.

Nguyen PQ, Huang X, Collins DS, Collins JJ and Lu T (2023) Harnessing synthetic biology to enhance ocean health. *Trends in Biotechnology* **41(7)**: 860–874, 10.1016/j.tibtech.2022.12.015.

Ni S, Li C, Zhang W *et al.* (2023) Immobilization of purified enzyme ereB in metalorganic framework (MOF) mesopores for erythromycin degradation. *Environmental Research* **237**, p. 117023, 10.1016/J.ENVRES.2023.117023.

Pande V, Pandey SC, Sati D, Pande V and Samant M (2020) Bioremediation: an emerging effective approach towards environment restoration. *Environmental Sustainability* **3(1)**: 91–103, 10.1007/s42398-020-00099-w.

Pandey B, Pandey AK, Bhardwaj L and Dubey SK (2025) Biodegradation of acetaminophen: current knowledge and future directions with mechanistic insights from omics. *Chemosphere* **372** p. 144096, 10.1016/j.chemosphere.2025.144096.

Prakash T and Taylor TD (2012) Functional assignment of metagenomic data: challenges and applications. *Briefings in Bioinformatics* **13(6)**: 711–727, 10.1093/BIB/BBS033.

Qu B, Zhao H and Zhou J (2010) Toxic effects of perfluorooctane sulfonate (PFOS) on wheat (Triticum aestivum L.) plant. *Chemosphere* **79(5)**: 555–560, 10.1016/j.chemosphere.2010.02.012.

Rafeeq H, Afsheen N, Rafique S *et al.* (2023) Genetically engineered microorganisms for environmental remediation. *Chemosphere* **310**: 136751, 10.1016/J.CHEMOSPHERE.2022.136751.

Rajpoot AS, Choudhary T, Chelladurai H, Nath Verma T and Shende V (2022) A comprehensive review on bioplastic production from microalgae. *Materials Today: Proceedings* **56**: 171–178, 10.1016/J.MATPR.2022.01.060.

Ramírez-Rendon D, Passari AK, Ruiz-Villafán B *et al.* (2022) Impact of novel microbial secondary metabolites on the pharma industry. *Applied Microbiology and Biotechnology* Springer **106(5)**: 1855–1878, 10.1007/S00253-022-11821-5.

Ren H, Qin M, Zhang L *et al.* (2025) Modular engineering of a synthetic biology-Based platform for sustainable bioremediation of residual antibiotics in aquatic environments. *Engineering* 10.1016/J.ENG.2025.03.033.

Robak K and Balcerek M (2018) Review of second generation bioethanol production from residual biomass. *Food Technology and Biotechnology*, University of Zagreb **56(2)**, p. 174, 10.17113/FTB.56.02.18.5428.

Rodriguez-Narvaez OM, Peralta-Hernandez JM, Goonetilleke A and Bandala ER (2017) Treatment technologies for emerging contaminants in water: a review. *Chemical Engineering Journal* **323**: 361–380, 10.1016/j.cej.2017.04.106.

Ruan Z, Chen K, Cao W *et al.* (2024) Engineering natural microbiomes toward enhanced bioremediation by microbiome modeling. *Nature Communications* **15(1)**, 10.1038/s41467-024-49098-z.

Rylott EL and Bruce NC (2020) How synthetic biology can help bioremediation. *Current Opinion in Chemical Biology*, October, 10.1016/j.cbpa.2020.07.004.

Saeed MU, Hussain N, Sumrin A *et al.* (2022) Microbial bioremediation strategies with wastewater treatment potentialities – a review. *Science of the Total Environment* **818**, 10.1016/j.scitotenv.2021.151754.

Santo É do E, Ishii M, Pinto UM, Matsudo MC and Carvalho JCM de (2022) Obtaining bioproducts from the studies of signals and interactions between microalgae and bacteria. *Microorganisms* **10(10)**: 1–17, doi: 10.3390/microorganisms10102029.

Sarma H, Islam NF, Prasad R *et al.* (2021) Enhancing phytoremediation of hazardous metal(loid)s using genome engineering CRISPR–Cas9 technology. *Journal of Hazardous Materials* **414**: 125493, 10.1016/J.JHAZMAT.2021.125493.

Saxena S (2015) Microbial enzymes and their industrial applications. *Applied Microbiology* 121–154, 10.1007/978-81-322-2259-0_9.

Sharma P, Singh SP, Iqbal HMN and Tong YW (2022) Omics approaches in bioremediation of environmental contaminants: an integrated approach for environmental safety and sustainability. *Environmental Research* **211**, 10.1016/j.envres.2022.113102.

Sims JL, Stroski KM, Kim S *et al.* (2022) Global occurrence and probabilistic environmental health hazard assessment of per- and polyfluoroalkyl substances (PFASs) in groundwater and surface waters. *Science of the Total Environment* **816**: 151535, 10.1016/j.scitotenv.2021.151535.

Singh P, Rani P, Pandit A and Kumar R (2025) Microbial genetic engineering for the mitigation of petrol pollution in affected areas. In *Biotechnology for Environmental Sustainability*, Springer, Singapore: 245–260, 10.1007/978-981-97-7221-6_9.

Singha LP and Shukla P (2023) Microbiome engineering for bioremediation of emerging pollutants. *Bioprocess and Biosystems Engineering* **46(3)**: 323–339, 10.1007/s00449-022-02777-x.

Somu P, Narayanasamy S, Gomez LA *et al.* (2022) Immobilization of enzymes for bioremediation: a future remedial and mitigating strategy. *Environmental Research* **212**: 113411, 10.1016/J.ENVRES.2022.113411.

de Souza FM and Gupta RK (2024) Bacteria for bioplastics: progress, applications, and challenges. *ACS Omega*, American Chemical Society **9(8)**: 8666–8686, 10.1021/ACSOMEGA.3C07372/ASSET/IMAGES/lARGE/AO3C07372_0009.JPEG.

Srain HS, Beazley KF and Walker TR (2021) Pharmaceuticals and personal care products and their sublethal and lethal effects in aquatic organisms. *Environmental Reviews* **29(2)**: 142–181, 10.1139/er-2020-0054.

Srikanth S, Kumar M, Singh MP and Das BP (2016) Bioelectro chemical systems: a sustainable and potential platform for treating waste. *Procedia Environmental Sciences* **35**: 853–859, 10.1016/J.PROENV.2016.07.102.

Srimathi R (2017) Isolation, identification, optimization of prodigiosin pigment produced by serratia marcescens and its applications. *International Journal of Latest Engineering and Management Research* **02(09)**: 11–21.

Swain PR, Parida PK, Majhi PJ, Behera BK and Das BK (2025) Microplastics as emerging contaminants: challenges in inland aquatic food web. *Water (Switzerland)* **17(2)**: 1–14, 10.3390/w17020201.

Tariq A, Guo S, Farhat F and Shen X (2025) Engineering synthetic microbial communities: diversity and applications in soil for plant resilience. *Agronomy* **15(3)**: 1–21, 10.3390/agronomy15030513.

Thelusmond JR, Strathmann TJ and Cupples AM (2019) Carbamazepine, triclocarban and triclosan biodegradation and the phylotypes and functional genes associated with xenobiotic degradation in four agricultural soils. *Science of The Total Environment* **657**: 1138–1149, 10.1016/J.SCITOTENV.2018.12.145.

Ugbogu O (2016) A review of microbial protein production: prospects and challenges. *FUW Trends in Science and Technology Journal.*

Vaksmaa A, Guerrero-Cruz S, Ghosh P *et al.* (2023) Role of fungi in bioremediation of emerging pollutants. *Frontiers in Marine Science* **10**, p. 1070905, 10.3389/FMARS.2023.1070905/XML/NLM.

Vázquez-Núñez E, Molina-Guerrero CE, Peña-Castro JM, Fernández-Luqueño F and de la Rosa-Álvarez MG (2020) Use of nanotechnology for the bioremediation of contaminants: a review. *Processes*, July, 10.3390/pr8070826.

Verma N, Jujjavarapu SE and Mahapatra C (2023) Green sustainable biocomposites: substitute to plastics with innovative fungal mycelium based biomaterial. *Journal of Environmental Chemical Engineering* **11(5)**, p. 110396, 10.1016/J.JECE.2023.110396.

Visvanathan P, Kirubakaran D, Selvam K *et al.* (2024) Recent strategies for natural bioremediation of emerging pollutants: development of a green and sustainable environment. *ACS Symposium Series*: 1–19, 10.1021/BK-2024-1476.CH001.

Wang H, Xi H, Xu L *et al.* (2021) Ecotoxicological effects, environmental fate and risks of pharmaceutical and personal care products in the water environment: a review. *Science of the Total Environment* **788**, p. 147819, 10.1016/j.scitotenv.2021.147819.

Wang J, Guleria S, Koffas MAG and Yan Y (2016) Microbial production of value-added nutraceuticals. *Current Opinion in Biotechnology* **37**: 97–104, 10.1016/J.COPBIO.2015.11.003.

Wang Z, Ye H, Li X *et al.* (2025) Enhancing tetracycline removal: performance and mechanisms of interspecies electron transfer in microbial consortia. *Journal of Hazardous Materials* **492**, p. 138302, 10.1016/J.JHAZMAT.2025.138302.

Weeks JA, Guiney PD and Nikiforovz AI (2012) Assessment of the environmental fate and ecotoxicity of N,N-diethyl-m-toluamide (DEET). *Integrated Environmental Assessment and Management* **8(1)**: 120–134, 10.1002/ieam.1246.

Wilberforce T, Sayed ET, Abdelkareem MA, Elsaid K and Olabi AG (2021) Value added products from wastewater using bioelectrochemical systems: current trends and perspectives. *Journal of Water Process Engineering* **39**, p. 101737, 10.1016/J.JWPE.2020.101737.

Xiang W, Wei X, Tang H, Li L and Huang R (2020) Complete genome sequence and biodegradation characteristics of benzoic acid-Degrading bacterium pseudomonas sp. SCB32. *BioMed Research International* **2020(1)**: 6146104, 10.1155/2020/6146104.

Xu X and Jiang J (2024) Engineering microbiomes for enhanced bioremediation. *PLOS Biology* **22(12)**: e3002951, 10.1371/JOURNAL.PBIO.3002951.

Xue W, Zhou Q and Li F (2019) Bacterial community changes and antibiotic resistance gene quantification in microbial electrolysis cells during long-term sulfamethoxazole treatment. *Bioresource Technology* **294**: 122170, 10.1016/J.BIORTECH.2019.122170.

Yamini V and Devi Rajeswari V (2023) Effective bio-mediated nanoparticles for bioremediation of toxic metal ions from wastewater – a review. *Journal of Environmental Nanotechnology*, Institute for Environmental Nanotechnology **12(2)**: 12–33, 10.13074/jent.2023.06.232467.

Yuzbasioglu D, Mahmoud JH, Mamur S and Unal F (2022) Cytogenetic effects of antidiabetic drug metformin. *Drug and Chemical Toxicology* **45(2)**: 955–962, 10.1080/01480545.2020.1844226.

Zalán Z, Hudáček J, Štětina J, Chumchalová J and Halász A (2010) Production of organic acids by lactobacillus strains in three different media. *European Food Research and Technology* **230(3)**: 395–404, 10.1007/S00217-009-1179-9/TABLES/1.

Zhao Q, Yue S, Bilal M *et al.* (2017) Comparative genomic analysis of 26 Sphingomonas and sphingobium strains: dissemination of bioremediation capabilities, biodegradation potential and horizontal gene transfer. *Science of The Total Environment* **609**: 1238–1247, 10.1016/J.SCITOTENV.2017.07.249.

Microbial Processes in Environmental Engineering: From Theory to Practice

Rajneesh Kumar, Ram Sharan Singh and Maulin P Shah
ISBN 978-1-83708-327-5
https://doi.org/10.1108/978-1-83708-326-820261008

Chapter 8
Emerging technologies in microbial environmental engineering

Aparna Pandey and Sourish Bhattacharya

Process Design and Engineering Cell, CSIR–Central Salt and Marine Chemicals Research Institute, Bhavnagar, India

Abstract

Utilising the capacities of various microorganisms, such as bacteria, microalgae and fungi, microbial environmental engineering tackles urgent global issues related to sustainability and climate change mitigation. This chapter examines new microbial technologies that offer integrated solutions for resource recovery, pollution reduction and energy transition in the fields of biofuels, bioelectricity, bioremediation, and carbon capture, utilisation and storage (CCUS). Utilising bacteria and microalgae to effectively produce biodiesel, bioethanol and advanced biofuels, microorganisms are at the forefront of the biofuel production process. These developments provide scalable renewable energy alternatives while lowering dependency on fossil fuels. In CCUS, bacteria aid in biomineralisation, which transforms CO_2 into stable carbonates, while microalgae use photosynthesis to absorb atmospheric carbon dioxide (CO_2) Also, methanotrophs help mitigate the effects of methane, another greenhouse gas. Microbial fuel cells and microbial electrolysis cells are examples of bioelectrochemical devices that use bacteria to transform organic waste into hydrogen or electricity. Microalgal-bacterial consortia show promise for working together to remediate wastewater and improve carbon fixation and bioenergy production. Microalgae, fungi and bacteria are used in bioremediation to break down organic contaminants and extract heavy metals from contaminated areas. The efficiency and specificity of these procedures are being improved by innovations like nanobiotechnology and genetically engineered microorganisms. These applications are further supported by real-time microbial biosensors, which make it possible to precisely monitor and optimise microbial systems.

Keywords: Synthetic biology; energy; bioremediation; microbial fuel cells; carbon sequestration.

8.1. Introduction

The growing global problems of resource depletion, climate change and environmental pollution demand creative and long-term solutions. Utilising the potential of microorganisms to create environmentally friendly technology for resource recovery, pollution reduction and energy generation, microbial environmental engineering is an area that shows great promise. The necessity for renewable energy sources is emphasised by the increased worldwide energy demands brought on by urbanisation and industrialisation (Karthick and Haribabu, 2020). Renewable sources are becoming more and more popular because of their favourable environmental effects as a means of addressing energy issues and creating sustainable alternatives to fossil fuels (Hardt *et al.*, 2018).

Approximately 3% of global electricity consumption goes towards wastewater treatment; of the treatment costs, roughly 50% are associated with sludge handling/disposal. The biggest challenges facing modern wastewater treatment methods are their high energy consumption and operating costs (Saba *et al.*, 2017). Technical limitations that standard wastewater treatment systems encounter, in addition to energy consumption, include the ineffective treatment of the generated greenhouse gases (GHGs) and dissolved components like ammonia and phosphates (Li *et al.*, 2013).

Recent years have seen the rise of fuel cells, particularly biological fuel cells (BFCs), like microbial fuel cells (MFCs), as a promising sustainable energy source. They use electrochemical processes to transform chemical energy into electrical power (Zhang *et al.*, 2020). Recent research has focused on recovering high-value products from MFCs in the form of heavy metals, methane (CH_4), hydrogen (H_2) and hydrogen peroxide (H_2O_2) (Jadhav *et al.*, 2017). As a result, MFCs are now viable bioelectricity generation systems with numerous advantages, including efficiency, cleanliness, recyclability and the production of less hazardous pollutants (Mansoorian *et al.*, 2013).

Since MFCs can treat wastewater and produce energy, they have swiftly evolved from an environmentally beneficial electrical gadget to a sustainable system. To generate electricity, the MFC requires anode and cathode electrodes. Typically, a conductive wire connects these electrodes and a proton exchange membrane (PEM) separates them (Oliveira *et al.*, 2013). In contrast to chemical fuel cells (such as hydrogen fuel cells), MFCs rely significantly on microorganisms for the oxidation reactions that take place at the anode side. Protons and electrons are created during these reactions, which also break down biological stuff. Microorganisms in a single or mixed culture are utilised in the anode chamber, and the substrate can be either real wastewater or synthetic wastewater.

Electric current is produced when the anode electrode gathers electrons and sends them to the cathode by way of an external wire, while the proton is released to the anolyte and diffuses across the proton-selective membrane. When the last electron acceptor, usually oxygen, is reduced at the cathode, water is created (Hernández-Fernández *et al.*, 2015).

The majority of microorganisms have the ability to function as biocatalysts in MFCs. Potter demonstrated the first MFC concept in 1910 (Ieropoulos *et al.*, 2005). Living cultures of *Escherichia coli* and *Saccharomyces* were utilised to produce electrical energy using platinum electrodes (Potter, 1911). Until the 1980s, when it was discovered that adding electron mediators may greatly enhance the power output and current density, this attracted little attention. Organisms cannot transmit electrons directly to the anode in the anodic chamber unless they are anodophiles.

The outer layers of most microbial species are composed of non-conductive lipid membranes, lipopolysaccharides and peptididoglycans, which inhibit direct electron transfer to the anode. Transmission is accelerated by electron mediators (Davis and Higson, 2007). Oxidised mediators can be easily reduced by taking the electrons out of the membrane. In the anodic chamber, the mediators subsequently penetrate the membrane, transfer the electrons to the anode and undergo further oxidation in the bulk solution.

Power output is increased by this cyclic process, which speeds up the electron transfer rate. Although a mediator with the lowest redox would theoretically have the lowest anodic redox and maximise the redox difference between the anode and cathode (i.e. the largest voltage difference), it may not be the best at removing electrons from the microorganisms' reduced intracellular systems (NADH, NADPH or reduced cytochromes). An E° mediator would have greater overall power than a mediator with the lowest redox (Ieropoulos *et al.*, 2005).

Common examples of synthetic exogenous mediators are metalorganics and dyes, such as thionine, methylene blue and neutral red (NR). Unfortunately, synthetic mediators are not widely used in MFCs due to their instability and toxicity. By employing naturally existing molecules like microbial metabolites (endogenous mediators), some bacteria can function as mediators. Humic acids, anthraquinone and sulfur oxyanions (sulfate and thiosulfate) can all transfer electrons from the cell membrane to the anode.

The discovery that certain microorganisms could transfer electrons straight to the anode was a significant advance. These microorganisms have a high Coulombic efficiency and are operationally stable. The anode serves as the last electron acceptor in the dissimilatory respiratory chain of the bacteria that are using the biofilm. The movement of electrons between the electrodes and the microorganisms may also depend on the development of biofilms on a cathode surface. In wastewater treatment and power production, mediator-less MFCs are beneficial since they do not require a mediator (Du *et al.*, 2007).

The electrochemical processes that occur between the organic matter and the ultimate acceptor, such as oxygen, determine how effectively an MFC operates. The real/actual cell potential is never as high as it should be because of three irreversible losses that take place in an MFC: ohmic losses, concentration polarisation and activation polarisation. The activation polarisation is more common at low current densities because the reacting species must overcome an activation energy. Factors including electron transit, reactant species adsorption or desorption and electrode surface properties all affect activation polarisation.

If microorganisms are unwilling to donate electrons to the anode, activation polarisation in MFCs increases. By adding mediators, the energy barrier created by this can be lowered. Ohmic loss, which results from ionic and electronic conduction, is the most common loss in the centre of the operational range. This loss can be avoided by shortening the distance between the two electrodes, using membranes with lower resistance and improving the electrolytes' ionic conductivity. At very high current densities, concentration polarisation is the main loss because of mass transport limitations and the incapacity to maintain the initial substrate concentration in the bulk fluid. In addition to the MFC design, stirring and/or bubbling aid in minimising this loss.

To get closer to the optimal potential, polarisation curves of an MFC show the different losses and their magnitudes, highlighting viable mitigation strategies. These actions could involve choosing different microorganisms, mediators and substrates as well as changing the MFC's architecture (short electrode spacing) and configurations (new catalyst electrolytes, improved electrode structure) (Oliveira *et al.*, 2013). Furthermore, MFC performance is impacted by biological elements such as pH and nutrient availability, in contrast to conventional fuel cells that have a consistent power output because of clearly defined chemical processes (Ali *et al.*, 2019).

It has been investigated how wastewater treated using MFCs may remove and recover contaminants like ammonia (NH_3), heavy metals and chemical oxygen demand (COD). This method produces electricity by using the biological decomposition of organic substances (Trapero *et al.*, 2017). MFCs are unable to remove most high-performance heavy metals, including thorium (Th), lead (Pb), cadmium (Cd), chromium (Cr), mercury (Hg), copper (Cu), arsenic (As) and nickel (Ni). MFCs are capable of oxidising simple carbonates to carbon dioxide (CO_2) and transporting electrons through biological reduction. Since the substrates on a battery's anode side act as the fuel needed for metabolic conversion, electron transport is similar to that occurring in conventional fuel cell systems (Rahimnejad *et al.*, 2012). MFCs can use organic resources, like wastewater, as fuel through microbial breakdown, in contrast to conventional fuel cells which rely on refined fuels and strictly controlled chemical processes. The main advantages of MFC technology include high substrate to energy conversion efficiency, sludge volume reduction and the recovery of high-value products (Do *et al.*, 2018; Rabaey and Verstraete, 2005).

8.2. General features, types and designs of MFCs

8.2.1 General features

MFCs produce electricity by oxidising both organic and inorganic materials using microorganisms as catalysts. Microorganisms in an MFC anodic chamber oxidise additional substrates, producing protons and electrons in the process. One of the results of oxidation is carbon dioxide. Since photosynthesis uses carbon dioxide from the atmosphere, the regenerated biomass does not contain any net carbon. H^+ ions and electrons are released when the bacteria on the anode decompose organic molecules. The electrons generated by these substrates can either travel to the anode (negative terminal) and be connected to the cathode (positive terminal) by a resistor and a conductive substance, or they can be operated under a load to power a device (Hernández-Flores *et al.*, 2015).

The H^+ ions are able to reach the cathode through the semi-permeable membrane. This process is driven by the electrochemical gradient caused by the high concentration of H^+ ions near the anode. When the electrons from the cathode combine with dissolved oxygen and H^+ ions, pure water is produced (Brown, 2012).

There are numerous materials and layouts being used in the construction of microbial fuel cells. In Figure 8.1, an MFC schematic diagram for generating electricity is displayed. It consists of cathodic and anodic chambers divided by a membrane for proton exchange.

The electrode reaction in the following example uses acetate as the substrate.

Anodic reaction:

$$CH_3COO^- + 2H_2O \rightarrow CO_2 + 7H^+ + 8e^- \tag{8.1}$$

Cathodic reaction:

$$O_2 + 4e^- + 4H^+ \rightarrow 2H_2O \tag{8.2}$$

Power is generated as a byproduct of the overall reaction, which breaks down the substrate into carbon dioxide and water. According to the electrode reaction pair previously indicated, an MFC bioreactor can be powered by the electron flow that begins in the anode and travels to the cathode in the external circuit. Depending on how actively microorganisms transfer electrons from the medium to the anode electrode, MFCs can be divided into two categories: mediator-less MFCs, which do not need a mediator, and mediator-containing MFCs, which do (Rahimnejad *et al.*, 2015).

The main factors that affect MFC performance and energy production in wastewater treatment are the microorganisms used (substrates that can act as an electron donor source), operating parameters like pH, electrode surface area and temperature, as well as the composition

Figure 8.1 Diagrammatic representation of a typical MFC

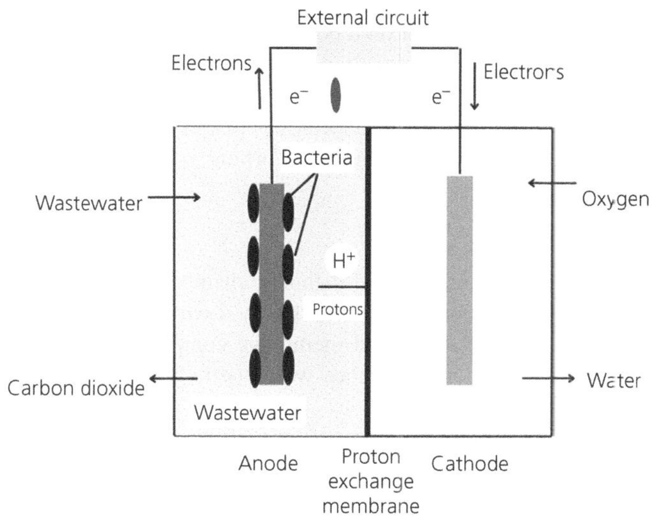

and design of the anode, cathode and membrane. Thus, energy production can be greatly impacted by the factors outlined above.

8.2.2 Types of MFC design

Based on their chamber design, MFCs are classified as single-, double-chamber, up-flow, stacked and paper MFCs (Kouam Ida and Mandal, 2023). MFC performance is influenced not only by biocatalysts but also by reactor configuration. Electrodes, a PEM or salt bridge and an external circuit are the main components of a fuel cell. The efficiency, generated power and budget cost of the cell are directly impacted by the MFC design. For an MFC to be designed efficiently, its size, shape, kind of electrode and stacking orientation are all crucial (Nawaz *et al.*, 2022).

8.2.2.1 Single-chamber MFCs

The cathode and anode are housed in the same chamber, in contrast to other kinds of MFC. Park and Zeikus were the first to use it (Park and Zeikus, 2003). Although the MFC lacks a clearly defined cathodic chamber, the PEM has a chamber with a porous cathode on one side that allows protons to flow through while preventing oxygen from reaching the anode (Liu and Logan, 2004). The circuit is completed on the other side by the absorption of oxygen from the atmosphere. Ordinary carbon electrodes serve as the anodes, whereas PEM-bonded flexible carbon cloth or porous carbon electrodes serve as the cathodes. That being stated, graphite coatings are commonly applied to cathodes. As a catholyte, electrolytes are frequently pumped into the cathode in a constant condition to prevent the cathode and membrane from drying out (Cheng *et al.*, 2006).

Single-chamber MFCs enable a direct reaction between the porous cathode electrode and oxygen-enriched air. The mode of operation is air-cathode. Reduced internal resistance, enhanced proton diffusion, reduced cost, increased cathode oxygen reduction rate, ease of use and aeration that increases adaptability are all advantages of single-chamber MFCs. However, a number of limitations, including evaporation, high-rate oxygen diffusion and liquid outflow, restrict their application on a commercial scale (Liu and Logan, 2004). The anode and cathode must remain closed for single MFCs to function at high power densities.

A large working volume is required for large-scale applications in order to provide adequate power density for wastewater treatment, but the working volume of each MFC unit must remain modest.

8.2.2.2 Double-chamber MFCs

In a double-chamber MFC, the anodic and cathodic chambers are anaerobic and aerobic, respectively. Both regions are connected externally by a wire and internally by the proton exchange membrane. The entire frame and membrane complete the circuit by decreasing the diffusion of oxygen in the anodic chamber, which permits protons to enter the cathodic chamber (Du *et al.*, 2007).

Any non-corrosive conducting substance can be used to create the submerged electrodes. This produces power in batch mode using a chemically specified medium. Because a significant distance between the electrode and cathode would result in a decreased output, an external resistance directly affects the output. Choi *et al.* (2013) found that the internal resistance decreased from 672 Ω (103 mW/m^2) to 93 Ω (57 mW/m^2) when the electrodes were placed closer together.

A double-chamber MFC is mostly used for treating wastewater to generate power on a large scale. To supply oxygen, however, the cathode requires regular solution replenishment with aeration. However, it is still possible to increase the power production by adjusting the cathode's pH, removing new oxygen, boosting flow rate, introducing more electron mediators and keeping internal resistance constant. A double-chamber microbial fuel cell typically operates in the water cathode mode. This special design can be altered to produce MFCs in cylindrical, stacked and single-chamber modes. For industrial applications, MFC setup is essential in determining power output to achieve greater efficiency and lower costs.

8.2.2.3 Up-flow MFCs

The shape of the up-flow MFC is cylinder-like. The MFC comprises a cathode chamber at the top and an anode at the bottom. Glass wool and glass bead layers separate the two sections. The substrate enters the anode at the bottom and moves upward to the cathode before leaving at the top. Between the electrodes, a gradient is created that contributes to the fuel cell's beneficial activity (Cheng *et al.*, 2006). Anolyte and catholyte are not well defined in this concept. Furthermore, there is no actual separating and proton transmission-related issues are quite rare (Venkata Mohan *et al.* 2014; Zhou *et al.* 2013).

The up-flow mode MFC designs are appealing for wastewater treatment since they are easily scalable in comparison to other systems. However, the fuel cell's main drawback is that the energy needed to pump the substrate is far greater than the electricity generated (Venkata Mohan *et al.* 2014; Zhou *et al.* 2013). Therefore, the up-flow MFC's main purpose is to treat wastewater rather than generate energy (Brutinel and Gralnick, 2012). The complex design, high internal resistance and electrode-based losses of these MFCs, which are commonly employed in basic research, are thought to be the causes of their low power densities.

8.2.2.4 Stacked MFCs

MFCs are essential for increasing power generation and can be stacked in series or parallel. When used in parallel, the stack design produces higher current output; when used in series, it produces higher voltage output. The design has little effect on the Coulombic efficiency, but it amplifies the battery's overall output (Flimban *et al.*, 2019). Using the Coulombic efficiency (CE) in MFCs, one may estimate the number of electrons transferred from the anode that contribute to the current production based on the total electrons created theoretically from substrate oxidation. By connecting the cells in stacked MFCs in different ways, different rates of CE can be achieved. Voltage reversal is the main barrier to higher voltage outputs in stacked MFCs. The depletion of the cell's substrate may be the cause of the voltage reversal, indicating the bacteria's decreased capacity to generate higher voltage (Kumar *et al.*, 2017).

Certain factors must be considered while treating wastewater with stacked MFCs that have the capacity to generate enough power. Some of the criteria include checking the organic concentrations at the end of MFC operation, keeping the reactor capacity, maintaining the substrate flow rate and height if the MFC is vertically stacked. Stacking MFCs can be divided into two subcategories: vertical stack and horizontal stack.

8.2.2.5 Paper MFCs

Paper-based microbial fuel cells provide a new and affordable way to produce bioelectricity by utilising the unique properties of paper as a substrate for microbial growth and electron transfer. Without the use of external pumps, these MFCs enable the passive movement of nutrients and electrolytes by utilising the capillary action of paper.

Because paper is porous, bacteria can adhere to it and create biofilms on its large surface area, increasing the efficiency of electron transport. Carbon-based anodes, like graphene or carbon cloth, and a proton exchange membrane (PEM) or paper layers coated with ion-conducting polymers are typically used in paper-based MFCs to isolate the anode and cathode compartments (Hashemi *et al.*, 2016). Paper-based MFCs are ideal for low-power applications including environmental biosensors and point-of-care diagnostic devices due to their simplicity, lightweight design and biodegradability. Nonetheless, issues like low power output and poor long-term stability continue to be important research topics for future advancement.

8.2.2.5 Summary of advantages and disadvantages of various MFCs

Table 8.1 summarises the advantages and disadvantages of the various MFC designs currently available.

8.2.3 Substrates and microorganisms used in MFCs

8.2.3.1 Substrates

Substrates are important because they give the bacteria in MFCs nourishment. Therefore, substrates are essential to the growth of biofilms. The substrate feed can be optimised to modify the biofilm and MFCs have employed a variety of substrates. There are a number of

Table 8.1 Advantages and disadvantages of various MFC designs

MFC type	Advantages	Disadvantages
Single-chamber MFCs	Simple, cost-effective, high power density, compact	Low efficiency, cathode biofouling, oxygen supply limitations
Double-chamber MFCs	Higher efficiency, stable performance, less biofouling	Complex, expensive, higher internal resistance
Up-flow MFCs	Efficient wastewater treatment, good mixing, high power output	Risk of clogging, requires hydraulic control, complex operation
Stacked MFCs	Higher voltage output, scalable, optimised space usage	Internal resistance, maintenance challenges, electrode interference
Paper MFCs	Low-cost, lightweight, ideal for biosensors	Low power output, short lifespan, not scalable

ways to alter the microbiome of MFCs, but using certain startup and operating circumstances would be a more successful approach than feed optimisation (Vilajeliu-Pons *et al.*, 2016).

Several factors affect the efficiency of MFCs, including the electrode material, the microorganisms that oxidise the substrate molecules and the type of substrates (complex or simple) (Khan *et al.*, 2016). MFCs can utilise substrates from a variety of wastewater sources, including industrial, animal, pharmaceutical, agricultural and residential (Angosto *et al.*, 2015; Min *et al.*, 2005b; Miran *et al.*, 2016a; Sonawane *et al.*, 2014).

A substrate is defined as a feature that supports microbial metabolism by acting as an electron donor and has a known chemical makeup, such as acetate, glycerol or synthetic wastewater. Research on MFC performance under controlled settings has been conducted using these substrates.

The most common substrate for MFCs is acetate, which electroactive bacteria use as their main carbon supply. Acetate is a desirable substrate in MFCs as, at normal temperature, it is inert to microbial transformations like fermentation or methanogenesis (Khater *et al.*, 2017).

Glycerol is a significant byproduct of the manufacturing of biodiesel and a necessary component of the food, beverage, tobacco and pharmaceutical sectors. Glycerol is chosen as an energy source for producing bioelectrochemical energy in order to make the production process more cost-effective (Tremouli *et al.*, 2016). It is a good feedstock since it is inexpensive, widely accessible and has a simplified chemical structure. Furthermore, glycerol can be naturally metabolised by bacteria and converted into a variety of different products and intermediates. A variety of alcohols, including ethanol, n-butanol, 2,3-butanediol and 1,2-propanediol, are produced during glycerol fermentation. However, there is still a great need for improved technology for efficient glycerol conversion.

In the search for practical and repeatable substrate conditions, synthetic wastewater (also referred to as simulated wastewater) is an efficient substrate for MFCs. The mixture usually resembles wastewater in one way or another. The three carbohydrate molecules that are often abundant in synthetic wastewater are sucrose, soluble starch and glucose. Furthermore, a lot of proteobacteria are often present in synthetic wastewater (Aldrovandi *et al.*, 2009).

Microbial communities are frequently found in complex substrates, which are a blend of several chemical substances. There are several sources of complex substrate, including

- food waste and household wastewater from municipal kitchens
- industrial wastewater from sources include refineries, petroleum facilities, distilleries, palm oil factories and dark fermentation systems
- agricultural sources such as cellulose, soybean, molasses and other waste products
- animal waste from swine, cattle, fish markets, seafood processing and so on
- fruit waste including peels and liquids
- chemical waste such as azole dye, ethanolamine, sulfide, nitrate, isopropanol and medicinal ingredients
- dairy waste such as cheese whey and yoghurt.

Based on the concentration of dissolved chemicals, domestic wastewater – a subset of municipal wastewater – is divided into three categories: weak, medium and strong. Weak, medium and strong wastewater are indicated by dissolved organic material concentrations of 250 m/L, 500 m/L and 1000 m/L, respectively (Sincero and Sincero, 2002) .

Household activities like washing, bathing and flushing generate wastewater. Municipal wastewater-activated sludge is used as an inoculum in mediator-less single-chamber MFCs that treat synthetic wastewater. Biohydrogen fermentation of food waste produces acetate and butyrate, which are promising substrates for MFCs. Food waste leachates, containing xenobiotic and dissolved organic pollutants, can be treated using MFCs, which simultaneously generate electricity and degrade pollutants (Choi and Ahn, 2015).

Industrial wastewater contains organic biopolymers like chitin and cellulose, which serve as good substrates for electricity generation. Winery and distillery wastewater are high in organic load but lack inorganic nutrients like nitrogen and phosphorus. Dark fermentation of biowaste releases volatile pollutants like acetic acid. Heavy metals, hydrocarbons, sulfides and phenols are among the organic and inorganic contaminants found in petrochemical and refinery wastewater. Petroleum wastewater is rarely utilised in MFCs because of its complicated nature (Varjani, 2017).

Lignocellulose, derived from agricultural and forestry residues, is the most abundant biomass and a rich source of organic polymers. Its treatment releases polyphenols, potential substrates for exoelectrogenic bacteria. Cellulose, a key component, is a glucose-based homopolymer.

Edible soybean oil refining generates high organic wastewater, with impurities removed through deacidification, degumming, bleaching and neutralisation. Molasses, a byproduct of sugar production, serves as a raw material for yeast and ethanol fermentation, with its wastewater supporting microbial growth due to nutrients like phosphate, polysaccharide and nitrate (Zeng *et al.*, 2009). Fish processing and market activities generate wastewater from washing, cutting and cleaning.

Animal manure wastewater from breeding industries is rich in phosphorus, organics and nitrates. Swine wastewater, high in organic content, has been used in MFCs for bioelectricity generation (Ogugbue *et al.*, 2015).

Seafood processing wastewater serves as a substrate in a tubular up-flow MFC. Lemon waste, including seeds, pulp and peels from juice extraction, can be fermented to produce volatile fatty acids and used in MFCs for bioelectricity generation. The potential of lemon peel as a carbon source for MFCs has been studied (Miran *et al.*, 2016b). Orange peel waste, a byproduct of juice consumption, is used to produce ethanol, methane and bioelectricity in MFCs without pretreatment (Miran *et al.*, 2016a).

Industrial wastewater contains organic-inorganic chemicals and heavy metals. In MFCs, nitrates act as electron acceptors in the cathode, while sulfides serve as electron donors in the anode, enabling their removal (Li *et al.*, 2014). Ethanolamine (ETA), used in intermediates and detergents, has been studied as an MFC substrate for wastewater treatment (Shin *et al.*,

2015). Dairy wastewater, rich in carbohydrates, proteins and fats, is easily biodegradable and serves as a substrate in MFCs (Faria *et al.*, 2017).

8.2.3.2 Microorganisms used in MFCs

The theory of MFC operation depends on an understanding of the anodic electron transport mechanism. The electron transport system used by bacteria to move electrons to the electrode is made up of a number of extracellular matrix components and electron shuttles that are dissolved in the bulk solution (Du *et al.*, 2007).

The metabolism of organic materials can provide electrons to the anode through a variety of microorganisms. Rich sources of these microorganisms include soil, freshwater sediment, wastewater, marine sediment and activated sludge (Niessen *et al.*, 2006). Microorganisms can transfer electrons to the anode electrode in three different ways: first, directly from the respiratory enzymes (i.e. cytochromes) to the electrode; second, using mediators produced by the bacteria; or third, through the use of exogenous mediators (external to the cell) such as potassium ferricyanide, thionines or neutral red.

Table 8.2 summarises the microorganisms used within MFCs.

Table 8.2 Microorganisms that are used in MFCs (continued on next page)

Microorganism	Electron transfer mechanism	Substrate utilised	Application	Reference
Actinobacillius succinogenes	MET (mediated electron transfer) by way of succinate and formate oxidation	Glucose	Succinic acid production for biopolymers	Park *et al.*, 1999
Aeromonas hydrophila	MET by way of quinone electron shuttles	Acetate	Biohydrogen production from wastewater	Pham *et al.*, 2006
Alcaligenes faecalis, Pseudomonas aeruginosa	MET by way of ubiquinone derivatives and menaquinone-based electron transfer	Glucose	Nitrate removal and bioremediation of nitrate and phosphate in wastewater	Holzman, 2005
Clostridium beijerinckii, Clostridium butyricum	Fermentative electron transfer by way of hydrogenases and butyrate production	Starch, lactate, glucose, molasses	Fermentative bacterium and biobutanol and biohydrogen production	Niessen *et al.*, 2004
Erwinia dissolven	MET by way of extracellular polysaccharides	Glucose	Enzyme production for industrial bioprocesses	Vega and Fernández, 1987

Table 8.2 Continued

Microorganism	Electron transfer mechanism	Substrate utilised	Application	Reference
Escherichia coli	MET by way of flavins and quinones	Glucose	Genetically engineered strains for enhanced bioelectricity production	Ieropoulos *et al.*, 2005b
Desulfovibrio desulfuricans	DET (direct election transfer) by way of periplasmic hydrogenases and c-type cytochromes	Sucrose	Sulfate/sulfide as mediator	Ieropoulos *et al.*, 2005b
Geobacter metallireducens, *Geobacter sulfurreducens*	DET by way of c-type cytochromes and conductive pili	Acetate	Mediator-less MFC and bioelectricity generation	Bond *et al.*, 2002; Min *et al.*, 2005a
Klebsiella pneumoniae	MET by way of flavin secretion	Glucose	Biohydrogen and ethanol production	Rhoads *et al.*, 2005
Lactobacillus plantarum	MET by way of excreted flavins	Glucose	Fermentative biohydrogen production	Vega and Fernández, 1987
Pseudomonas aeruginosa	MET by way of phenazine compounds (pyocyanin, phenazine-1-carboxylate)	Glucose	Bioelectricity generation using phenazines	Choi *et al.*, 2003
Shewanella oneidensis	DET by way of cytochrome complex	Lactate	Biocorrosion prevention	Min *et al.*, 2005a
Shewanella putrefaciens	DET by way of outer membrane cytochromes	Lactate, acetate, glucose	Mediator-less MFC and biosensors for environmental monitoring	Cydzik-Kwiatkowska and Nosek, 2022
Streptococcus lactis	Fermentative electron transfer by way of lactate oxidation	Glucose	Ferric chelate complex as mediators and dairy industry applications in fermentation	Vega and Fernández, 1987
Chlorella vulgaris	Oxygenic photosynthesis	CO_2, sunlight, wastewater nutrients	Bioelectricity, CO_2 sequestration, wastewater treatment	Flimban *et al.*, 2019
Synechocystis sp.	Biophotovoltaic electron transfer	CO_2, light, organic waste	Photo-bioelectricity generation	Slate *et al.*, 2019
Oscillatoria sp.	Photosynthetic electron transfer	CO_2, wastewater	Sustainable bioelectricity production	Min *et al.*, 2005a

8.3. Working principles/treatability of MFCs and generation of energy

8.3.1 Working principles

Electrogenic microorganisms are used in MFCs to oxidise organic or inorganic substrates, producing protons and electrons as byproducts of microbial metabolism. Bacteria in the anaerobic anode chamber break down organic materials, producing electrons that are either directly transported to the anode by way of cytochromes and conductive nanowires, or indirectly through redox mediators.

While protons move to the cathode chamber by way of a proton exchange membrane (PEM) or salt bridge, the electrons move to the cathode by way of an external circuit, generating an electric current. The redox reaction is completed at the cathode when protons, electrons and an electron acceptor, usually oxygen, combine to generate water. The entire electrochemical process is similar to microbial respiration, except it is used to generate bioelectricity.

The efficiency of MFCs depends on several variables, such as temperature, pH, substrate type, microbial community makeup and electrode material. Microbial desalination cells (MDCs) for water purification and microbial electrolysis cells (MECs) for hydrogen production are examples of advanced designs that increase the possible uses of MFC technology. The production of bioelectricity combined with carbon capture, resource recovery and wastewater treatment make MFCs an environmentally responsible and sustainable answer to energy and environmental problems.

8.3.2 Generation of energy

The chemical energy contained in the chemical components of biomass can be converted into electrical energy by microorganisms using microbial fuel cells. The anode chamber's microorganisms consume substrates as fuel to generate electricity; consequently, they overcome a resistance to transport electrons to the anode. The electrons diminish the electron acceptors (O_2) once they arrive at the cathode. Protons are transferred from the anode chamber to the anolyte for reduction by way of a membrane. Equations 8.3 and 8.4 employ acetate as the electrode reaction substrate, whereas Equations 8.5 and 8.6 use sucrose.

$$\text{Anode: } CH_3COO^- + H_2O \xrightarrow{\textit{microorganism}} 2CO_2 + 2H^+ + 8e^- \quad (8.3)$$

$$\text{Cathode: } O_2 + 4e^- + 4H^+ \rightarrow 2H_2O \quad (8.4)$$

$$\text{Anode: } C_{12}H_{22}O_{11} + 13H_2O \xrightarrow{\textit{microorganism}} 12CO_2 + 48H^+ + 48e^- \quad (8.5)$$

$$\text{Cathode: } O_2 + 4e^- + 4H^+ \rightarrow 2H_2O \quad (8.6)$$

In MFCs, the output voltage, which is calculated by multiplying the external resistance by the current in accordance with Equation 8.7, has a direct effect on the cathode and anode potentials. This equation explains the relationship between activation overpotential and current density (Xia *et al.*, 2018).

$$V_{cell} = V_c - V_a - \eta_{ohm} \tag{8.7}$$

The cathode potential is denoted by V_c, the anode potential by V_a and the ohm overpotential by η_{ohm}. However, the majority of voltage readings have been made using voltmeters and other quantitative tools. Using Equation 8.8, the power density may be determined.

$$P = \frac{IV}{A} \tag{8.8}$$

where A (m^2) is the anode's surface area, V (V) is the voltage that can be measured with a multimeter and I (A) is the current (Min *et al.*, 2005b).

Key operational and design parameters that influence electricity generation in MFCs are summarised in Figure 8.2.

8.4. Application and performance of MFCs in wastewater treatment

8.4.1 Applications

Several MFC applications have emerged in the last few decades. These are as follows.

8.4.1.1 Generation of bioelectricity

It is evident that most MFC research is conducted towards the technology's core use, which is the generation of electricity. In the anode chamber of the MFC, the microorganisms oxidise the substrate, generating protons and electrons that are transferred to the cathode by way of PEM and electrical connection, respectively. The two chambers of the MFC can be electri-

Figure 8.2 Important parameters affecting electricity generation in MFCs

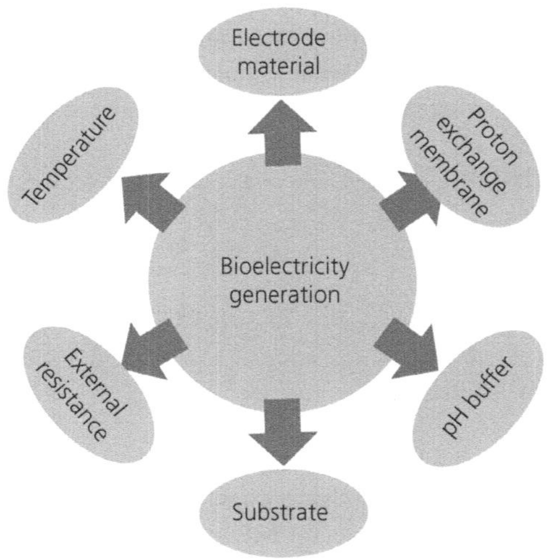

cally connected to a multimeter and an external resistor box to measure the voltage. Ohm's law can then be used to determine the power.

MFCs rely heavily on substrates that can be fully oxidised into electrons in order to increase their Coulombic efficiency and, consequently, their power production. According to a study (Kumar *et al.*, 2017), *Geobacter sulfurreducens* is capable of fully reducing acetate to protons and electrons. The main factors influencing the MFC's electrical output are its design, electrode composition, inoculum (pure or mixed culture), proton exchange membrane and operational environment.

8.4.1.2 Biohydrogen production

MFCs can be easily modified to produce hydrogen instead of electricity. Under normal operating conditions, protons generated by the anodic process travel to the cathode, where they combine with oxygen to form water.

In an MFC, producing hydrogen from the protons and electrons produced by microbial metabolism is thermodynamically unfavourable. Overcoming the thermodynamic barrier can be achieved by using an external potential to raise the cathode potential in an MFC circuit. In this mode, hydrogen is created at the cathode by combining protons and electrons from the anodic reaction. Since some energy originates from the biomass oxidation process in the anodic chamber, the theoretically required external potential for an MFC is 110 mV, which is significantly less than the 1210 mV needed for direct electrolysis of water at neutral pH.

In contrast to conventional fermentation, which typically yields 4 mol H_2/mol glucose, MFCs have the potential to produce 8–9 mol H_2/mol glucose. The cathodic chamber no longer requires oxygen when producing biohydrogen with MFCs. Because there is no longer a problem with oxygen leaking to the anodic chamber, MFC efficiency increases. Another benefit is that hydrogen may be stored and accumulated for later use, which gets around the MFC's low power feature. Therefore, in a hydrogen economy, MFCs provide a sustainable source of hydrogen that can assist in meeting the entire demand for hydrogen (Holzman, 2005).

8.4.1.3 Wastewater treatment

Early in the 1990s, MFC technology was investigated for the treatment of wastewater (Habermann and Pommer, 1991). According to recent research, MFCs are a potential technology for treating many types of wastewater in industrial, agricultural and municipal settings (He *et al.*, 2017). Low material costs and great operating sustainability are desirable attributes for an effective treatment system in general.

MFCs are also utilised at a field level to treat wastewater in combination with other treatment units and to monitor wastewater treatment facilities, water quality and so on in remote locations where power access is limited. Even though the highest power production efficiency of MFCs (±60%) is within acceptable bounds, this particular issue is currently a drawback for researchers trying to improve MFC performance (Surya Ramadan and Purwono, 2017). Scientists claim that biological treatment is frequently used as a reliable and very cost-effective way to remove organic compounds and nitrogen from leachate (Mehmood *et al.*, 2009).

A special ability of certain bacteria employed in MFCs is their ability to remove sulfides, which is essential for wastewater treatment. MFCs exhibit excellent operational stability by fostering the growth of bioelectrochemically active microorganisms during wastewater treatment. Reports have shown a Coulombic efficiency of up to 80% and in some cases, up to 90% of the COD can be removed (Kim *et al.*, 2005).

8.4.1.4 Bioremediation

For the (in situ) bioremediation of contaminated soils, sediments, groundwater and surface water, the MFC technique has been studied. The microbial community that naturally inhabits streams is responsible for this biological process, which converts pollutants into less harmful forms. By using specific contaminants in the system to generate electricity and boost bioremediation rates, bioremediation now known as bioelectroremediation offers cost advantages over more conventional methods such as mycoremediation, phytoremediation and bacterial bioremediation in MFCs (Pisciotta and Dolceamore Jr, 2016). MFCs offer both reduction and oxidation reactions, functioning as a comprehensive treatment system that results in a thorough pollutant breakdown process.

8.4.1.5 Desalination

MFCs may perform desalination with minimal modifications to their fundamental operating principle since they are microbial desalination cells (MDCs). These MDCs can perform desalination in addition to wastewater treatment and electricity production. This is accomplished by incorporating an additional chamber between the anodic and cathodic chambers on a typical DC MFC design. The chambers are separated by cation exchange membranes (CEM) and anion exchange membranes (AEM) (Saeed *et al.*, 2015).

8.4.1.6 Biosensors

In addition to the applications previously mentioned, MFC technology can be used as a sensor for in situ process monitoring and control, as well as for pollution analysis. MFCs can be used as biological oxygen demand (BOD) sensors because of the proportionate relationship between their Coulombic yield and wastewater strength. Calculating a liquid stream's Coulombic yield is a precise way to determine its BOD value. Several studies (Chang *et al.*, 2004; Kim *et al.*, 2003) demonstrated a strong linear correlation between the wastewater's strength and Coulombic yield over a significant range of BOD concentrations. However, a high BOD concentration requires a long response time since, in the absence of a dilution mechanism, the Coulombic yield cannot be calculated until the BOD has been depleted. There has been an effort to improve the dynamic responses of MFCs used as sensors.

8.4.2 Performance of microbial fuel cells

8.4.2.1 Ideal performance

Optimised performance is determined by the electrochemical reactions between the organic substrate (such as glucose) and the final electron acceptor (such as oxygen), which has a low potential. But its optimal cell voltage is unknown since the electrons are moved from the organic substrate to the anode by way of a complicated respiratory chain that differs for each microorganism, including the same bacteria, depending on the growth conditions. Despite the fact that the respiratory chain is still poorly understood, the main anodic reactions that control the voltage are the anode, the mediator (if one is used) or the last cytochrome in the system for the electrophile/anodophile if this has conducting pili. Bacterial species that cannot release electrons directly to the anode need a redox mediator to move them there (Rabaey and Verstraete, 2005).

8.4.2.2 Actual MFC performance

The real cell potential is always lower than the equilibrium potential because of irreversible losses. Equation 8.9 represents the irreversible losses in a genuine MFC.

$$V_{cell} = E_{cathode} - |\eta_{act,c} + h_{conc,c}| - E_{anode} - |\eta_{act,a} + \eta_{conc,a}| - iR_{i} \tag{8.9}$$

The cathode and anode activation polarisation losses are represented by $\eta_{act,c}$ and $\eta_{act,a}$, respectively. $\eta_{conc,c}$ and $\eta_{conc,a}$ in cathodic and anodic chambers, respectively, represent concentration polarisation. Ohmic loss, or η_{ohm}, is the result of resistance to the passage of ions in the electrolyte and to the flow of electrons across the electrode. Thanks to Ohm's law, it may be expressed as iR_i, where i is the current flowing through the MFC and R_i is the total cell internal resistance of the MFC.

The responding species must overcome an activation energy, which is the cause of activation polarisation. It is a limiting step in electrochemical reactions where slow reaction kinetics control the rate of the reaction at the electrode surface. The physical characteristics of the electrode surface, the adsorption of reactant species, the desorption of product species and the movement of electrons across the double-layer cell membrane all play a part in activation polarisation. The addition of mediators can help overcome activation polarisation in microorganisms that do not easily transfer electrons to the anode surface (Appleby and Foulkes, 1989).

8.4.3 Factors affecting performance of MFCs during wastewater treatment

Current performance of laboratory MFCs is still much below optimal. There might be a number of causes. Several factors influence an MFC's power generation, including the following.

8.4.3.1 Electrode properties

An MFC's performance can be enhanced by using better-performing electrode materials, as different anode materials produce varying activation polarisation losses. For both anode and cathode structures, platinum (Pt) and Pt black electrodes are superior to graphite, graphite felt and carbon-cloth electrodes, despite their somewhat higher cost. Schröder *et al.* (2007) state that a platinumised carbon-cloth anode may produce a current of 2–4 mA in an agitated anaerobic culture of *E. coli* using a conventional glucose solution at 0.55 mmol/L. However, with the unmodified carbon cloth, no microbially aided current flow was seen under the same working circumstances. Pt shows higher catalytic activity with regard to oxygen than graphite materials. MFCs with Pt or Pt-coated cathodes generated higher power densities than those with graphite or graphite felt cathodes (Jang *et al.*, 2004).

8.4.3.2 pH

Protons are encouraged to migrate from the anode to the cathode when the pH levels at the anode and cathode differ, which is ideal for MFC operation. This natural behaviour is usually limited by the membrane's slow rate of proton diffusion, which leads to the accumulation of protons generated by the bacteria on the anode and a drop in the anode pH (Kumar *et al.*, 2020).

The pH reduction may negatively impact microorganisms, reducing their activity and, consequently, the rate at which the substrate oxidation reaction proceeds, as the reaction in the anodic compartment involves a biological process. Additionally, the pH will rise on the cathode side due to the lack of protons, slowing down the rate of the oxygen reduction reaction (Margaria *et al.*, 2017; Sasaki *et al.*, 2016). Several strategies, including DC design and the use of chemical buffers like phosphate, have been proposed to stop that. However, due to the high expense of buffers and environmental concerns, this is not appealing in real-world applications, especially in wastewater treatment. To avoid this, saline solutions and CO_2 have been used on the cathode side of an MFC to control the pH, and they appear to be a suitable and reasonably priced alternative to chemical buffers (Goswami and Mishra, 2018).

The CO_2 is used in the cathode, where it forms buffering compounds such as carbonate and bicarbonate by combining with hydroxide anions. It is also advantageous to use saline solution since it allows for pH control and improves ionic conductivity, both of which increase power output. Although it is possible, using the saline solution at the anode needs caution since it may change the microbial activity (Tremouli *et al.*, 2017).

8.4.3.3 Temperature

In MFCs, temperature is one of the most important factors affecting the system's overall dynamics. Any notable temperature change during operation could have a big effect on the MFC's performance. Mass transfer, microbial metabolism and thermodynamics (electrode potentials and Gibbs free energy) are its main areas of influence.

The MFC's ability to generate electricity and efficiently treat wastewater depends on a number of critical factors, including temperature (Amend and Shock, 2001). Usually, the MFCs are operated between 25 and 30°C. Nevertheless, MFCs have also shown promising results at higher temperatures (both in terms of COD removal and power density). The boost in power density may be due to an increase in bacterial metabolism and membrane permeability. Furthermore, the high temperature can increase the conductivity of the anolyte and catholyte, lowering the ohmic resistance of the MFC and increasing its power density (Phung *et al.*, 2004).

A small rise in temperature has been found to have no effect on membrane permeability. Temperature has an exponential effect on bacterial activity, according to the findings. As a result, the higher bacterial activity is directly responsible for the greater power output brought on by the temperature rise. Biofilm growth at the anode, which affects current output, can be used to measure bacterial activity (Wang *et al.*, 2014).

8.4.3.4 Aeration

Aeration has a major effect on the anode and cathode processes, which in turn affects power output and overall efficiency in MFCs. The reduction reaction in the cathode chamber is improved by the presence of oxygen, which raises electron acceptance and the cell voltage overall. The enhanced oxygen reduction reaction (ORR) kinetics of aerated cathodes often result in better power densities than those of non-aerated ones. However, too much aeration might cause oxygen to diffuse into the anode chamber, which lowers Coulombic efficiency by upsetting the anaerobic conditions necessary for electrogenic bacteria to transfer electrons efficiently.

Aeration can also affect the nature of the microbial community, the formation of biofilms and the rates at which substrates are utilised, all of which affect MFC performance. In MFC systems, optimising aeration parameters like oxygen content and airflow rate is essential for maximising bioelectricity generation while balancing energy input.

8.5. Different product recovery from wastewater using MFCs

Bioelectrochemical systems called microbial fuel cells (MFCs) turn organic waste into electrical energy while also recovering valuable goods from wastewater. MFCs' dual-purpose qualities make them a desired sustainable solution for resource recovery and wastewater treatment. MFC-based systems are more economically and environmentally viable since they can recover a variety of products, such as bioelectricity, biohydrogen, biochemicals, heavy metals and nutrients.

8.5.1 Bioelectricity generation

Bioelectricity is one of the main products that MFCs recover from wastewater. Electrons are released when electroactive microorganisms in the anode chamber break down organic materials. An electric current is then created when these electrons go to the cathode through an external circuit. The efficiency of bioelectricity generation is influenced by the microbial community composition, electrode materials, substrate availability and system design. Improvements in electrode materials, system architecture and operational optimisation can greatly increase power density and scalability, even while MFC power output is still low when compared to traditional energy sources (Mohan *et al.*, 2008).

8.5.2 Production of hydrogen

As microbial electrolysis cells (MECs) function under applied voltage settings, MFCs can also be altered to aid in hydrogen recovery. In this process, protons and electrons from the anode's microbial metabolism mix to form hydrogen gas at the cathode. The utilisation of organic waste as a fuel and the reduction of greenhouse gas emissions make this technology a sustainable substitute for traditional hydrogen generation procedures like steam methane reforming (Holzman, 2005). However, for practical implementation, issues like diffusion-induced hydrogen loss and the requirement for extra energy input must be resolved.

8.5.3 Biochemicals and value-added compounds

MFCs provide a special platform for the recovery of high-value metabolites from wastewater, including polyhydroxyalkanoates (PHAs) and volatile fatty acids (VFAs). VFAs are produced by specific microbial species in the anode biofilm and are used as building blocks for biofuels and bioplastics. Furthermore, certain bacterial strains are capable of producing PHAs, biodegradable polymers that find use in the medical and sustainable packaging sectors. By cleaning wastewater and allowing the selective recovery of these substances, MFC-based systems combine environmental protection with waste valorisation.

8.5.4 Heavy metal removal and recovery

MFCs are necessary for both bioremediation and the recovery of heavy metals from industrial wastewater. Microbial activity at the anode facilitates the reduction of metal ions at the cathode, which leads to the precipitation and recovery of metals such as copper, silver and chromium. This process not only detoxifies wastewater but also recovers valuable metals that can be used in a range of industrial applications (Mathuriya and Yakhmi, 2014). The efficiency of metal recovery depends on several factors, such as pH, electrode material and microbial metabolism, in order to increase metal precipitation and minimise toxicity effects on microbial populations.

8.5.5 Nutrient recovery (nitrogen and phosphorus)

For agricultural applications, MFCs make it possible to recover vital nutrients like phosphate and nitrogen (Bhattacharya *et al.*, 2024). Nitrification-denitrification processes can transform ammonium nitrogen (NH_4^+) into nitrogen gas (N_2), lowering the risk of eutrophication in water bodies. The slow-release fertiliser struvite (magnesium ammonium phosphate), which increases soil fertility, is another way that phosphorus can precipitate. By integrating MFCs with nutrient recovery systems, wastewater treatment efficiency is increased and nutrient cycles are closed in an environmentally beneficial manner (Baby and Ahammed, 2022).

8.6. Recent advancements in MFC technologies

As a sustainable technology for wastewater treatment, bioelectricity generation and resource recovery, MFCs have attracted a lot of attention. The development of MFC technologies over time has concentrated on enhancing power output, electrode composition, microbial community optimisation, system scalability and integration with other treatment procedures.

Optimising the anode chamber's substrate concentration is one method of raising the MFC's efficiency because it directly affects the device's energy output and substrate removal efficiency. Abrupt shifts in the substrate concentration have the potential to decrease microbial diversity and prevent energy production. Highly adaptable microorganisms that are carefully cultured have been shown to improve the stability and removal effectiveness of MFCs.

Choosing the right external resistance is another crucial operational factor for enhancing the MFC's performance. Kloch and Toczyłowska-Mamińska (2020) investigated the effects of changes in external resistance on the removal of COD and energy production in MFCs that were powered by wood hydrothermal wastewater. Peak power produced in the MFC did not always correspond with the maximum COD removal efficiency. The external resistance when the power density was not maximised yielded the maximum wastewater treatment efficiency.

According to Potrykus *et al.* (2021), the population distribution in the MFC is greatly influenced by external resistance. Higher external resistances cause the population to shift from being more electrogenic to being more fermentative. A change in external resistance significantly affects the production of electricity but has a less significant effect on the elimination of COD. As a result, after running with high external resistance, the MFC shows decreased electrical efficiency.

Recently, changing the anode has been one of the most common ways to boost the power produced in the MFC. Heat modification with acid can enhance the anodic carbon fabric's surface structure and characteristics. According to research by Ni *et al.* (2020), when compared to untreated carbon cloth, thermo-acid anode modification increased the power density by 350% and significantly altered the structure of the MFC microbial population and energy output. Anode changes that improve electron transit in the cell and microorganism deposition can also change the anode's performance. Metal oxides containing conductive polymers such as polypyrrole (PPy) are suitable materials for this use because they reduce the internal resistance of the cell and boost the kinetic activity of the anode reaction.

The efficiency of the MFC is greatly influenced by the kind of membrane that is used in the cell. The anode chamber and cathode chamber are kept anaerobic and are separated by the membrane. Studies on other membrane types have been carried out as Nafion 117 is the most often used membrane, although it is somewhat expensive. In comparison to the Nafion membrane, graphene oxide, sulfonated polyether ketone, sulfonated silicon dioxide and sulfonated polystyrene ethylene butylene polystyrene membranes ensured a 55% and 107% increase in energy generation, respectively (Cydzik-Kwiatkowska and Nosek, 2022).

8.7. Challenges and future prospects

The increased demand for energy, wastewater treatment, heavy metal removal, bioremediation, environmental monitoring and even water supply problems, like desalination, can all be addressed by MFCs, demonstrating their flexibility as a technology. Despite their promise, the large-scale implementation of MFCs faces several technological and economic barriers. Key challenges include minimising energy losses during scale-up, reducing material costs and improving system stability for long-term operation.

One of the primary challenges in MFC scale-up is minimising energy losses while maintaining efficiency. Stacked MFC configurations have demonstrated significant potential in improving power output. However, achieving this power enhancement requires an optimal balance between increased efficiency and cost management. The expenses associated with upscaling MFCs rise considerably due to high material costs, particularly for electrodes and separators. Developing cost-effective designs that maintain high performance without significantly increasing operational expenses is essential for making MFCs a viable large-scale technology.

Catalysts are essential for improving the performance of the cathodic reaction in MFCs, especially the oxygen reduction reaction (ORR). Although Pt is the most effective catalyst for ORR, its high price and scarcity prevent it from being widely used in MFCs. To address this issue, researchers are exploring alternatives such as non-precious metal catalysts, metal-organic frameworks (MOFs), transition metal oxides and carbon-based catalysts. Additionally, improving catalyst stability by enhancing metal-support interactions and carbon support durability is necessary for long-term operation. Despite these advancements, further research is required to optimise alternative catalysts for real-world, continuous operation conditions.

The majority of MFCs use Nafion membranes as separators because of their great chemical stability and proton conductivity. However, these membranes are expensive and present several operational limitations, including biofouling and cation crossover which decrease system efficiency over prolonged use. Biofouling on the membrane increases internal resistance, reducing proton transfer rates and overall performance. Although operating MFCs without a separator could mitigate biofouling-related inefficiencies, it introduces new issues such as cathode contamination, further degrading system efficiency. Developing cost-effective and durable membrane alternatives, such as ion-exchange membranes, ceramic membranes or functionalised polymer membranes, is crucial for long-term MFC optimisation. Additionally, incorporating effective binders and advanced self-cleaning strategies at the cathode could help mitigate biofouling effects.

Another significant challenge in MFC operation is maintaining stable power output over extended periods. Thick biofilms at the anode can lead to diffusion limitations, restricting substrate availability and reducing electron transfer efficiency. While biofilm formation is essential for microbial electron transfer, excessive growth can hinder system performance. Strategies such as optimising electrode surface properties, implementing periodic biofilm control mechanisms and integrating self-regulating biofilm architectures are being explored to enhance long-term stability. Furthermore, research into microbial community dynamics can help identify robust electrogenic bacteria capable of sustaining stable energy generation under varying environmental conditions. To improve strain selection, biofilm conductivity and operating conditions, future studies on MFCs should concentrate on comprehending the metabolic processes of electrochemically active bacteria.

Improving MFC efficiency in wastewater treatment requires optimising reactor architecture and design. Practical scalability in MFC stacks requires addressing issues like voltage reversal and ionic short circuits brought on by biocatalysed electrode processes. The commercial viability of MFC technology will increase with the development of power gathering and utilisation systems. Furthermore, combining MFCs with other wastewater treatment techniques may improve overall treatment effectiveness and increase the range of real-world uses for them (He *et al.*, 2017).

8.8. Conclusions

A promising technology in the field of microbial environmental engineering, microbial fuel cells (MFCs) offer a sustainable way to clean wastewater and generate energy. The exploration of different types of MFCs, including single-chamber, double-chamber, up-flow, stacked and paper-based designs, showcases their versatility and adaptability for various applications. A key factor in maximising MFC performance is the choice of microorganisms and substrates, which directly influence the efficiency of bioelectricity generation and the effectiveness of pollutant degradation.

The fundamental working principles of MFCs emphasise their ability to treat wastewater while simultaneously producing electrical energy. Their integration into wastewater treatment facilities provides a dual advantage: mitigating environmental pollution and harnessing bioelectricity as a renewable energy source. The efficiency of MFCs is influenced by several factors, including electrode properties, pH, temperature and aeration, all of which must be carefully managed to maximise output and reliability.

Beyond energy generation and wastewater treatment, MFCs also facilitate the recovery of valuable byproducts such as hydrogen, biochemicals and nutrients, enhancing their economic viability. The efficiency and scalability of MFC technology have been greatly increased by recent developments, such as the use of innovative electrode materials, better reactor designs and genetic alterations of microbial consortia. However, despite these advancements, challenges such as high operational costs, electrode fouling, low power output and system scalability remain key barriers to widespread industrial implementation.

8.9. Recommendations

To enhance the practical application of MFCs in environmental engineering, research and development should focus on cost-effective and high-performance electrode materials, as well as novel proton exchange membranes to improve energy conversion efficiency while reducing costs.

Engineering robust microbial consortia with enhanced electron transfer capabilities through genetic modifications or selective adaptation can significantly improve MFC performance and stability. Integrating MFCs with existing wastewater treatment and renewable energy technologies, such as anaerobic digestion, constructed wetlands or solar energy systems, can further enhance their efficiency and energy recovery potential.

Additionally, pilot-scale studies and industrial trials are essential for assessing real-world feasibility and ensuring successful commercialisation. Advanced control strategies, automation and real-time monitoring systems should be implemented to maintain optimal operational conditions, reduce maintenance requirements and enhance long-term performance. Research, commercialisation and public awareness campaigns highlighting the economic and environmental advantages of MFCs should be supported by government policies and financing programmes in order to promote their wider use in municipalities and industry.

Lastly, comprehensive lifecycle assessments should be conducted to evaluate the environmental footprint and economic feasibility of MFCs, ensuring their long-term sustainability and competitiveness compared to conventional treatment technologies. MFC technology will develop if these issues are resolved, making it a practical option for producing renewable energy and treating wastewater sustainably.

REFERENCES

Aldrovandi A, Marsili E, Stante L *et al.* (2009) Sustainable power production in a membrane-less and mediator-less synthetic wastewater microbial fuel cell. *Bioresource Technology* **100(13)**: 3252–3260, 10.1016/j.biortech.2009.01.041.

Ali AH, Al-Mussawy HA, Hussein MJ and Hamadi NJ (2019) Comparison between conventional and modified microbial fuel cell for wastewater treatment and electricity generation. *International Journal of Environmental Science and Technology* **16(12)**: 8141–8150, 10.1007/s13762-019-02355-x.

Amend JP and Shock EL (2001) Energetics of overall metabolic reactions of thermophilic and hyperthermophilic Archaea and Bacteria. *FEMS Microbiology Reviews* **25(2)**: 175–243, 10.1111/j.1574-6976.2001.tb00576.x.

Angosto JM, Fernández-López JA and Godínez C (2015) Brewery and liquid manure wastewaters as potential feedstocks for microbial fuel cells: a performance study. *Environmental Technology* **36(1)**: 68–78, 10.1080/09593330.2014.937769.

Appleby AJ and Foulkes FR (1989) *Fuel Cell Handbook*. Van Nostrand Reinhold, New York, NY, USA.

Baby MG and Ahammed MM (2022) Nutrient removal and recovery from wastewater by microbial fuel cell-based systems – a review. *Water Science and Technology* **86(1)**: 29–55, 10.2166/wst.2022.196.

Bhattacharya A, Neena M and Chatterjee P (2024) Microbial nutrient recovery cell as an efficient and sustainable nutrient recovery option in sewage treatment. *Journal of Environmental Management* **366**: 121753, 10.1016/j.jenvman.2024.121753.

Bond DR, Holmes DE, Tender LM and Lovley DR (2002) Electrode-reducing microorganisms that harvest energy from marine sediments. *Science* **295(5554)**: 483–485, 10.1126/science.1066771.

Brown JL (2012) Fuel cells treat wastewater, generate electricity. *Civil Engineering Magazine* **82(7)**: 34–35, 10.1061/ciegag.0000643.

Brutinel ED and Gralnick JA (2012) Shuttling happens: soluble flavin mediators of extracellular electron transfer in Shewanella. *Applied Microbiology and Biotechnology* **93(1)**: 41–48, 10.1007/s00253-011-3653-0.

Chang IS, Jang JK, Gil GC *at al.* (2004) Continuous determination of biochemical oxygen demand using microbial fuel cell type biosensor. *Biosensors and Bioelectronics* **19(6)**: 607-613.

Cheng S, Liu H and Logan BE (2006) Increased performance of single-chamber microbial fuel cells using an improved cathode structure. *Electrochemistry Communications* **8(3)**: 489–494, 10.1016/j.elecom.2006.01.010.

Choi J and Ahn Y (2015) Enhanced bioelectricity harvesting in microbial fuel cells treating food waste leachate produced from biohydrogen fermentation. *Bioresource Technology* **183**: 53–60, 10.1016/j.biortech.2015.01.109.

Choi S, Kim JR, Cha J *et al.* (2013) Enhanced power production of a membrane electrode assembly microbial fuel cell (MFC) using a cost effective poly [2,5-benzimidazole] (ABPBI) impregnated non-woven fabric filter. *Bioresource Technology* **128**: 14–21, 10.1016/j.biortech.2012.10.013.

Choi Y, Jung E, Kim S and Jung S (2003) Membrane fluidity sensoring microbial fuel cell. *Bioelectrochemistry* **59(1–2)**: 121–127, 10.1016/S1567-5394(03)00018-5.

Cydzik-Kwiatkowska A and Nosek D (2022) Advances in microbial fuel cell technologies. *Energies* **15(16)**, p. 5958, 10.3390/en15165958.

Davis F and Higson SPJ (2007) Biofuel cells—recent advances and applications. *Biosensors and Bioelectronics* **22(7)**: 1224–1235, 10.1016/j.bios.2006.04.029.

Do MH, Ngo HH, Guo WS *et al.* (2018) Challenges in the application of microbial fuel cells to wastewater treatment and energy production: a mini review. *Science of The Total Environment* **639**: 910–920, 10.1016/j.scitotenv.2018.05.136.

Du Z, Li H and Gu T (2007) A state of the art review on microbial fuel cells: a promising technology for wastewater treatment and bioenergy. *Biotechnology Advances* **25(5)**: 464–482, 10.1016/j.biotechadv.2007.05.004.

Faria A, Gonçalves L, Peixoto JM *et al.* (2017) Resources recovery in the dairy industry: bioelectricity production using a continuous microbial fuel cell. *Journal of Cleaner Production* **140**: 971–976, 10.1016/j.jclepro.2016.04.027.

Flimban SGA, Ismail IMI, Kim T and Oh S-E (2019) Overview of recent advancements in microbial fuel cells from fundamentals to applications: design, major elements, and scalability. *Energies* **12(17)**: 3390. https://doi.org/10.3390/en12173390.

Goswami R and Mishra VK (2018) A review of design, operational conditions and applications of microbial fuel cells. *Biofuels* **9(2)**: 203–220, 10.1080/17597269.2017.1302682.

Habermann W and Pommer EH (1991) Biological fuel cells with sulphide storage capacity. *Applied Microbiology and Biotechnology* **35(1)**, 10.1007/BF00180650.

Hardt L, Owen A, Brockway P *et al.* (2018) Untangling the drivers of energy reduction in the UK productive sectors: efficiency or offshoring? *Applied Energy* **223**: 124–133, 10.1016/j.apenergy.2018.03.127.

Hashemi N, Lackore JM, Sharifi F *et al.* (2016) A paper-based microbial fuel cell operating under continuous flow condition. *Technology* **04(02)**: 98–103, 10.1142/S2339547816400124.

He L, Du P, Chen Y *et al.* (2017) Advances in microbial fuel cells for wastewater treatment. *Renewable Sustainable Energy Reviews* **71**: 388–403. https://doi.org/10.1016/j.rser.2016.12.069

Hernández-Fernández FJ, Pérez de los Ríos A, Salar-García MJ *et al.* (2015) Recent progress and perspectives in microbial fuel cells for bioenergy generation and wastewater treatment. *Fuel Processing Technology* **138**: 284–297, 10.1016/j.fuproc.2015.05.022.

Hernández-Flores G, Poggi-Varaldo HM, Solorza-Feria O *et al.* (2015) Characteristics of a single chamber microbial fuel cell equipped with a low cost membrane. *International Journal of Hydrogen Energy* **40(48)**: 17380–17387, 10.1016/j.ijhydene.2015.10.024.

Holzman DC (2005) Microbe Power! *Environmental Health Perspectives* **113(11)**, 10.1289/ehp.113-a754.

Ieropoulos IA, Greenman J, Melhuish C and Hart J (2005) Comparative study of three types of microbial fuel cell. *Enzyme and Microbial Technology* **37(2)**: 238–245, 10.1016/j.enzmictec.2005.03.006.

Jadhav DA, Ghosh Ray S and Ghangrekar MM (2017) Third generation in bio-electrochemical system research – a systematic review on mechanisms for recovery of valuable by-products from wastewater. *Renewable and Sustainable Energy Reviews* **76**: 1022–1031, 10.1016/j.rser.2017.03.096.

Jang JK, Pham TH, Chang IS *et al.* (2004) Construction and operation of a novel mediator- and membrane-less microbial fuel cell. *Process Biochemistry* **39(8)**: 1007–1012, 10.1016/S0032-9592(03)00203-6.

Karthick S and Haribabu K (2020) Bioelectricity generation in a microbial fuel cell using polypyrrole-molybdenum oxide composite as an effective cathode catalyst. *Fuel* **275**: 117994, 10.1016/j.fuel.2020.117994.

Khan MR, Baranitharan E, Prasad DMR and Cheng CK (2016) Fast biofilm formation and its role on power generation in palm oil mill effluent fed microbial fuel cell. *MATEC Web of Conferences* **62**: 04002, 10.1051/matecconf/20166204002.

Khater DZ, El-Khatib KM and Hassan HM (2017) Microbial diversity structure in acetate single chamber microbial fuel cell for electricity generation. *Journal of Genetic Engineering and Biotechnology* **15(1)**: 127–137, 10.1016/j.jgeb.2017.01.008

Kim BH, Chang IS, Gil GC, Park HS and BOD HKN (2003) Novel BOD (biological oxygen demand) sensor using mediator-less microbial fuel cell. *Biotechnology Letters* **25(7)**: 541–545.

Kim JR, Min B and Logan BE (2005) Evaluation of procedures to acclimate a microbial fuel cell for electricity production. *Applied Microbiology and Biotechnology* **68(1)**: 23–30, 10.1007/s00253-004-1845-6.

Kouam Ida T and Mandal B (2023) Microbial fuel cell design, application and performance: a review. *Materials Today: Proceedings* **76**: 88–94, 10.1016/j.matpr.2022.10.131.

Kumar R, Singh L and Zularisam AW (2017) Microbial fuel cells: types and applications. In *Waste Biomass Management – A Holistic Approach*. Springer, Cham, Switzerland, pp. 367–384, 10.1007/978-3-319-49595-8_16.

Kumar SS, Kumar V, Gnaneswar Gude V, Malyan SK and Pugazhendhi A (2020) Alkalinity and salinity favor bioelectricity generation potential of Clostridium, Tetrathiobacter and Desulfovibrio consortium in Microbial Fuel Cells (MFC) treating sulfate-laden wastewater. *Bioresource Technology* **306**, p. 123110, 10.1016/j.biortech.2020.123110.

Li W, Zhang S, Chen G and Hua Y (2014) Simultaneous electricity generation and pollutant removal in microbial fuel cell with denitrifying biocathode over nitrite. *Applied Energy* **126**: 136–141, 10.1016/j.apenergy.2014.04.015.

Li W-W, Yu H-Q and He Z (2013) Towards sustainable wastewater treatment by using microbial fuel cells-centered technologies. *Energy & Environmental Science* **7(3)**: 911–924, 10.1039/C3EE43106A.

Liu H and Logan BE (2004) Electricity generation using an air-cathode single chamber microbial fuel cell in the presence and absence of a proton exchange membrane. *Environmental Science & Technology* **38(14)**: 4040–4046, 10.1021/es0499344.

Pisciotta JM and Dolceamore Jr JJ (2016) Bioelectrochemical and Conventional Bioremediation of Environmental Pollutants. *Journal of Microbial and Biochemical Technology* **8(4)**: 327–343. https://doi.org/10.4172/1948-5948.1000306

Potter MC (1911) Electrical effects accompanying the decomposition of organic compounds. *Proceedings of the Royal Society of London. Series B, Containing Papers of a Biological Character* **84(571)**: 260–276, 10.1098/rspb.1911.0073.

Mansoorian HJ, Mahvi AH, Jafari AJ *et al.* (2013) Bioelectricity generation using two chamber microbial fuel cell treating wastewater from food processing. *Enzyme and Microbial Technology* **52(6–7)**: 352–357, 10.1016/j.enzmictec.2013.03.004.

Margaria V, Tommasi T, Pentassuglia S *et al.* (2017) Effects of pH variations on anodic marine consortia in a dual chamber microbial fuel cell. *International Journal of Hydrogen Energy* **42(3)**: 1820–1829, 10.1016/j.ijhydene.2016.07.250.

Mathuriya AS and Yakhmi JV (2014) Microbial fuel cells to recover heavy metals. *Environmental Chemistry Letters* **12(4)**: 483–494, 10.1007/s10311-014-0474-2.

Mehmood MK, Adetutu E, Nedwell DB and Ball AS (2009) In situ microbial treatment of landfill leachate using aerated lagoons. *Bioresource Technology* **100(10)**: 2741–2744, 10.1016/j.biortech.2008.11.031.

Min B, Cheng S and Logan BE (2005a) Electricity generation using membrane and salt bridge microbial fuel cells. *Water Research* **39(9)**: 1675–1686, 10.1016/j.watres.2005.02.002.

Min B, Kim J, Oh S, Regan JM and Logan BE (2005b) Electricity generation from swine wastewater using microbial fuel cells. *Water Research* **39(20)**: 4961–4968, 10.1016/j.watres.2005.09.039.

Miran W, Nawaz M, Jang J and Lee DS (2016a) Conversion of orange peel waste biomass to bioelectricity using a mediator-less microbial fuel cell. *Science of The Total Environment* **547**: 197–205, 10.1016/j.scitotenv.2016.01.004.

Miran W, Nawaz M, Jang J and Lee DS (2016b) Sustainable electricity generation by biodegradation of low-cost lemon peel biomass in a dual chamber microbial fuel cell. *International Biodeterioration & Biodegradation* **106**: 75–79, 10.1016/j.ibiod.2015.10.009.

Mohan Y, Manojmuthukumar S and Das D (2008) Electricity generation using microbial fuel cells. *International Journal of Hydrogen Energy* **33(1)**: 423–426, 10.1016/j.ijhydene.2007.07.027.

Nawaz A, ul Haq I, Qaisar K *et al.* (2022) Microbial fuel cells: insight into simultaneous wastewater treatment and bioelectricity generation. *Process Safety and Environmental Protection* **161**: 357–373, 10.1016/j.psep.2022.03.039.

Niessen J, Harnisch F, Rosenbaum M, Schroder U and Scholz F (2006) Heat treated soil as convenient and versatile source of bacterial communities for microbial electricity generation. *Electrochemistry Communications* **8(5)**: 869–873, 10.1016/j.elecom.2006.03.025.

Niessen J, Schroder U and Scholz F (2004) Exploiting complex carbohydrates for microbial electricity generation: a bacterial fuel cell operating on starch. *Electrochemistry Communications* **6(9)**: 955–958, 10.1016/j.elecom.2004.07.010.

Ogugbue C, Ebode E and Leera S (2015) Electricity generation from swine wastewater using microbial fuel cell. *Journal of Ecological Engineering* **16**: 26–33, 10.12911/22998993/60450.

Oliveira VB, Simões M, Melo LF and Pinto AMFR (2013) Overview on the developments of microbial fuel cells. *Biochemical Engineering Journal* **73**: 53–64, 10.1016/j.bej.2013.01.012.

Park DH, Laivenieks M, Guettler MV, Jain MK and Zeikus JG (1999) Microbial utilization of electrically reduced neutral red as the sole electron donor for growth and metabolite production. *Applied and Environmental Microbiology* **65(7)**: 2912–2917, 10.1128/AEM.65.7.2912-2917.1999.

Park DH and Zeikus JG (2003) Improved fuel cell and electrode designs for producing electricity from microbial degradation. *Biotechnology and Bioengineering* **81(3)**: 348–355, 10.1002/bit.10501.

Pham TH, Rabaey K, Aelterman P *et al.* (2006) Microbial fuel cells in relation to conventional anaerobic digestion technology. *Engineering in Life Sciences* **6(3)**: 285–292, 10.1002/elsc.200620121.

Phung NT, Lee J, Kang KH *et al.* (2004) Analysis of microbial diversity in oligotrophic microbial fuel cells using 16S rDNA sequences. *FEMS Microbiology Letters* **233(1)**: 77–82, 10.1016/j.femsle.2004.01.041.

Pisciotta JM and Dolceamore Jr JJ (2016) Bioelectrochemical and conventional bioremediation of environmental pollutants. *Journal of Microbial & Biochemical Technology* **8(4)**, 10.4172/1948-5948.1000306.

Potrykus S, León-Fernández LF, Nieznański J, Karkosiński D and Fernandez-Morales FJ (2021) The influence of external load on the performance of microbial fuel cells. *Energies* **14(3)**: 612.

Rabaey K and Verstraete W (2005) Microbial fuel cells: novel biotechnology for energy generation. *Trends in Biotechnology* **23(6)**: 291–298, 10.1016/j.tibtech.2005.04.008.

Rahimnejad M, Adhami A, Darvari S, Zirepour A and Oh S-E (2015) Microbial fuel cell as new technology for bioelectricity generation: a review. *Alexandria Engineering Journal* **54(3)**: 745–756, 10.1016/j.aej.2015.03.031.

Rahimnejad M, Ghoreyshi AA, Najafpour GD, Younesi H and Shakeri M (2012) A novel microbial fuel cell stack for continuous production of clean energy. *International Journal of Hydrogen Energy* **37(7)**: 5992–6000, 10.1016/j.ijhydene.2011.12.154.

Rhoads A, Beyenal H and Lewandowski Z (2005) Microbial fuel cell using anaerobic respiration as an anodic reaction and biomineralized manganese as a cathodic reactant. *Environmental Science & Technology* **39(12)**: 4666–4671, 10.1021/es048386r.

Saba B, Christy AD, Yu Z and Co AC (2017) Sustainable power generation from bacterio-algal microbial fuel cells (MFCs): an overview. *Renewable and Sustainable Energy Reviews* **73**: 75–84, 10.1016/j.rser.2017.01.115.

Saeed HM, Husseini GA, Yousef S *et al.* (2015) Microbial desalination cell technology: a review and a case study. *Desalination* **359**: 1–13, 10.1016/j.desal.2014.12.024.

Sasaki D, Sasaki K, Tsuge Y and Kondo A (2016) Comparative metabolic state of microflora on the surface of the anode electrode in a microbial fuel cell operated at different pH conditions. *AMB Express* **6(1)**: 125, 10.1186/s13568-016-0299-4.

Schröder U (2007) Anodic electron transfer mechanisms in microbial fuel cells and their energy efficiency. *Physical Chemistry Chemical Physics* **9(21)**: 2619–2629.

Shin J-W, Seo S-J, Maitlo HA and Park J-Y (2015) The enhancement of ammonium removal from ethanolamine wastewater using air-cathode microbial fuel cells coupled to ferric reduction. *Bioresource Technology* **190**: 466–473, 10.1016/j.biortech.2015.03.048.

Sincero AP and Sincero GA (2002) *Physical-Chemical Treatment of Water and Wastewater*, CRC Press, Boca Raton, FL, USA, 10.1201/9781420031904.

Slate AJ, Whitehead KA, Brownson DAC and Banks CE (2019) Microbial fuel cells: an overview of current technology. *Renewable and Sustainable Energy Reviews* **101**: 60–81, 10.1016/j.rser.2018.09.044.

Sonawane JM, Marsili E and Chandra Ghosh P (2014) Treatment of domestic and distillery wastewater in high surface microbial fuel cells. *International Journal of Hydrogen Energy* **39(36)**: 21819–21827, 10.1016/j.ijhydene.2014.07.085.

Surya Ramadan B and Purwono (2017) Challenges and opportunities of microbial fuel cells (MFCs) technology development in Indonesia. *MATEC Web of Conferences* **101**: 02018, 10.1051/matecconf/201710102018.

Trapero JR, Horcajada L, Linares JJ and Lobato J (2017) Is microbial fuel cell technology ready? An economic answer towards industrial commercialization. *Applied Energy* **185**: 698–707, 10.1016/j.apenergy.2016.10.109.

Tremouli A, Martinos M and Lyberatos G (2017) The effects of salinity, pH and temperature on the performance of a microbial fuel cell. *Waste and Biomass Valorization* **8(6)**: 2037–2043, 10.1007/s12649-016-9712-0.

Tremouli A, Vlassis T, Antonopoulou G and Lyberatos G (2016) Anaerobic degradation of pure glycerol for electricity generation using a MFC: the effect of substrate concentration. *Waste and Biomass Valorization* **7(6)**: 1339–1347, 10.1007/s12649-016-9498-0.

Varjani SJ (2017) Microbial degradation of petroleum hydrocarbons. *Bioresource Technology* **223**: 277–286, 10.1016/j.biortech.2016.10.037.

Vega CA and Fernández I (1987) Mediating effect of ferric chelate compounds in microbial fuel cells with Lactobacillus plantarum, Streptococcus lactis and Erwinia dissolvens. *Bioelectrochemistry and Bioenergetics* **17(2)**: 217–222, 10.1016/0302-4598(87)80026-0.

Venkata Mohan S, Velvizhi G, Annie Modestra J and Srikanth S (2014) Microbial fuel cell: critical factors regulating bio-catalyzed electrochemical process and recent advancements. *Renewable and Sustainable Energy Reviews* **40**: 779–797, 10.1016/j.rser.2014.07.109.

Vilajeliu-Pons A, Bañeras L, Puig S *et al.* (2016) External resistances applied to MFC affect core microbiome and swine manure treatment efficiencies. *PLOS ONE* **11(10)**: e0164044, 10.1371/journal.pone.0164044.

Wang Y, Wang A, Zhou A *et al.* (2014) Electrode as sole electrons donor for enhancing decolorization of azo dye by an isolated Pseudomonas sp. WYZ-2. *Bioresource Technology* **152**: 530–533, 10.1016/j.biortech.2013.11.001.

Xia C, Zhang D, Pedrycz W, Zhu Y and Guo Y (2018) Models for microbial fuel cells: a critical review. *Journal of Power Sources* **373**: 119–131, 10.1016/j.jpowsour.2017.11.001.

Zeng Y-F, Liu Z-L and Qin Z-Z (2009) Decolorization of molasses fermentation wastewater by SnO2-catalyzed ozonation. *Journal of Hazardous Materials* **162(2–3)**: 682–687, 10.1016/j.jhazmat.2008.05.094.

Zhang L, Wang J, Fu G and Zhang Z (2020) Simultaneous electricity generation and nitrogen and carbon removal in single-chamber microbial fuel cell for high-salinity wastewater treatment. *Journal of Cleaner Production* **276**, p. 123203, 10.1016/j.jclepro.2020.123203.

Zhou M, Wang H, Hassett DJ and Gu T (2013) Recent advances in microbial fuel cells (MFCs) and microbial electrolysis cells (MECs) for wastewater treatment, bioenergy and bioproducts. *Journal of Chemical Technology & Biotechnology* **88(4)**: 508–518, 10.1002/jctb.4004.

Microbial Processes in Environmental Engineering: From Theory to Practice

Rajneesh Kumar, Ram Sharan Singh and Maulin P Shah
ISBN 978-1-83708-327-5
https://doi.org/10.1108/978-1-83708-326-820261009

Chapter 9

Microbial biosensors for environmental monitoring and sustainability

Padmapriya Subbiah[1], Selvarani Subbiah[2] and Sasikala Subbiah[3]

[1]*Department of Marine and Coastal Studies, Madurai Kamaraj University, Madurai, India*
[2]*Thiagarajar College (Autonomous), Affiliated to Madurai Kamaraj University, Madurai, India*
[3]*Sree Krishna College of Engineering Unai, Anaicut, Vellore, India*

Abstract

Microbial biosensors, which utilise the biological responses of microorganisms to detect pollutants with high specificity and sensitivity, have emerged as indispensable tools for environmental biomonitoring. These are analytical devices based on the integration of biological components with electrochemical, optical or microbial fuel cell transducers for detecting environmental contaminants such as heavy metals, pesticides, toxins and organic waste. Providing real-time, onsite detection, biosensors play essential roles in numerous fields such as environmental monitoring, food safety and clinical diagnostics. Based on this versatility, different organisms, from bacteria, plants and lichens to small mammals, can be used to detect pollutants and measure their ecological impact. Apart from agricultural chemicals, these devices can also quantify mycotoxins, phenols and cyanides, ideally suited for measuring agricultural chemicals along with environmental contaminants. Challenges remain in the scaling from laboratory to field applications, regulatory acceptance and commercialisation of biosensor technologies, despite many of the advantages biosensors afford. This chapter will describe the functioning, application and limitations of microbial biosensors, focusing on the environmental field through monitoring of soil contamination and ensuring ecological safety. This chapter explores the literature through case studies that showcase biosensor performance for quantifying water quality, soil health and air pollution, while addressing opportunities for biosensors harnessed with smart technologies and data analytics in the near future. This chapter emphasises what functions biosensors are capable of playing, to what extent they provide information on their functionality and their potential importance in attaining environmental sustainability.

Keywords: Biosensors; biomonitoring; environmental contaminants; sustainability; pollution.

9.1. Introduction

Microbial biosensors are sophisticated analytical instruments that integrate biological components, typically microorganisms or their derivatives, to detect and quantify environmental contaminants. These sensors translate biological interactions into quantifiable electrical signals by way of transducer mechanisms, rendering them indispensable tools for environmental biomonitoring aligned with sustainable development goals (SDGs) (Huang *et al.*, 2023; Naresh and Lee, 2021).

The significance of environmental biomonitoring is paramount, as it underpins pollution assessment, ecosystem protection and sustainable resource utilisation in alignment with SDGs 6 (Clean Water and Sanitation), 12 (Responsible Consumption and Production), 13 (Climate Action), 14 (Life Below Water) and 15 (Life on Land) (Huang *et al.*, 2023; Yun *et al.*, 2022). The integration of advanced biosensor technologies empowers stakeholders to proactively track pollutant levels across diverse environmental matrices, thereby preserving biodiversity and mitigating detrimental ecological consequences.

Biosensors comprise essential components, including biological elements such as enzymes, antibodies or whole cells that selectively interact with target analytes, and transducers that convert these interactions into interpretable signals. Various transducer types, including electrochemical, optical and mass-sensitive systems, offer versatility in detecting and assessing environmental pollutants (Naresh and Lee, 2021). The effectiveness of these biosensors is contingent on their structural design, the selection of biological recognition elements and the transduction methodology employed.

Microbial biosensors have been extensively developed to target key environmental pollutants, such as heavy metals including zinc and cadmium, organic pollutants and pathogens (Mishra *et al.*, 2018; Yun *et al.*, 2022). The application of transgenic organisms, such as zebrafish, used within biosensor platforms has demonstrated significant potential for real-time environmental health assessments (Liu *et al.*, 2016). Furthermore, innovations in biosensor technology, including whole-cell biosensor systems and nanostructured materials, have expanded their applicability, enabling the detection of a broad spectrum of contaminants, while offering cost-efficient solutions for continuous environmental monitoring (Lu *et al.*, 2015; Tankus *et al.*, 2024).

As illustrated in Figure 9.1, biosensors consist of key components, including analytes derived from food, environmental and human samples; bioreceptors such as nucleic acids, whole cells, enzymes and antibodies; and transducers including electrodes, nanowire arrays and nanoparticles. These components work synergistically to facilitate efficient and accurate detection, reinforcing the critical role of microbial biosensors in environmental monitoring and sustainability initiatives. Microbial biosensors constitute a pivotal advancement in environmental monitoring, contributing substantially to realising SDGs. Their sophisticated design and operational efficacy enhance our capacity to detect hazardous pollutants, ensuring that sustainable development initiatives uphold water quality, ecological integrity and climate stability.

Figure 9.1 Schematic representation of biosensor components and applications in environmental monitoring, emphasising their role in achieving SDGs

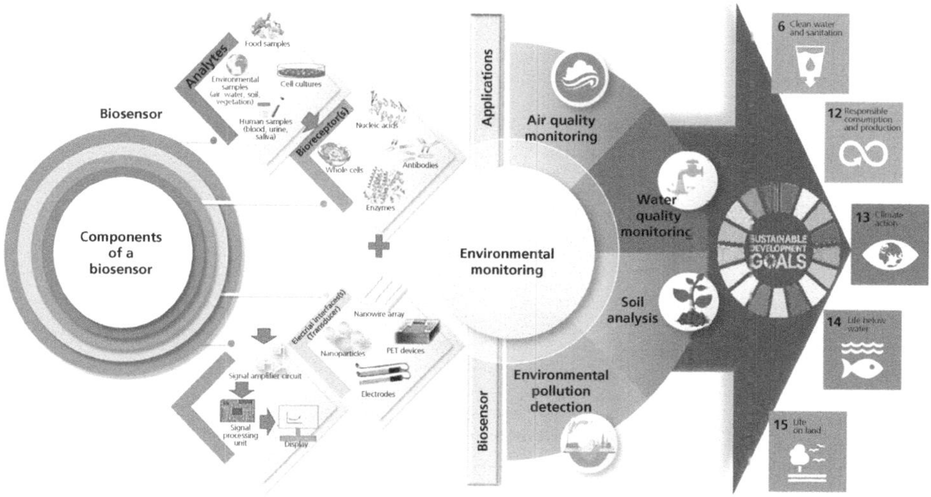

9.2. Principles of microbial biosensors

Microbial biosensors operate on fundamental principles that integrate biological recognition with transduction technologies, enabling the detection of specific environmental contaminants. Understanding the structure and working mechanisms of microbial biosensors is essential for their effective application in biomonitoring. As illustrated in Figure 9.2, biosensors can be classified based on their bioreceptors, transducers, detection systems and technology, providing a comprehensive framework for diverse applications.

9.2.1 Basic structure and working mechanism

The core structure of a microbial biosensor comprises two essential components: biological recognition elements and transducers. The biological elements, primarily microorganisms, selectively respond to targeted pollutants, while the transducer converts these biological responses into quantifiable signals, allowing for real-time environmental monitoring and assessment (Qi *et al.*, 2014).

9.2.1.1 Microorganisms as biological recognition elements

Microorganisms function as highly efficient biological recognition elements in biosensors due to their ability to specifically interact with environmental contaminants. As living entities, they metabolise or bind to various pollutants, generating detectable outputs such as electrical signals (Qi *et al.*, 2014). The use of whole cells, microbial communities or genetically modified strains enhances biosensor sensitivity and specificity, enabling the detection of a

Figure 9.2 Classification of biosensors based on bioreceptors, transducers, detection systems and technology (Concept adapted from Padmapriya *et al.*, 2024)

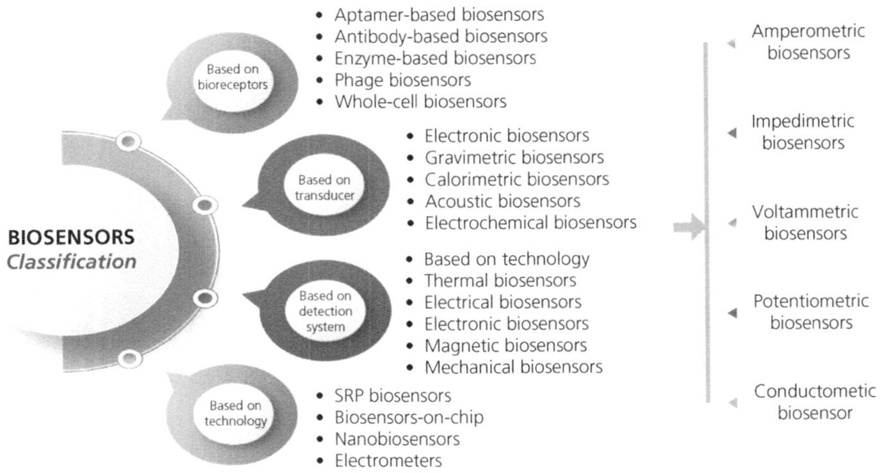

broad spectrum of pollutants, including heavy metals and organic compounds (Andriukonis *et al.*, 2021).

9.2.1.2 Types of transducers

Microbial biosensors employ diverse transduction mechanisms to convert biological responses into measurable signals. The primary transducer types include electrochemical, optical and microbial fuel cell-based transducers, each offering distinct advantages depending on the application. Figure 9.2 shows classification based on the transducer and further categorises electrochemical transducers into amperometric, potentiometric, voltammetric, impedimetric and conductometric biosensors (Padmapriya *et al.*, 2024), and they are detailed below.

(*a*) *Electrochemical transducers*. These transducers detect variations in electrical properties, such as current, potential or impedance, resulting from microbial activity. Their high sensitivity and capability to deliver real-time quantitative data make them particularly valuable in environmental monitoring – for example, amperometric biosensors utilising microbial cells effectively measure sulfide concentrations, demonstrating their potential for monitoring sulfate-reducing bacteria in environmental samples (Qi *et al.*, 2014).

(*b*) *Optical transducers*. Optical biosensors rely on changes in light properties, such as absorbance, fluorescence or luminescence, induced by microbial interactions. Their primary advantage lies in delivering high-resolution detection, which is especially useful for analysing complex environmental samples with multiple pollutants (Cui *et al.*, 2019).

(*c*) *Microbial fuel cell-based transducers*. Microbial fuel cell (MFC) biosensors leverage microbial metabolism to generate electricity, which serves as the output signal.

This approach is particularly advantageous for in situ water quality monitoring, where real-time detection of organic pollutants is critical. MFC biosensors facilitate electron transfer from microorganisms to an external circuit, directly correlating biological activity with electrical current generation (Peixoto *et al.*, 2011).

The principles governing microbial biosensors underscore the innovative integration of biological recognition and technological transduction methods for environmental monitoring. The strategic selection of microorganisms as recognition elements, coupled with diverse transducer types, ensures the effective detection of pollutants vital for sustaining ecological health.

9.2.2 Specificity and sensitivity in pollutant detection

The efficacy of microbial biosensors is predominantly contingent on their capacity to exhibit high specificity and sensitivity in pollutant detection. Specificity pertains to the biosensor's ability to selectively identify target contaminants without interference from extraneous substances, whereas sensitivity denotes its capability to detect minimal concentrations of these analytes (Long *et al.*, 2013). Attaining both high specificity and sensitivity is imperative for robust environmental monitoring, particularly within complex matrices where numerous potential interferents may be present.

Microbial biosensors, particularly those employing whole cells as recognition elements, can be genetically or biochemically engineered to detect specific pollutants by leveraging intrinsic affinity mechanisms, enzymatic activities or metabolic pathways – for example, biosensors based on cyanobacteria have been developed to detect heavy metals and pesticides with remarkable specificity, utilising the organism's inherent biochemical pathways (Wong *et al.*, 2013). This exploitation of biological interactions substantially enhances detection limits while minimising false positives, thereby ensuring precise assessment of pollution levels.

Moreover, advancements in biosensing technologies, such as the incorporation of nanomaterials and other enhancement strategies, have significantly augmented sensitivity. The integration of electrochemical transducers, for example, has facilitated the development of biosensors capable of detecting pollutants at nanomolar concentrations, paving the way for portable devices capable of real-time environmental monitoring (Goradel *et al.*, 2017). Ultimately, the continuous evolution of biosensor designs and their analytical capabilities will be instrumental in addressing the escalating environmental challenges posed by pollution.

9.2.3 Role of genetically engineered microorganisms in enhancing biosensor performance

The advent of genetically engineered microorganisms has revolutionised microbial biosensors, augmenting their performance and broadening their applicability for environmental monitoring. Through the application of synthetic biology techniques, researchers can modify microbial strains to enhance their responsiveness to specific pollutants, thereby enabling more precise and reliable detection methodologies. These engineered microorganisms can be designed to express reporter genes that generate detectable outputs such as fluorescence or luminescence on interaction with target contaminants (He *et al.*, 2015) – for instance, genetically modified bacterial strains have been developed to detect environmental contaminants like arsenic by incorporating genetic circuits that elicit a measurable signal on exposure (Li *et al.*, 2015).

This approach allows for customised specificity and sensitivity, facilitating the detection of a diverse spectrum of pollutants across varying environmental conditions. Biosensor performance can be further optimised through promoter engineering, which enables the fine-tuning of gene expression levels in response to specific pollutant concentrations (Wang *et al.*, 2023), thereby refining the detection dynamic range with greater precision.

The synergy between genetic engineering and cutting-edge biosensing technologies fosters the development of highly sophisticated biosensors capable of functioning effectively in intricate environmental scenarios. By catering to the specific demands of diverse pollutant detection, these engineered microorganisms represent a pivotal advancement in the pursuit of efficient and sustainable environmental monitoring solutions.

9.3. Applications in environmental monitoring

Microbial biosensors have become indispensable in environmental surveillance, particularly in detecting a wide range of contaminants, including heavy metals. These biosensing technologies harness the intrinsic biochemical pathways of microorganisms to enable rapid, highly sensitive and specific pollutant detection across various environmental matrices, thereby contributing to public health protection and ecosystem sustainability.

9.3.1 Detection of environmental contaminants

9.3.1.1 Heavy metals

The contamination of ecosystems with heavy metals poses a profound environmental threat due to their toxic effects on both human health and biodiversity. Microbial biosensors have been designed to efficiently detect hazardous metals such as lead (Pb), cadmium (Cd), zinc (Zn) and mercury (Hg) in environmental samples. Liu *et al.* (2016) described the development of a transgenic zebrafish biosensor engineered to detect zinc and cadmium in aquatic environments using enhanced green fluorescent protein (eGFP) technology. This biosensor facilitates real-time monitoring of these metals, demonstrating the feasibility of employing genetically modified organisms to assess water quality and potential toxicological impacts.

A biosensor platform utilising eGFP was introduced by Lee *et al.* (2019) for cadmium detection in environmental samples. Through genetic modification of *Escherichia coli*, they constructed a biosensor capable of responding to cadmium exposure, thereby illustrating the effectiveness of split-protein systems as advanced biosensing tools. These systems not only exhibit high sensitivity but also offer adaptability for various detection applications, rendering them particularly suitable for environmental monitoring.

Furthermore, Hui *et al.* (2021) reported the development of indigoidine-based biosensors designed to detect the presence of heavy metals. These whole-cell biosensors provide a visual indication of metal toxicity, representing a pioneering approach for monitoring environmental water samples, particularly for bioavailable metal fractions that pose the greatest risk to living organisms. The ability to selectively monitor toxic metal concentrations is instrumental in evaluating environmental hazards and implementing necessary safety measures.

Microbial biosensors present substantial advantages in the detection of heavy metals by integrating genetic engineering strategies and biochemical pathways to enhance specificity, sensitivity and real-time analytical capabilities. These biosensing tools play a crucial role in assessing environmental resource quality, ensuring adherence to health and safety regulations.

9.3.1.2 Pesticides and agricultural chemicals

Pesticides, widely employed in agriculture to manage insects, weeds and fungi, pose significant environmental and health concerns due to their extensive application (Febbraio *et al.*, 2011). The integration of microbial biosensors into environmental monitoring systems plays a pivotal role in detecting these chemical agents within ecological matrices, facilitating effective surveillance and sustainable agricultural management (Zamora-Sequeira *et al.*, 2019).

Febbraio *et al.* (2011) engineered a thermostable esterase derived from *Alicyclobacillus acidocaldarius*, functioning as a biosensor for organophosphate pesticide detection. This biosensor capitalises on the enzyme's specificity toward these compounds, enabling rapid and precise quantification. The development of such enzymatic biosensors highlights the potential of microbial components in advancing highly sensitive detection methodologies applicable across diverse agricultural settings.

Nguyen and Jang (2021) introduced a cationic surfactant-decorated liquid crystal-based aptasensor for malathion, a commonly used organophosphate pesticide. This biosensor employed a label-free detection mechanism, ensuring swift and reliable results, which are particularly advantageous for onsite agricultural assessments. Its pronounced specificity for malathion and its ability to differentiate target compounds from potential interferents underscore the necessity of developing selective biosensing technologies for pesticide analysis.

Zamora-Sequeira *et al.* (2019) conducted a comprehensive review of sensor-based methodologies for pesticide quantification in agricultural applications, emphasising the effectiveness of colorimetric and electrochemical detection techniques. This review accentuated the adaptability of biosensors in monitoring a broad spectrum of pesticides, reinforcing their significance in environmental surveillance strategies.

Despite these advancements, mitigating pesticide residues in the environment remains a critical challenge due to their toxicity and persistence. Biopesticides, derived from natural biological sources, present a sustainable alternative. By addressing the detrimental effects of synthetic pesticides, microbial biopesticides contribute to restoring ecological balance within agricultural ecosystems (Samada and Tamburan, 2020). The role of microbial biosensors in monitoring both conventional pesticides and emerging biopesticides will be instrumental in promoting sustainable agricultural practices while minimising ecological harm.

Microbial biosensors represent a paradigm shift in the detection of pesticides and agrochemicals within environmental monitoring frameworks. Their exceptional specificity and sensitivity support regulatory compliance and best practices, ultimately fostering safer agricultural environments and safeguarding public health.

9.3.1.3 Organic pollutants: phenols and cyanides

Microbial biosensors are emerging as pivotal tools in environmental surveillance, particularly for identifying organic pollutants such as phenols and cyanides, which are major contributors to water and soil contamination. The hazardous nature of these pollutants necessitates efficient detection strategies to evaluate and mitigate their environmental impact.

Phenolic compounds and cyanides are well-documented for their toxicity and detrimental effects on living organisms. Phenols, extensively utilised in industrial processes, tend to accumulate in the environment, leading to persistent contamination of soil and water. Microbial biosensors have been engineered to facilitate the rapid detection of phenolic pollutants, enabling real-time monitoring of contamination levels. Recent research has focused on designing biosensors incorporating microbial strains capable of degrading phenolic compounds (Ren *et al.*, 2022).

Cyanides, frequently encountered in industrial activities such as mining and electroplating, pose critical environmental hazards due to their acute toxicity to aquatic ecosystems and severe health implications for humans. Biosensors tailored for cyanide detection typically utilise microorganisms exhibiting selective responses to cyanide ions. Genetically modified strains of *Escherichia coli* and other microbial species have been investigated for their potential in developing highly sensitive biosensing platforms that generate quantifiable responses on cyanide exposure. Nonetheless, the references provided do not sufficiently substantiate these claims (Abdel-Razzaq *et al.*, 2023).

Recent advancements in biosensing technologies have explored the integration of advanced materials, including magnetic nanoparticles and carbon-based nanomaterials, to enhance biosensor performance in detecting organic pollutants. However, the cited studies do not specifically address the application of these innovations for phenol and cyanide detection (Mirasole *et al.*, 2016; Qu *et al.*, 2017).

Additionally, passive sampling methodologies have gained attention as an emerging approach for detecting organic pollutants. Although research has investigated environmental monitoring techniques for organic contaminants, the reference cited does not provide concrete support for the role of passive sampling in detecting phenols and cyanides (Sun *et al.*, 2019).

While microbial biosensors hold substantial promise for monitoring organic pollutants, the current body of literature does not consistently support the specific claims regarding their application and advancements in phenol and cyanide detection. Further targeted research is required to validate these assertions and strengthen the scientific foundation for their environmental deployment.

9.3.1.4 Mycotoxins

Microbial biosensors are gaining prominence as highly effective instruments for detecting mycotoxins, which are toxic secondary metabolites produced by fungi such as *Aspergillus*, *Penicillium* and *Fusarium*. The presence of mycotoxins in food and animal feed presents serious threats to public health and food security, necessitating stringent monitoring protocols to ensure agricultural safety (Fiorbelli *et al.*, 2024).

The identification of mycotoxins in food, feed and environmental samples is critical due to their severe toxicological effects. Among the most concerning mycotoxins are aflatoxins, ochratoxin A, fumonisins and deoxynivalenol (DON), each of which poses distinct health risks to humans and animals (Gámiz-Gracia *et al.*, 2020). Their capacity to induce both acute and chronic toxicity underscores the urgency of developing reliable detection and mitigation systems (Greeff-Laubscher *et al.*, 2019; Rodríguez *et al.*, 2012).

Advancements in microbial biosensor technology have focused on engineering biosensors capable of targeting specific mycotoxins across diverse matrices. *Bacillus subtilis* strains, for example, have demonstrated potential in mycotoxin detoxification and can be integrated into biosensing applications (Greco *et al.*, 2025). Harnessing the natural biochemical interactions of microorganisms with these toxic compounds offers a novel approach to both detection and remediation.

In parallel, sophisticated analytical techniques such as liquid chromatography-tandem mass spectrometry (LC-MS/MS) have been explored for quantifying mycotoxin levels in environmental samples (Gámiz-Gracia *et al.*, 2020; Kappenberg and Juraschek, 2021). These techniques enhance microbial biosensing by providing precise quantitative data, enabling comprehensive assessments of food safety and contamination levels.

Environmental factors, particularly temperature variations, significantly influence mycotoxin prevalence. Studies highlight the impact of rising temperatures and humidity fluctuations on fungal mycotoxin production, increasing contamination risks in agricultural settings (Fels-Klerx *et al.*, 2016). Developing biosensors capable of adapting to such environmental shifts will be instrumental in ensuring effective monitoring and early detection of mycotoxin outbreaks.

Microbial biosensors represent a transformative solution for mycotoxin surveillance across food and environmental sectors. Their rapid detection capabilities not only facilitate timely intervention but also contribute to the formulation of strategic measures aimed at minimising health hazards associated with mycotoxin contamination.

9.3.2 Environmental quality assessment

Microbial biosensors and advanced monitoring technologies play a crucial role in assessing environmental quality, particularly in water, soil and air. Evaluating these parameters is fundamental to maintaining ecological balance, safeguarding public health and promoting sustainable agricultural systems.

9.3.2.1 Water quality assessment: pollutants and toxins

Ensuring water quality through the detection of pollutants and toxins is essential, given the hazards posed by contaminants such as heavy metals, pesticides and microbial pathogens. Microbial biosensors present a viable approach for real-time identification of these hazardous substances across different aquatic environments (Bruyn and Andrews, 2016). Genetically engineered microorganisms, for instance, offer highly sensitive and selective detection mechanisms, which are critical for securing potable water supplies and preserving aquatic biodiversity.

Advancements in sensor technology, particularly biosensors incorporating nanotechnology, have significantly improved pollutant monitoring capabilities (Wilhelm *et al.*, 2022). Studies emphasise the integration of sophisticated sensor platforms for detecting organic pollutants, phenols and mycotoxins in surface and groundwater systems (Head *et al.*, 2020). Consistent and precise monitoring of these contaminants supports effective pollution control strategies and enables swift intervention in contamination incidents.

9.3.2.2 Soil health evaluation: contaminants and nutrient dynamics

Assessing soil health is integral to sustainable agricultural management and environmental conservation. Monitoring soil contaminants and nutrient composition allows farmers and land managers to implement strategies that enhance productivity while maintaining ecological stability. Essential soil health parameters include nutrient concentrations, organic matter content and contamination levels from agrochemicals and heavy metals (Blair *et al.*, 2024; Giuliani *et al.*, 2024).

Recent technological advancements have introduced cutting-edge methodologies such as multispectral imaging and deep learning-based analysis, which improve the precision and efficiency of soil assessments (Dhanalakshmi, 2024). These innovations facilitate real-time identification of nutrient imbalances and contamination, ensuring proactive soil management. Establishing standardised soil health metrics, as highlighted in recent research, is essential for adapting agricultural practices to evolving environmental conditions (Griffiths *et al.*, 2018).

Furthermore, a comprehensive understanding of the impact of soil health on agricultural productivity is vital for fostering long-term sustainability. Findings by Blair *et al.* (2024) illustrate the influence of different land management strategies on soil quality indicators, underscoring the necessity of tailored monitoring systems for effective soil conservation.

9.3.2.3 Air pollution surveillance: volatile organic compounds and toxins

Monitoring air quality, particularly the presence of volatile organic compounds (VOCs) and other airborne toxins, is essential for mitigating public health risks and environmental deterioration. Indoor air pollutants, including VOCs, such as formaldehyde and toluene, present significant health hazards, reinforcing the urgency for efficient detection methodologies (Chi *et al.*, 2016). Emerging studies have established the detrimental effects of VOCs on air quality and their association with various respiratory and systemic conditions, highlighting the importance of continuous air quality assessment (Panagos *et al.*, 2024).

The integration of biosensors and chemical sensing technologies has been instrumental in tracking VOC concentrations in urban and industrial settings (Babür and Dindaroğlu, 2020). The development of cost-effective, user-friendly air monitoring devices has empowered communities to conduct localised air quality evaluations, fostering a deeper understanding of pollution dynamics (Seaton *et al.*, 2020). Additionally, the incorporation of citizen science initiatives has significantly enhanced data collection efforts and public engagement in addressing air pollution concerns (Head *et al.*, 2020).

Comprehensive monitoring of water quality, soil health and air pollution is fundamental to addressing contemporary environmental challenges. The synergy between microbial biosensors and advanced sensing technologies provides innovative solutions for real-time environmental assessment, ensuring that regulatory frameworks and conservation efforts are guided by precise and reliable data.

9.4. Performance of microbial biosensors in environmental monitoring

Microbial biosensors have emerged as essential tools in environmental monitoring, offering real-time contaminant detection and facilitating data-driven decision-making. Their application across diverse environmental matrices, particularly in soil contamination analysis and water quality assessment, underscores their adaptability and effectiveness. Table 9.1 provides an overview of microbial biosensors used in environmental monitoring, highlighting their detection methods, sensitivity and microbial sources.

9.4.1 Soil contamination analysis

Microbial biosensors provide an efficient and practical approach to evaluating soil contamination, enabling precise detection of pollutants and assessment of soil health. One of the most notable applications involves genetically engineered bacterial strains designed to respond to specific soil contaminants. Liu *et al.* (2010) investigated the use of genetically modified *Pseudomonas* strains as whole-cell microbial biosensors for detecting polychlorinated biphenyls (PCBs) and chlorobenzoate bioavailability in polluted soils. This study highlights the capacity of engineered biosensors to identify and track the degradation of hazardous soil pollutants with high accuracy.

Enhancing microbial biosensor performance in soil environments can be achieved through the development of artificial soils that simulate natural conditions. Valle *et al.* (2022) introduced biosensors under controlled experimental conditions replicating real soil osmotic potentials, thereby improving the accuracy of microbial activity assessments and their responses to environmental stressors. This advancement enhances the reliability of biosensors in evaluating soil contamination under diverse ecological scenarios.

Beyond contamination detection, microbial biosensors contribute to broader soil health assessments. Cheng *et al.* (2016) explored microbial gene expression relevant to soil greenhouse gas fluxes, offering insights into microbial community dynamics and the influence of contaminants on ecological processes. Collectively, microbial biosensors represent a powerful approach for soil monitoring, enabling rapid pollutant identification and providing valuable data for remediation strategies.

9.4.2 Water quality assessment

Microbial biosensors play a pivotal role in water quality assessment by enabling rapid and precise pollutant detection. Recent advancements in biosensing technology have led to the integration of MFC-based systems for continuous monitoring of toxic substances in aquatic environments. Chen *et al.* (2022) introduced a closed-loop microbial fuel cell control system for monitoring total organic carbon (TOC) fluctuations in public swimming pools, demonstrating the effectiveness of MFCs in real-time water quality surveillance. The correlation between microbial electricity generation and toxin concentration marks a significant innovation in continuous water monitoring.

Table 9.1 Microbial biosensors for environmental monitoring

Pollutant category/ analyte	Microbial biosensor example/ microorganism	Detection method/transducer type	Detection range/detection limit (LOD), EC50 or IC50	Reference
Silver nanoparticles	*E. coli, P. fluorescens, S. cerevisiae, P. subcapitata*	Optical	EC50: 0.01–8.17 mg/L	Ivask *et al.*, 2014
MMS, 4-NQO, phlomycin, hydrogen peroxide, tert-butyl hydroperoxide methyl viologen, chlorambucil and cisplatin	*S. cerevisiae* BY4741	Optical	4-NQO: 0.12 ng/mL MMS: 0.36 μg/mL	Wei *et al.*, 2013
Copper	Heterogeneous microbial populations	MFC	LOD = 5 ppm	Shen *et al.*, 2013
Copper, cadmium, zinc, chromium, mercury, lead	*Psychrobacter*	Electrochemical	EC50: 0.8–110.1 mg/L	Wang *et al.*, 2013
Isothiocyanate, cinnamaldehyde	Recombinant *Acinetobacter baylyi* ADP1	Optical	NA	Chan *et al.*, 2013
Chromium, iron, nitrate, sodium acetate	Heterogeneous microbial populations	MFC	Detection range: 1–50 μg/L	Di Lorenzo *et al.*, 2014
Zinc, cadmium	Recombinant *E. coli* DH5	Optical	NA	Kim *et al.*, 2015
Arsenic, selenite	Recombinant *E. coli* DH5α	MFC	Arsenite: LOD = 2.0×10^{-4} mg/L selenite: LOD = 5.8×10^{-6} mg/L	Ooi *et al.* 2015
Lindane	Recombinant *E. coli* BL21	Electrochemical	LOD = 2 ppt	Anu Prathap *et al.*, 2012
Methyl parathion, paraoxon	Recombinant *Moraxella*	Electrochemical	Methyl parathion: LOD = 1 μM, paraoxon: LOD = 0.2 μM	Mulchandani and Rajesh, 2011
Mercury	Light-up biosensor	Micro-spin column (Bio-Rad)	60.7 nM	Yang *et al.*, 2017
Copper, zinc, chromium	Whole-cell bacterial biosensors	Graphite sheet	1.42×10^{-4}– 4.7 h	Naik and Jujjavarapu, 2021

The development of compact biosensing devices has further expanded the potential for online water quality assessment. Yagur-Kroll *et al.* (2015) engineered a flow-cell biosensor incorporating bacterial sensor cells, demonstrating the efficacy of aluminium oxide-based platforms for real-time pollutant detection in water. These systems offer sustained monitoring capabilities, which are crucial for detecting abrupt pollution events and ensuring water safety.

Additionally, microbial biosensors have been extensively investigated for their role in detecting specific contaminants, such as heavy metals. Research conducted by Bose *et al.* (2021) highlights the effectiveness of microbial biosensors in monitoring mercury levels in aquatic systems, offering a cost-effective and highly sensitive solution for addressing significant environmental and public health concerns.

Microbial biosensors exhibit exceptional performance in environmental monitoring, particularly in soil contamination analysis and water quality assessment. Their capacity to deliver real-time, highly specific and sensitive data underscores their potential for enhancing environmental quality evaluations and guiding effective responses to contamination incidents.

9.4.3 Detection of airborne toxins

Microbial biosensors serve as indispensable tools in environmental monitoring, offering precise and real-time detection of airborne contaminants. These technologies hold particular significance in detecting volatile organic compounds (VOCs) and airborne pathogens, presenting a compelling alternative to conventional analytical methods. Airborne pollutants pose substantial risks to public health and environmental stability, necessitating effective detection strategies in urban spaces, occupational settings and natural ecosystems. Recent advancements in microbial biosensing have led to the development of highly sensitive and selective systems for monitoring airborne toxins.

Electrochemical biosensors have demonstrated remarkable efficacy in detecting airborne pathogens and VOCs (Sivakumar and Lee, 2022). The portability and ease of fabrication of these biosensors make them particularly suitable for point-of-care testing (POCT) in dynamic environments. Additionally, a specialised nanofibre bioreceptor has been engineered for detecting airborne *Escherichia coli* and *Staphylococcus aureus*, providing an advanced method for pathogen surveillance in air samples (Varvařovská *et al.*, 2024). This innovation underscores the expanding role of biosensors in air quality monitoring and contamination control.

Further innovations include the application of bioluminescent sensors for indoor air toxicity assessments. A biosensor utilising bacteria immobilised on optical fibres enables direct pollutant detection in the air (Eltzov *et al.*, 2015). By integrating bioluminescence with fibre-optic technology, this system enhances sensitivity and enables the detection of airborne toxins at significantly lower concentrations.

9.4.4 Comparative evaluation of microbial biosensors and conventional analytical techniques

The integration of microbial biosensors within environmental monitoring frameworks presents distinct advantages over traditional analytical methods such as chromatography and mass spectrometry. While conventional techniques are renowned for their precision and reliability, they often necessitate labour-intensive sample preparation, prolonged processing times and sophisticated laboratory instrumentation. Conversely, microbial biosensors offer several advantages, including rapid detection, operational simplicity and direct analysis capability (Ahuekwe *et al.*, 2024). Moreover, the miniaturisation of biosensors enables field-based applications, significantly enhancing adaptability in diverse monitoring environments.

Despite these benefits, microbial biosensors face limitations concerning selectivity, stability and interference from complex sample matrices, which may affect their analytical performance (Lim *et al.*, 2015). Traditional techniques, although more resource-intensive, generally provide superior specificity and quantification across a wider range of environmental pollutants.

A multi-analyte biosensor has demonstrated a strong correlation with conventional photometric analysis for organic acids, highlighting the potential for biosensors to match the accuracy of established methods while simultaneously detecting multiple analytes (Pilas *et al.*, 2018). This finding underscores the viability of microbial biosensors in comprehensive environmental monitoring while also emphasising the need for continued refinement and the potential for hybrid approaches that integrate biosensors with conventional methodologies.

Microbial biosensors are poised to transform environmental monitoring, particularly in airborne toxin detection and comparative analysis with traditional analytical methods. Their rapid response times and ease of deployment render them a promising alternative, yet further research and optimisation remain imperative to address existing challenges and enhance their overall efficacy.

9.5. Advantages and limitations

Microbial biosensors possess distinct advantages and limitations that influence their effectiveness in environmental monitoring. A comprehensive understanding of these factors is crucial for optimising their applications and mitigating potential challenges that may hinder their broader implementation.

9.5.1 Advantages

9.5.1.1 High specificity and sensitivity

Microbial biosensors are engineered to detect specific contaminants, including pathogens and toxins, while minimising interference from extraneous substances. Their high specificity is attributed to the biological recognition elements employed, such as genetically modified microorganisms or specialised enzymes that identify unique molecular markers associated with pollutants. Studies have demonstrated that these biosensors can achieve sensitivities comparable to or surpassing those of conventional laboratory-based methods, allowing for the rapid and precise detection of even trace levels of environmental pollutants (Zhang *et al.*, 2019).

9.5.1.2 Real-time and onsite detection

A key advantage of microbial biosensors is their capacity for real-time monitoring and onsite detection, facilitating immediate assessments of environmental conditions. This capability is essential for the prompt management of contamination events and the continuous evaluation of environmental health. The portability and rapid analytical functionality of microbial biosensors significantly enhance their utility in field-based assessments of water quality and soil health, addressing limitations associated with traditional monitoring techniques (Kivirand *et al.*, 2019).

9.5.1.3 Eco-friendly and cost-effective approach

Microbial biosensors primarily employ naturally occurring microorganisms or require minimal use of hazardous reagents, rendering them a more environmentally sustainable alternative to conventional detection methods, which often rely on toxic chemicals (Garcia-Junior, 2025). Additionally, advancements in biosensor design, including the incorporation of nanomaterials, have contributed to enhanced cost-effectiveness. The reduction in extensive sample preparation requirements and the potential for miniaturisation further drive down operational costs, fostering broader accessibility and adoption (Al-Gethami *et al.*, 2022).

9.5.2 Limitations

9.5.2.1 Laboratory-to-field scalability challenges

Despite their advantages, microbial biosensors encounter significant obstacles when transitioning from controlled laboratory environments to real-world field applications. Variability in environmental parameters, including temperature fluctuations, humidity levels and the presence of interfering compounds, can compromise sensor stability and performance (Restrepo *et al.*, 2021). Ensuring consistency in biosensor functionality across diverse conditions necessitates extensive field validation and iterative optimisation of biosensor configurations.

9.5.2.2 Regulatory and standardisation barriers

The widespread adoption of microbial biosensors in environmental monitoring is hindered by the absence of comprehensive regulatory frameworks and standardised validation procedures. Regulatory agencies often require rigorous verification protocols comparable to those established for conventional analytical methodologies, creating barriers to biosensor approval and implementation (Walter *et al.*, 2020). Furthermore, inconsistencies in biosensor performance across different studies and applications highlight the need for standardised protocols to enhance reproducibility and reliability.

9.5.2.3 Challenges in commercialisation and industry adoption

The commercialisation of microbial biosensors is impeded by factors such as high developmental costs, limitations in scalable manufacturing and the necessity for ongoing technical support to ensure proper deployment and maintenance (Yojo *et al.*, 2020). Moreover, resistance to adopting new biosensing technologies within established industries stems from entrenched practices and reliance on traditional monitoring techniques. Raising awareness of the advantages of microbial biosensors and fostering educational initiatives will be instrumental in promoting their integration into environmental monitoring frameworks (Prasanna *et al.*, 2024).

Microbial biosensors offer significant benefits, including high specificity, real-time analytical capabilities and environmental sustainability. However, addressing their limitations, particularly in scalability, regulatory acceptance and industry adoption, is imperative to unlocking their full potential in environmental monitoring. Continued research and development efforts aimed at overcoming these challenges will be essential in advancing the role of microbial biosensors in environmental science.

9.6. Integration of SMART technologies

The incorporation of smart technologies into microbial biosensors has significantly expanded their capabilities, optimising their application in environmental monitoring. The utilisation of data analytics, the Internet of Things (IoT) and wireless communication technologies has transformed biosensors into advanced tools for real-time environmental assessment. As illustrated in Figure 9.3, IoT-enabled biosensors facilitate real-time monitoring and data-driven decision-making, significantly improving water quality assessments (Essamlali *et al.*, 2024).

9.6.1 Role of data analytics and machine learning in biosensor applications

Data analytics and machine learning (ML) serve as fundamental components in enhancing the functionality of microbial biosensors. Machine learning algorithms facilitate the efficient processing of extensive datasets generated by biosensors, thereby improving accuracy and reliability. Advanced computational techniques are indispensable for interpreting biosensor-derived data, particularly in detecting healthcare-related markers (Singh *et al.*, 2021).

The application of ML enables the identification of patterns and correlations in sensor data, thereby augmenting sensitivity and specificity in pollutant detection, especially within complex environmental matrices. Moreover, ML models can be trained to recognise distinct signatures of toxins and pathogens, reducing false positives and refining biosensor performance (Zhang *et al.*, 2021). The integration of machine learning with biosensor technology enables

Figure 9.3 Integrated smart biosensor system for environmental monitoring (Concept adapted from Essamlali *et al.*, 2024)

the development of predictive models that facilitate timely interventions in environmental monitoring and public health, thereby enhancing the practical utility of these tools.

9.6.2 Biosensors integrated with IoT for real-time monitoring

The convergence of biosensors with IoT technology has facilitated substantial advancements in real-time environmental monitoring. IoT-enabled biosensors allow for the continuous and remote transmission of environmental data, ensuring immediate access to critical information. This functionality is particularly advantageous in scenarios requiring rapid responses to hazardous fluctuations in air or water quality, as exemplified in the development of smart healthcare monitoring systems (Catarinucci *et al.*, 2015).

The integration of IoT with biosensors enhances automated monitoring and information dissemination, significantly contributing to urban environmental management and public safety. These systems provide real-time alerts to authorities on the detection of toxic substances at critical levels, enabling prompt remedial measures to mitigate associated health risks.

Furthermore, IoT-integrated biosensors facilitate advanced data collection strategies, allowing environmental monitoring data to be incorporated into sustainability initiatives and policy development frameworks (Yao *et al.*, 2022). This interconnected approach is instrumental in achieving a holistic understanding of environmental dynamics and implementing effective sustainability measures.

9.6.3 Potential of wireless biosensors for remote environmental assessments

Wireless biosensors present a transformative approach to remote environmental assessments, enabling the deployment of detection systems in inaccessible or hazardous locations without necessitating constant human intervention. These biosensors often utilise battery-free operation and energy-harvesting technologies, facilitating prolonged monitoring in remote settings (Tomichan *et al.*, 2023). Wireless capabilities enhance data collection convenience while allowing continuous monitoring of environmental conditions.

Advancements in wireless communication technologies have expedited biosensor data transmission, enhancing the responsiveness of environmental monitoring systems (Yao *et al.*, 2022). Wireless biosensors integrated with machine learning algorithms further refine real-time pollutant analysis, fostering swift decision-making in environmental management.

Furthermore, integrating wireless communication into biosensors significantly enhances accessibility for onsite monitoring across various applications, including environmental pollution surveillance and healthcare diagnostics (Zhang *et al.*, 2021). By streamlining data acquisition and analysis, wireless biosensors contribute substantially to the expansion of remote sensing technologies, thereby improving the efficiency and effectiveness of environmental assessments.

The incorporation of smart technologies into microbial biosensors holds immense potential for advancing environmental monitoring. The convergence of data analytics, IoT and wireless technologies is set to augment the functionality and impact of biosensing systems, thereby paving the way for more sophisticated and efficient monitoring solutions.

9.7. Conclusions

The use of microbial biosensors has become essential in the field of environmental monitoring, as they provide accurate and efficient methods for detecting pollutants such as metals, pesticides, organic pollutants and pathogens. Their ability for real-time, highly sensitive, and specific detection, as well as their effectiveness in assessing environmental health and safety, has significantly improved.

The integration of advanced technologies, such as the IoT and machine learning, has further expanded their capabilities, enabling comprehensive data acquisition and analysis for informed decision-making. Their ability to perform onsite, real-time analysis ensures immediate responses to environmental hazards, thereby safeguarding public health and ecosystems. Beyond their monitoring functions, microbial biosensors contribute significantly to environmental sustainability. By quickly assessing soil, water and air quality, they facilitate more efficient resource management and help mitigate the adverse effects of pollution. Their accuracy, cost-effectiveness and environmentally friendly nature make them particularly suitable for large-scale use, especially in remote or resource-poor areas. Furthermore, their integration with sustainable strategies such as bioremediation and precision agriculture emphasise their potential in fostering long-term environmental sustainability.

Microbial biosensors represent an exciting advance in environmental monitoring, combining cutting-edge detection methods with standard applications. As technological innovations continue to refine their capabilities, these biosensors are poised to play an increasingly important role in protecting environmental health and supporting global sustainability efforts.

REFERENCES

Abdel-Razzaq FA-W, Dhaigham AAR and Yaseen SK (2023) Design and construction of a dust detection system using infrared laser: The case of dust storms in Baghdad in the summer 2022. *Iraqi Journal of Physics* **21(1)**: 34–41, 10.30723/ijp.v21i1.1077.

Ahuekwe C, Smith J and Patel R (2024) A review of fabrication techniques and optimization strategies for microbial biosensors. *IOP Conference Series: Earth and Environmental Science* **1342(1)**: 012015. doi:10.1088/1755-1315/1342/1/012015.

Al-Gethami W, Alhashmialameer D, Al-Qasmi N, Ismail SH and Sadek AH (2022) Design of a novel nanosensors based on green synthesized $CoFe_2O_4$/Ca-alginate nanocomposite-coated QCM for rapid detection of Pb(II) ions. *Nanomaterials* **12(20)**: 3620, 10.3390/nano12203620.

Andriukonis E, Celiešiūtė-Germanienė R, Ramanavičius S, Viter R and Ramanavičius A (2021) From microorganism-based amperometric biosensors towards microbial fuel cells. *Sensors* **21(7)**: 2442, 10.3390/s21072442.

Anu Prathap MU, Chaurasia AK, Sawant SN and Apte SK (2012) Polyaniline-based highly sensitive microbial biosensor for selective detection of lindane. *Analytical Chemistry* **84(15)**: 6672–6678, 10.1021/ac301077d.

Blair HK, Gutknecht JL, Jelinski NA *et al.* (2024) Nature versus nurture: quantifying the effects of management, region and hillslope position on soil health indicators in an on-farm survey in Minnesota. *Soil Science Society of America* **88(6)**: 2135–2155, 10.1002/saj2.20739.

Bose S, Maity S and Sarkar A (2021) Review of microbial biosensor for the detection of mercury in water. *Environmental Quality Management* [Preprint], (tqem.21742), 10.1002/tqem.21742.

Catarinucci L, De Donno D, Mainetti L *et al.* (2015) An IoT-Aware Architecture for Smart Healthcare Systems. *IEEE Internet of Things Journal* **2(6)**: 515–526, 10.1109/jiot.2015.2417684.

Chan AC, Ager D and Thompson IP (2013) Resolving the mechanism of bacterial inhibition by plant secondary metabolites employing a combination of whole-cell biosensors. *Journal of Microbiological Methods* **93(3)**: 209–217, 10.1016/j.mimet.2013.03.021.

Chen H, Meng X, Liu D *et al.* (2022) Closed-loop microbial fuel cell control system designed for online monitoring of TOC dynamic characteristics in public swimming pool. *International Journal of Environmental Research and Public Health* **19(20)**: 13024, 10.3390/ijerph192013024.

Cheng H-Y, Masiello CA, Bennett GN and Silberg JJ (2016) Volatile gas production by methyl Halide transferase: an in situ reporter of microbial gene expression in soil. *Environmental Science and Technology* **50(16)**: 8750–8759, 10.1021/acs.est.6b01415.

Chi C, Chen W, Guo M *et al.* (2016) Law and features of TVOC and Formaldehyde pollution in urban indoor air. *Atmospheric Environment* 132: 85–90, 10.1016/j.atmosenv.2016.02.043.

Cui Y, Lai B and Tang X (2019) Microbial fuel cell-based biosensors. *Biosensors* **9(3)**: 92, 10.3390/bios9030092.

Dhanalakshmi B (2024) Soil health intelligence system using multispectral imaging and advanced deep learning techniques (SHIDS-ADLT). *Communications on Applied Nonlinear Analysis* 31(6s): 417–426, 10.52783/cana.v31.1233.

Di Lorenzo M, Thomson AR, Schneider K, Cameron PJ and Ieropoulos I (2014) A small-scale air-cathode microbial fuel cell for on-line monitoring of water quality. *Biosensors and Bioelectronics* 62, pp. 182–188, 10.1016/j.bios.2014.06.050.

Eltzov E, Cohen A and Marks RS (2015) Bioluminescent liquid light guide pad biosensor for indoor air toxicity monitoring. *Analytical Chemistry* **87(7)**: 3655–3661, 10.1021/ac5038208.

Essamlali I, Nhaila H and El Khaili M (2024) Advances in machine learning and IoT for water quality monitoring: A comprehensive review. *Heliyon* **10(6)**: e27920, 10.1016/j.heliyon.2024.e27920.

Febbraio F, Merone L, Cetrangolo GP *et al.* (2011) Thermostable esterase 2 from Alicyclobacillus acidocaldarius as biosensor for the detection of organophosphate pesticides. *Analytical Chemistry* **83(5)**: 1530–1536, 10.1021/ac102025z.

Fiorbelli E, Lapris M, Errico M *et al.* (2024) Mycotoxin challenge in dairy cows: Assessment of the efficacy of an anti-mycotoxin agent by adopting an in vitro Rumen simulation method. *Toxins* **16(11)**, 10.3390/toxins16110490.

Gámiz-Gracia L, García-Campaña AM and Arroyo-Manzanares N (2020) Application of LC-MS/MS in the mycotoxins studies. *Toxins* **12(4)**: 272, 10.3390/toxins12040272.

Garcia-Junior MA, Andrade BS, Lima AP *et al.* (2025) Artificial-intelligence bio-inspired peptide for salivary detection of SARS-CoV-2 in electrochemical biosensor integrated with Machine Learning algorithms. *Biosensors* **15(2)**: 75, 10.3390/bios15020075.

Greco D, D'Ascanio V, Santovito E *et al.* (2025) Unlocking the potential of Bacillus subtilis: a comprehensive study on mycotoxin decontamination, mechanistic insights and efficacy assessment in a liquid food model. *Foods* **14(3)**, 10.3390/foods14030360.

Greeff-Laubscher MR, Beukes, Marais GJ and Jacobs K (2019) Mycotoxin production by three different toxigenic fungi genera on formulated abalone feed and the effect of an aquatic environment on fumonisins. *Mycology* **11(2)**: 105–117, 10.1080/21501203.2019.1604575.

Griffiths BS, Faber J and Bloem J (2018) Applying soil health indicators to encourage sustainable soil use: The transition from scientific study to practical application. *Sustainability* **10(9)**: 3021, 10.3390/su10093021.

Hashemi Goradel N, Mirzaei H, Sahebkar A *et al.* (2018) Biosensors for the detection of environmental and urban pollutions. *Journal of Cellular Biochemistry* **119(1)**: 207–212, 10.1002/jcb.26030.

He W, Yuan S, Zhong W-H, Siddikee Md A and Dai C-C (2016) Application of genetically engineered microbial whole-cell biosensors for combined chemosensing. *Applied Microbiology and Biotechnology* **100(3)**: 1109–1119, 10.1007/s00253-015-7160-6.

Head JS, Crockatt ME, Didarali Z, Woodward M-J and Emmett BA (2020) The role of citizen science in meeting SDG targets around soil health. *Sustainability* **12(24)**: 10254, 10.3390/su122410254.

Huang C-W, Lin C, Nguyen MK *et al.* (2023) A review of biosensor for environmental monitoring: principle, application and corresponding achievement of sustainable development goals. *Bioengineered* **14(1)**: 58–80, 10.1080/21655979.2022.2095089.

Hui C-Y, Guo Y, Li L *et al.* (2021) Indigoidine biosynthesis triggered by the heavy metal-responsive transcription regulator: a visual whole-cell biosensor. *Applied Microbiology and Biotechnology* **105(14–15)**: 6087–6102, 10.1007/s00253-021-11441-5.

Hui Y, Huang Z, Alahi MEE *et al.* (2022) Recent advancements in electrochemical biosensors for monitoring the water quality. *Biosensors* **12(7)**: 551, 10.3390/bios12070551.

Ivask A, Kurvet I, Kasemets K *et al.* (2014) Size-dependent toxicity of silver nanoparticles to bacteria, yeast, algae, crustaceans and mammalian cells in vitro. *PloS one* **9(7)**: e102108, 10.1371/journal.pone.0102108.

Kappenberg A and Juraschek LM (2021) Development of a LC-MS/MS method for the simultaneous determination of the mycotoxins deoxynivalenol (DON) and zearalenone (ZEA) in soil matrix. *Toxins* **13(7)**: 470, 10.3390/toxins13070470.

Kim M, Lim JW, Kim HJ *et al.* (2015) Chemostat-like microfluidic platform for highly sensitive detection of heavy metal ions using microbial biosensors. *Biosensors and Bioelectronics* **65**: 257–264, 10.1016/j.bios.2014.10.028.

Kivirand K, Min M and Rinken T (2019) Challenges and applications of impedance-based biosensors in water analysis. In *Biosensors for Environmental Monitoring* (Rinken T and Kivirand K (eds)). InTechOpen, London, UK.

Lee SH, Lee JH, Jung WC *et al.* (2020) Changes in soil health with remediation of petroleum hydrocarbon contaminated soils using two different remediation technologies. *Sustainability* **12(23)**: 10078, 10.3390/su122310078.

Lee W, Kim H, Kang Y, Lee Y and Yoon Y (2019) A biosensor platform for metal detection based on enhanced green fluorescent protein. *Sensors* **19(8)**: 1846, 10.3390/s19081846.

Li L, Liang J, Hong W *et al.* (2015) Evolved bacterial biosensor for arsenite detection in environmental water. *Environmental Science and Technology* **49(10)**: 6149–6155, 10.1021/acs.est.5b00832.

Lim JW, Ha D, Lee J, Lee SK and Kim T (2015) Review of micro/nanotechnologies for microbial biosensors. *Frontiers in Bioengineering and Biotechnology* **3**: 61, 10.3389/fbioe.2015.00061.

Liu L, Yan Y, Wang J, Wu W and Xu L (2016) Generation of mt:egfp transgenic zebrafish biosensor for the detection of aquatic zinc and cadmium. *Environmental Toxicology and Chemistry* **35(8)**: 2066–2073, 10.1002/etc.3362.

Liu X, Germaine KJ, Ryan D and Dowling DN (2010) Genetically modified Pseudomonas biosensing biodegraders to detect PCB and chlorobenzoate bioavailability and biodegradation in contaminated soils. *Bioengineered Bugs* **1(3)**: 198–206, 10.4161/bbug.1.3.12443.

Lobry de Bruyn L and Andrews S (2016) Are Australian and United States farmers using soil information for soil health management?. *Sustainability* **8(4)**: 304, 10.3390/su8040304.

Logue JM, McKone TE, Sherman MH and Singer BC (2011) Hazard assessment of chemical air contaminants measured in residences: Hazard assessment of indoor air contaminants. *Indoor Air* **21(2)**: 92–109, 10.1111/j.1600-0668.2010.00683.x.

Long F, Zhu A and Shi H (2013) Recent advances in optical biosensors for environmental monitoring and early warning. *Sensors* **13(10)**: 13928–13948, 10.3390/s131013928.

Lu Y, Macias D, Dean ZS, Kreger NR and Wong PK (2015) A UAV-mounted whole cell biosensor system for environmental monitoring applications. *IEEE Transactions on Nanobioscience* **14(8)**: 811–817, 10.1109/TNB.2015.2478481.

Mirasole C, Di Carro M, Tanwar S and Magi E (2016) Liquid chromatography-tandem mass spectrometry and passive sampling: powerful tools for the determination of emerging pollutants in water for human consumption. *Journal of Mass Spectrometry* **51(9)**: 814–820, 10.1002/jms.3813.

Mishra GK, Sharma V and Mishra RK (2018) Electrochemical aptasensors for food and environmental safeguarding: a review. *Biosensors* **8(2)**, 10.3390/bios8020028.

Mulchandani A and Rajesh (2011) Microbial biosensors for organophosphate pesticides. *Applied Biochemistry and Biotechnology* **165(2)**: 687–699, 10.1007/s12010-011-9288-x.

Naik S and Jujjavarapu SE (2021) Self-powered and reusable microbial fuel cell biosensor for toxicity detection in heavy metal polluted water. *Journal of Environmental Chemical Engineering* **9(4)**: 105318, 10.1016/j.jece.2021.105318.

Naresh V and Lee N (2021) A review on biosensors and recent development of nanostructured materials-enabled biosensors. *Sensors* **21(4)**: 1109, 10.3390/s21041109.

Nguyen DK and Jang C-H. (2021) A cationic surfactant-decorated liquid crystal-based aptasensor for label-free detection of malathion pesticides in environmental samples. *Biosensors* **11(3)**: 92, 10.3390/bios11030092.

Ooi L, Heng LY and Mori IC (2015) A high-throughput oxidative stress biosensor based on Escherichia coli roGFP2 cells immobilized in a k-carrageenan matrix. *Sensors* **15(2)**: 2354–2368, 10.3390/s150202354.

Padmapriya S, Anand M, Rangesh K and Maruthupandy M (2024) Metal nanoparticle-based impedimetric biosensors for rapid detection of bacterial pathogen in aquaculture. In *Nanoscience and Nanotechnology for Smart Prevention Diagnostics and Therapeutics*. Wiley, Hoboken, NJ, USA, pp. 183–215, 10.1002/9781394175222.ch8.

Panagos P, Broothaerts N, Ballabio C *et al.* (2024) How the EU Soil observatory is providing solid science for healthy soils. *European Journal of Soil Science* **75(3)**, 10.1111/ejss.13507.

Peixoto L, Min B, Martins G *et al.* (2011) In situ microbial fuel cell-based biosensor for organic carbon. *Bioelectrochemistry* **81(2)**: 99–103, 10.1016/j.bioelechem.2011.02.002.

Prasanna NL, Choudhary S, Kumar S *et al.* (2024) Advances in plant disease diagnostics and surveillance: a review. *Plant Cell Biotechnology and Molecular Biology* **25(11–12)**: 137–150, 10.56557/pcbmb/2024/v25i11-128918.

Qi P, Zhang D and Wan Y (2014) Development of an amperometric microbial biosensor based on *Thiobacillus thioparus* cells for sulfide and its application to detection of sulfate-reducing bacteria. *Electroanalysis* **26(8)**: 1824–1830, 10.1002/elan.201400198.

Qu L, Wang N, Xu H *et al.* (2017) Gold nanoparticles and g-C 3 N 4 -intercalated graphene oxide membrane for recyclable surface enhanced Raman scattering. *Advanced Functional Materials* **27(31)**: 1701714, 10.1002/adfm.201701714.

Ren X, Breadmore MC and Maya F (2022) Biphasic magnetic levitation to detect organic pollutants on microplastics. *Analytical Chemistry* **94(25)**: 9033–9039, 10.1021/acs.analchem.2c01094.

Restrepo M, Huffenberger AM, Hanson CW, Draugelis M and Laudanski, K (2021) Remote monitoring of critically-ill post-surgical patients: lessons from a biosensor implementation trial. *Healthcare* **9(3)**: 343, 10.3390/healthcare9030343.

Rodríguez A, Rodríguez M, Martín A, Delgado J and Córdoba JJ (2012) Presence of ochratoxin A on the surface of dry-cured Iberian ham after initial fungal growth in the drying stage. *Meat Science* **92(4)**: 728–734, 10.1016/j.meatsci.2012.06.029.

Samada LH and Tambunan USF. (2020) Biopesticides as promising alternatives to chemical pesticides: a review of their current and future status. *OnLine Journal of Biological Sciences* **20(2)**: 66–76, 10.3844/ojbsci.2020.66.76.

Seaton FM, Barrett G, Burden A *et al.* (2021) Soil health cluster analysis based on national monitoring of soil indicators. *European Journal of Soil Science* **72(6)**: 2414–2429, 10.1111/ejss.12958.

Shen Y, Wang M, Chang IS and Ng HY (2013) Effect of shear rate on the response of microbial fuel cell toxicity sensor to Cu(II). *Bioresource Technology* 136: 707–710, 10.1016/j.biortech.2013.02.069.

Shing WL, Heng LY and Surif S (2013) Performance of a cyanobacteria whole cell-based fluorescence biosensor for heavy metal and pesticide detection. *Sensors* **13(5)**: 6394–6404, 10.3390/s130506394.

Singh A, Sharma A, Ahmed A *et al.* (2021) Recent advances in electrochemical biosensors: Applications, challenges and future scope. *Biosensors* **11(9)**: 336, 10.3390/bios11090336.

Sivakumar R and Lee NY (2022) Recent advances in airborne pathogen detection using optical and electrochemical biosensors. *Analytica Chimica Acta* **1234(340297)**: 340297, 10.1016/j.aca.2022.340297.

Sun Y, Zhou J, Zhou Y, Zeng Y and Chen Y (2019) Influencing factors of groundwater organic pollution around the Bosten Lake area of Xinjiang China. *E3S web of Conferences* 98: 09029, 10.1051/e3sconf/20199809029.

Tankus M, Olceroglu B, Balci G *et al.* (2024) Antibody DNA, and enzyme-based biosensing platforms as an on-site testing kit. In *Nanotechnology and Nanomaterials Annual Volume 2024*. InTechOpen, London, UK.

Tomichan R, Sharma A, Akash K *et al.* (2023) Insight of smart biosensors for COVID-19: a review. *Luminescence: the Journal of Biological and Chemical Luminescence* **38(7)**: 1102–1110, 10.1002/bio.4430.

Valle ID, Gao X, Ghezzehei TA, Silberg JJ and Masiello CA (2022) Artificial Soils Reveal Individual Factor Controls on Microbial Processes. *bioRxiv*, 10.1101/2022.02.01.478713.

Van der Fels-Klerx HJ, Liu C and Battilani P (2016) Modelling climate change impacts on mycotoxin contamination. *World Mycotoxin Journal* **9(5)**: 717–726, 10.3920/wmj2016.2066.

Varvařovská L, Gao X, Ghezzehei TA, Silberg JJ and Masiello CA (2024) The development of a specific nanofiber bioreceptor for detection of Escherichia coli and Staphylococcus aureus from air. *Biosensors* **14(5)**: 234, 10.3390/bios14050234.

Walter J-G, Eilers A, Alwis L, Roth B and Bremer K (2020) SPR biosensor based on polymer multi-mode optical waveguide and nanoparticle signal enhancement. *Sensors* **20(10)**: 2889, 10.3390/s20102889.

Wang X, Liu M, Wang X *et al.* (2013) p-Benzoquinone-mediated amperometric biosensor developed with Psychrobacter sp. for toxicity testing of heavy metals. *Biosensors and Bioelectronics* **41**: 557–562, 10.1016/j.bios.2012.09.020.

Wang Y, Xuejiang Li S, Xue N *et al.* (2023) Modulating sensitivity of an erythromycin biosensor for precise high-throughput screening of strains with different characteristics. *ACS Synthetic Biology* **12(6)**: 1761–1771, 10.1021/acssynbio.3c00059.

Wei T, Zhang C, Xu X *et al.* (2013) Construction and evaluation of two biosensors based on yeast transcriptional response to genotoxic chemicals. *Biosensors and Bioelectronics* **44**: 138–145, 10.1016/j.bios.2013.01.029.

Wilhelm RC, Amsili JP, Kurtz KSM *et al.* (2022) Ecological insights into soil health according to the life-history traits and environment-wide associations of bacteria in agricultural soils. *bioRxiv*, 10.1101/2022.02.03.479020.

Yagur-Kroll S, Schreuder E, Ingham CJ *et al.* (2015) A miniature porous aluminum oxide-based flow-cell for online water quality monitoring using bacterial sensor cells. *Biosensors and Bioelectronics* **64**: 625–632, 10.1016/j.bios.2014.09.076.

Yang X, He Y, Wang X and Yuan R *et al.* (2017) A SERS biosensor with magnetic substrate $CoFe_2O_4$ @Ag for sensitive detection of Hg^{2+}. *Applied Surface Science* **416**: 581–586, 10.1016/j.apsusc.2017.04.106.

Yao Z, Tan L and She K (2022) 5G-BSS: 5G-based universal blockchain smart sensors. *Sensors* **22(12)**: 4607, 10.3390/s22124607.

Yojo LS, Rangel RC, Sasaki KRA and Martino JA *et al.* (2020) Study of a charge-based biosensor and reconfigurability using BESOI MOSFET. *ECS Transactions* **97(5)**: 115–120, 10.1149/09705.0115ecst.

Zamora-Sequeira R, Starbird-Pérez R, Rojas-Carillo O and Vargas-Villalobos S (2019) What are the main sensor methods for quantifying pesticides in agricultural activities? A review. *Molecules* **24(14)**: 2659, 10.3390/molecules24142659.

Zhang K, Wang J, Liu T *et al.* (2021) Machine learning-reinforced noninvasive biosensors for healthcare. *Advanced Healthcare Materials* **10(17)**: e2100734, 10.1002/adhm.202100734.

Zhang Z, Zhou J and Du X (2019) Electrochemical biosensors for detection of foodborne pathogens. *Micromachines* **10(4)**: 222, 10.3390/mi10040222.

Microbial Processes in Environmental Engineering: From Theory to Practice

Rajneesh Kumar, Ram Sharan Singh and Maulin P Shah
ISBN 978-1-83708-327-5
https://doi.org/10.1108/978-1-83708-326-820261010

Chapter 10
Endophytes: diversity, ecological roles and applications in pollution mitigation

Sajitha Menon K

Department of Botany, Government Victoria College, Palakkad, Kerala, India

Abstract

Endophytes, the microorganisms residing asymptomatically within plant tissues, represent a vast array of unexplored biodiversity. They have a profound role in ecological processes and environmental sustainability. These microorganisms, which usually include bacteria and fungi, form intricate associations with their host plants, enhancing their growth, resilience and stress tolerance. The study of endophytes has divulged their pivotal roles in nutrient cycling, phytoremediation and detoxification of pollutants, making them natural allies in promoting environmental sustainability. Endophytes have diverse metabolic capabilities that can be utilised for pharmaceutical and industrial applications. Their potential in mitigating environmental challenges, such as heavy metal contamination and plastic degradation, highlights their importance in addressing pollution. Their symbiotic interactions and growth-promoting potential offer promising insights into sustainable agriculture, and they offer alternatives to chemical fertilisers and pesticides. Despite their diverse potential, challenges such as host specificity, field efficacy, environmental adaptability, large-scale isolation and cultivation need to be addressed. This chapter discusses the diversity of endophytes, emphasising their ecological roles and potential in mitigating pollution. It highlights innovative research and emerging applications on endophytes as key players in achieving environmental sustainability with their diverse potential.

Keywords: Endophytes; biodiversity; ecological role; pollution control; phytoremediation.

10.1. Introduction

Endophytes are microorganisms that inhabit plant tissues without exhibiting any symptoms in their host. Plant endophytes are primarily made up of fungi and bacteria, whereas algae, protozoa, archaebacteria and nematodes are also observed. Even though endophytes are primarily found in abundance in roots, they are also common in other plant parts like leaves, stem, petiole, fruits and so on. They have been discovered in all plants that have been examined (Aleynova and Kiselev, 2023).

They are found everywhere in nature and engage in intricate interactions with their hosts, including mutualism, antagonism and infrequently parasitism. They are recognised for boosting host development and nutrient acquisition. They assist their host in enduring different forms of abiotic and biotic stresses, while also boosting the plant's defence against numerous insects and pests. The endophyte population within a plant species varies significantly and is influenced by several factors, including plant species, developmental stage, environmental conditions and more (Gouda *et al.*, 2016).

Endophytes are said to generate several bioactive metabolites. These secondary metabolites consist of alkaloids, flavonoids, quinones, steroids, phenolic substances, saponins, tannins, terpenoids and various others. The main role of these secondary metabolites is to provide protection to plants against pathogens and stress-inducing factors. In addition to secondary metabolites, numerous endophytes are recognised for producing enzymes such as cellulase, laccase, amylase, lyase and so on, which have important industrial uses. The release of various plant hormones like auxins, gibberellins and cytokinin, which are known to promote plant growth and development, has also been documented (Singh and Kumar, 2023).

Numerous bioactive substances extracted from endophytes have demonstrated therapeutic promise and are under thorough examination for their antiarthritic, antimicrobial, anticancer, antidiabetic, anti-insect and immunosuppressant effects (Barkodia *et al.*, 2018). In addition to their various recognised functions, numerous endophytes have been effectively utilised in phytoremediation. The affordability and environmentally friendly characteristics of endophytes render endophyte-mediated phytoremediation a promising technology that can be well explored for a greener, pollution-free earth.

10.2. Diversity of endophytes

Endophytes, in most cases, are latent pathogens that reside within plant tissue without apparently damaging the plant, but they do have the ability to induce disease under specific circumstances. These microorganisms are drawn to their host plants by chemotaxis, reacting to chemical cues such as root exudates, by some signal molecules produced by plants or by unknown means, and they penetrate plant tissues through natural openings such as stomata, hydathodes or lenticels, or through wounds or enzymatic cell wall degradation. On entering the plant, they infect deeper tissues, avoiding the immune system of the plant by repressing defence reactions, impersonating host compounds, mimicking plant molecules or deploying effectors that antagonise immune signalling (Wippel, 2023).

Endophytes occupy particular niches, like intercellular space or vascular tissues, and can spread systemically or even vertically through seeds (Fadiji and Babalola, 2020). In spite of their latent pathogenicity, endophytes tend to offer considerable advantages to their hosts, such as increased resistance to pathogens through induced systemic resistance and production of antimicrobial compounds, tolerance to abiotic stresses such as drought, salinity and heavy metals, and plant growth promotion through phytohormone synthesis and nutrient uptake (Agbessenou *et al.*, 2020). In addition, they have the ability to cause epigenetic changes in the host that control gene expression and precondition the plant for subsequent stress events (Gupta *et al.*, 2020). This dynamic interaction, oscillating between mutualism and potential pathogenicity, emphasises the ecological and evolutionary importance of endophytes, showing their ability to increase plant resilience and indicating their potential use in many fields.

Endophytes have a high level of diversity within both bacterial and fungal communities. The functional complexity of endophytic species is closely tied to the habitual behaviours and lifestyles of their host plants. Comparative analyses of endophytic microbial communities across different plant species suggest that phylogenetically related plants tend to harbour more similar microbial communities than those that are evolutionarily distant. This highlights the strong connection between the behaviour of bacterial and fungal endophytes and the specific lifestyles of their host plants.

Interestingly, bacterial communities appear to be less influenced by the annual growth cycles of plants compared to fungal communities. This difference may be attributed to the faster reproductive rates of bacteria, which likely facilitates their colonisation and adaptation. As environmental conditions shift, new species emerge and dominant species fluctuate, leading to ongoing changes in the structure and function of these communities. This dynamic interplay underscores the adaptability and resilience of endophytic microorganisms in response to both host plant characteristics and external environmental pressures (Hernández-Tasco *et al.*, 2023).

Endophytes thus offer remarkable potential as a source of valuable bioactive compounds, including antibiotics, anticancer agents and immunosuppressants, and, agriculturally, they enhance crop resilience by improving stress tolerance to drought, salinity and pathogens. Environmentally, endophytes are screened for their potential in pollution control. They are being studied for their ability to degrade contaminants like heavy metals, pesticides and petroleum waste. Their unique plant – microorganism interactions make them invaluable across sectors, including human health, food security and environmental conservation.

10.3. Environmental pollution: an overview of challenges and solutions

Pollution is the introduction of contaminants into the natural environment. These contaminants are called pollutants. Introduction of pollutants to the environment can bring about serious impact on it. Polluting natural resources like air, water and soil brings about serious consequences on living organisms. Urbanisation, mining, burning of fossil fuels, industrial effluent discharge, agricultural activities, plastic and so on are some of the causes of pollution.

Environmental pollutants are broadly categorised into two types: elemental pollutants and organic pollutants. Elemental pollutants include radioisotopes such as uranium, tellurium and so on, as well as heavy metals and metalloids like mercury, cobalt and arsenic. These contaminants are typically immobile and accumulate in plant tissues. In contrast, organic pollutants consist of compounds like polycyclic aromatic hydrocarbons and halogenated hydrocarbons (Meagher, 2000).

Pollution has become a prominent issue and several remedial methods have been suggested, including biological, chemical and physical methods (Ukaogo *et al.*, 2020). Physical methods remove pollutants without altering the chemical nature, but not all types of pollutant can be handled in this way. Chemical methods transform pollutants into less harmful substances through reactions, which again have limitations. A combination of physical and chemical methods like adsorption, photocatalysis, membrane filtration and oxidation processes are also extensively used to control the impact of pollution. But these techniques have various limitations, like high cost, the production of toxic products and low efficiency causing secondary pollution (Wu *et al.*, 2023).

Biological processes are increasingly recognised as promising and sustainable solutions for environmental challenges. Among the most effective biological methods are bioremediation and phytoremediation. Bioremediation harnesses the power of microorganisms to break down pollutants, while phytoremediation utilises plants to absorb, filter and detoxify contaminants from soil and water. These natural approaches offer environmentally friendly alternatives to traditional remediation techniques.

However, biological remediation does face certain challenges, such as slow progress due to environmental adaptability issues, nutrient limitations, toxicity resistance and limited bioavailability of contaminants. These factors can hinder the efficiency and speed of the process. To address these limitations, innovative improvement strategies are being developed (Liu *et al.*, 2023).

10.4. Endophytes in pollution control

A recent trend in biological methods employs endophytes in pollution control. Endophytes, which include bacteria, fungi and actinomycetes, form mutually beneficial relationships with their host plants. These microorganisms not only enhance plant growth and stress tolerance but also play a pivotal role in mitigating environmental pollutants. Through mechanisms such as degradation of toxic compounds, metal sequestration and production of enzymes that break down contaminants, endophytes act as natural bioremediation agents. Their ability to thrive in polluted environments and interact with plants makes them a powerful tool for phytoremediation, a process where plants and their associated microorganisms work together to clean up contaminated soil and water.

The use of endophytes in pollution control offers several advantages, including cost-effectiveness, sustainability and minimal disruption to ecosystems. Moreover, their ability to enhance the bioavailability and degradation of pollutants makes them a promising solution for addressing complex contamination issues. As research in this field advances, endophytes are increasingly being recognised as key players in the development of innovative green technologies for environmental restoration.

In this scenario, bacterial endophytes have garnered significant interest. Beyond their ability to improve the growth and health of their host plants, they are also utilised in the bioremediation of polluted soils and waterways, aiding in the degradation of harmful substances and the recovery of affected ecosystems. Certain endophytes also demonstrate promise in boosting biofuel yields derived from plant biomass, contributing to advancements in bioenergy production (Muhammad *et al.*, 2024).

Heavy metal pollution is a pressing global issue, threatening ecosystems and human health due to its persistence, toxicity and non-biodegradable nature. These metals accumulate in crops, compromising food safety and security worldwide. Current remediation methods, such as physical and chemical techniques, are effective but resource-intensive and risk secondary pollution. Microbial remediation can reduce metal toxicity by altering valence states, while phytoremediation, although environmentally friendly, is limited by the slow growth of hyperaccumulators.

A promising alternative is the combined use of microbial–plant systems, which leverages the strengths of both for environmentally friendly and cost-effective remediation. Endophytic bacteria, having evolved resistance mechanisms like adsorption, methylation and redox reactions, play a key role in mitigating metal toxicity (Li *et al.*, 2017). Some of the bacteria that have been studied for heavy metal accumulation include *Bacillus thuringiensis* GDB-1 from *Alnus firma* (Babu *et al.*, 2013), *Sphingomonas* sp. *ZYG-4* from *Ageratina adenophora* (Li *et al.*, 2024), *Enterobacteriaceae genus* CBSB1 from *Brassica juncea* (Qiu *et al.*, 2014) and *Mesorhizobium loti* HZ76 from *Robinia pseudoacacia* (Fan *et al.*, 2018).

However, real-world application faces challenges, including mixed pollutants, soil variability and competition from native microorganisms. Additionally, endophytic bacteria's stability under harsh conditions complicates their consistent performance. Large-scale field trials are crucial to validate and refine this approach, bridging the gap between laboratory research and practical implementation. Addressing these challenges will advance sustainable solutions for heavy metal pollution, contributing to global environmental and food security goals (Liu *et al.*, 2024).

Wu *et al.* (2019) conducted a study on endophytic bacterium *Bacillus safensis* ZY16, isolated from the roots of the halophytic plant *Chloris virgata*, which has emerged as a promising candidate for bioremediation and plant growth promotion. This strain exhibits multiple beneficial traits, including the ability to degrade petroleum hydrocarbons such as n-alkanes and polycyclic aromatic hydrocarbons, produce biosurfactants to enhance hydrocarbon bioavailability and promote plant growth through different mechanisms. Its high salt tolerance further enables it to thrive in hypersaline environments, making it particularly suited for remediating petroleum-contaminated saline soils. These findings underscore the potential of plant – endophyte partnerships, particularly in challenging environments, for sustainable bioremediation.

However, translating laboratory success to field applications remains a challenge due to factors like nutrient availability, microbial competition and environmental variability. Future research should focus on large-scale field trials to validate and optimise the endophytes for the real-world remediation of hydrocarbon-contaminated soils, especially in saline ecosystems.

Fungal endophytes are also reported to have a role in bioremediation, especially in the absorption of heavy metals. In a study, a total of 253 fungal strains were isolated from different parts of *Hordeum vulgare* cv. Arna and *Glycine max* cv. Almaty. The highest fungal presence was noticed in roots. Among the isolates, 23 strains demonstrated cadmium (Cd) tolerance, with tolerance indices ranging from 0.24 to 1.12. Five strains, namely *Beauveria bassiana* T7, *Beauveria bassiana* T15, *Rhodotorula mucilaginosa* MK1, *Rhodotorula mucilaginosa* RH2 and *Metschnikowia pulcherrima* MP2, showed the highest Cd resistance, with minimum inhibitory concentrations (MICs) between 290 and 2400 μg/mL. These fungi exhibited significant Cd removal efficiency (up to 59%), along with excellent bioaccumulation capacity (Ignatova *et al.*, 2021; Nascimento *et al.*, 2023).

In addition to the cadmium-tolerant strains discussed earlier, several other ubiquitous fungi, including *Penicillium* spp., exhibit significant potential for bioremediation. These fungi, found both as free-living organisms and endophytes, have demonstrated remarkable capabilities in degrading environmental pollutants, such as synthetic dyes (e.g. azo dyes) and polycyclic aromatic hydrocarbons (Barman *et al.*, 2023). Studies suggest that fungal enzymes like laccases, peroxidases and cytochrome P450 monooxygenases play a crucial role in breaking down these contaminants through oxidative and metabolic pathways (AbuQamar *et al.*, 2023).

Studies have shown that fungal endophytes can be a sustainable solution for plastic waste. They have shown a four-stage biodegradation process passing through biodeterioration, biofragmentation, assimilation and mineralisation. These microorganisms have been shown to secrete enzymes, like cutinases, peroxidases, esterases and so on, that can degrade polyethylene and polyurethane. Additionally, their symbiotic relationships with plants enhance efficiency by modifying soil conditions while promising environmental adaptability. Harnessing endophytes could revolutionise plastic waste management, merging bioremediation with circular economy principles for a cleaner future (Sharma *et al.*, 2025).

Endophytes possess the ability to synthesise diverse enzymes and bioactive compounds that can break down or modify both organic and inorganic pollutants, contributing to environmental restoration making them good candidates for bioremediation to convert hazardous organic compounds into harmless substances, thereby promoting ecological health. Recent studies have shown endophytic fungi, including *Aspergillus fumigatus*, *Rhizopus* spp., *Penicillium radicum* and *Fusarium proliferatum*, have potential in remediating hexavalent chromium in soil, supporting safer agricultural practices (Bibi *et al.*, 2018).

A novel approach involving the combination of endophytes with nanoparticles has also shown promising results in bioremediation. A nanobiocomposite composed of magnetite nanoparticles (Fe_3O_4) and *Aspergillus flavus* (Af-CL-7) demonstrated remarkable efficiency in bioremediation, significantly degrading Reactive Black 5 (RB5) textile dye while reducing its toxicity. Additionally, this composite exhibited excellent regenerative properties, allowing for multiple reuse cycles without losing effectiveness (Polli *et al.*, 2023).

Recent research has extensively investigated the combined use of endophytic microorganisms and nanoparticles for water treatment, demonstrating significant potential in pollutant removal. While these hybrid systems offer promising advantages for water purification, several challenges must be addressed before large-scale implementation can be realised. A primary limitation is the inherently slow growth rate of endophytes, which currently restricts industrial-scale production and commercial viability. To overcome these barriers, focused research efforts are needed to optimise the efficiency and scalability of these bio-composite systems. Additionally, comprehensive environmental risk assessments are crucial to evaluate potential long-term ecological impacts before widespread adoption (Manganyi *et al.*, 2024).

10.5. Challenges

Endophytes, the microorganisms residing within plant tissues, have emerged as promising agents for environmental remediation due to their ability to degrade pollutants and enhance plant stress tolerance. However, several challenges limit their large-scale application in pollution control. One major constraint is environmental variability, as field conditions such as fluctuating temperatures, pH and moisture levels often reduce endophyte efficacy compared to controlled laboratory settings. Additionally, many endophytes exhibit host specificity, restricting their colonisation to particular plant species and limiting their broader applicability (Nandy *et al.*, 2020).

Competition with native soil microorganisms presents another hurdle, as indigenous microorganisms may outcompete or suppress introduced endophytes, diminishing their survival and pollutant-degrading capacity (Xia *et al.*, 2022). The complexity of mixed contaminants, such as coexisting heavy metals and organic pollutants, further complicates remediation efforts, as few endophytes possess the multifunctional traits needed to address diverse pollutants simultaneously. Moreover, natural degradation rates by endophytes are often slow compared to the rapid generation of industrial waste, making them less effective for immediate clean-up needs (Liu *et al.*, 2024).

Genetic instability is another concern, as engineered endophytes may lose their degradation traits over time under environmental stress (Lee *et al.*, 2025). Regulatory and safety issues, including potential ecological risks like gene transfer or invasiveness, also pose barriers to field application (Xia *et al.*, 2022). Finally, scalability and cost-effectiveness remain significant challenges, as mass production, storage and field deployment of endophytes are still economically unfeasible for widespread use.

While in vitro studies prove endophytes as effective warriors to degrade pollutants, real-world application faces hurdles. Variable field conditions, climates, complex ecosystems and unpredictable microbial interactions can limit their efficacy. Scaling up production, ensuring survival in contaminated sites and addressing regulatory barriers remain challenges. Bridging this gap requires tailored research and bioengineering solutions for practical implementation.

10.6. Conclusions

Endophytes represent a groundbreaking biological tool for environmental remediation, offering sustainable solutions to some of the most pressing pollution challenges, including plastic waste, heavy metal contamination and water pollution. These plant-associated microorganisms demonstrate remarkable capabilities – degrading synthetic plastics through enzymatic processes, detoxifying heavy metals by way of biosorption and sequestration, and breaking down organic pollutants in aquatic systems. Their symbiotic relationships with host plants further enhance phytoremediation efficiency, making them invaluable for environmentally friendly clean-up strategies.

Despite their potential, several challenges must be addressed to realise large-scale applications. Environmental variability, competition with native microorganisms and slower degradation rates compared to industrial pollution levels remain significant hurdles. Additionally, the host-specific nature of many endophytes limits their broad applicability, while regulatory and economic barriers complicate field deployment.

Future advancements hinge on multidisciplinary approaches. Genetic engineering can optimise endophyte strains for enhanced pollutant degradation and stress resilience, while microbial consortia could improve adaptability to complex contamination scenarios. Scaling up from laboratory to field requires rigorous testing under real-world conditions, integrating endophyte-based solutions with circular economy models such as converting degraded plastics into biodegradable materials. Collaboration between researchers, policymakers and industries will be essential to develop standardised protocols and ensure ecological safety.

In the long term, endophytes could transform pollution management by providing cost-effective, energy-efficient and scalable alternatives to conventional methods. By harnessing their natural symbiotic relationships and augmenting their capabilities through biotechnology, endophyte-mediated remediation promises to play a pivotal role in achieving a cleaner, more sustainable planet. Continued innovation and investment in this field will be critical to unlocking their full potential.

REFERENCES

AbuQamar SF, Abd El-Fattah HI, Nader MM *et al.* (2023) Exploiting fungi in bioremediation for cleaning-up emerging pollutants in aquatic ecosystems. *Marine Environmental Research* **190**: 106068.

Agbessenou A, Akutse KS, Yusuf AA *et al.* (2020) Endophytic fungi protect tomato and nightshade plants against Tuta absoluta (Lepidoptera: Gelechiidae) through a hidden friendship and cryptic battle. *Scientific Reports* **10(1)**: 22195.

Aleynova OA and Kiselev KV (2023) Interaction of plants and endophytic microorganisms: molecular aspects, biological functions, community composition, and practical applications. *Plants* **12(4)**: 714.

Babu AG, Kim JD and Oh BT (2013) Enhancement of heavy metal phytoremediation by Alnus firma with endophytic Bacillus thuringiensis GDB-1. *Journal of Hazardous Materials* **250**: 477–483.

Barkodia M, Joshi U, Wati L and Rami NV (2018) Endophytes: a hidden treasure inside plant. *International Journal of Chemical Studies* **6(5)**: 1660–1665.

Barman S, Chowdhury R and Bhattacharya SS (2023) Fungal-based land remediation. In *Bio-Inspired Land Remediation*. Springer, Cham, Switzerland, pp. 165–188.

Bibi S, Hussain A, Hamayun M *et al.* (2018) Bioremediation of hexavalent chromium by endophytic fungi; safe and improved production of Lactuca sativa L. *Chemosphere* **211**: 653–663.

Fadiji AE and Babalola OO (2020) Elucidating mechanisms of endophytes used in plant protection and other bioactivities with multifunctional prospects. *Frontiers in Bioengineering and Biotechnology* **8**: 467.

Fan M, Liu Z, Nan L *et al.* (2018) Isolation, characterization, and selection of heavy metal-resistant and plant growth-promoting endophytic bacteria from root nodules of Robinia pseudoacacia in a Pb/Zn mining area. *Microbiological Research* **217**: 51–59.

Gouda S, Das G, Sen SK, Shin HS and Patra JK (2016) Endophytes: a treasure house of bioactive compounds of medicinal importance. *Frontiers in Microbiology* **7**: 1538.

Gupta S, Kulkarni MG, White JF and Van Staden J (2020) Epigenetic-based developments in the field of plant endophytic fungi. *South African Journal of Botany* **134**: 394–400.

Hernández-Tasco AJ, Tronchini RA, Apaza-Castillo GA *et al.* (2023) Diversity of bacterial and fungal endophytic communities in the leaf blades of Sinningia magnifica, Sinningia schiffneri and Sinningia speciosa from different clades of Gesneriaceae family: a comparative analysis in three consecutive years. *Microbiological Research* **271**: 127365.

Ignatova L, Kistaubayeva A, Brazhnikova Y *et al.* (2021) Characterization of cadmium-tolerant endophytic fungi isolated from soybean (Glycine max) and barley (Hordeum vulgare). *Heliyon* **7(11)**: e08240.

Lee KC, Bentz PC, Vaughn J and Missaoui A (2025) genomic characterization of novel endophyte strains from tall fescue shows genome fragmentation post-hybridization. *Biorxiv*, pp. 2025–02.

Li JK, Yu H, Zeng WM, Wu XL and Yu RL (2017) Research progress of rhizosphere growth promoting bacteria to strengthen phytoremediation of heavy metal contaminated soil. *Life Sciences* **29**: 434–442.

Li Q, Yao S, Wen H *et al.* (2024) Improving lead phytoremediation using endophytic bacteria isolated from the pioneer plant Ageratina adenophora (spreng.) from a mining area. *Toxics* **12(4)**: 291.

Liu CJ, Deng SG, Hu CY *et al.* (2023) Applications of bioremediation and phytoremediation in contaminated soils and waters: CREST publications during 2018–2022. *Critical Reviews in Environmental Science and Technology* **53(6)**: 723–732.

Liu L, Quan S, Li L *et al.* (2024) Endophytic Bacteria Improve Bio-and Phytoremediation of Heavy Metals. *Microorganisms* **12(11)**: 2137.

Manganyi MC, Dikobe TB and Maseme MR (2024) Exploring the potential of endophytic microorganisms and nanoparticles for enhanced water remediation. *Molecules* **29(12)**: 2858.

Meagher RB (2000) Phytoremediation of toxic elemental and organic pollutants. *Current Opinion in Plant Biology* **3(2)**: 153–162.

Muhammad M, Wahab A, Waheed A *et al.* (2024) Harnessing bacterial endophytes for environmental resilience and agricultural sustainability. *Journal of Environmental Management* **368**: 122201.

Nandy S, Das T, Tudu CK *et al.* (2020) Fungal endophytes: Futuristic tool in recent research area of phytoremediation. *South African Journal of Botany* **134**: 285–295.

Nascimento JMD, Otaviano JJS, Sousa HSD and Oliveira JDD (2023) Biological method of heavy metal management: biosorption and bioaccumulation. In *Heavy Metals in the Environment: Management Strategies for Global Pollution*, American Chemical Society, Washington, DC, USA, pp.315–360.

Polli AD, de Oliveira Junior VA, dos Santos Ribeiro MA *et al.* (2023) Synthesis, characterization, and reusability of novel nanobiocomposite of endophytic fungus Aspergillus flavus and magnetic nanoparticles (Fe_3O_4) with dye bioremediation potential. *Chemosphere* **340**: 139956.

Qiu Z, Tan H, Zhou S and Cao L (2014) Enhanced phytoremediation of toxic metals by inoculating endophytic Enterobacter sp. CBSB1 expressing bifunctional glutathione synthase. *Journal of Hazardous Materials* **267**: 17–20.

Sharma D, Khan AA, Idris MG and Khan A (2025) Forest endophytes as a novel source of plastic degradation. In *Forest Fungi*. Academic Press, London, UK, pp. 311–318.

Singh VK and Kumar A (2023) Secondary metabolites from endophytic fungi: Production, methods of analysis, and diverse pharmaceutical potential. *Symbiosis* **90(2)**: 111–125.

Ukaogo PO, Ewuzie U and Onwuka CV (2020) Environmental pollution: causes, effects, and the remedies. In *Microorganisms for Sustainable Environment and Health*. Elsevier, Amsterdam, the Netherlands, pp. 419–429.

Wippel K (2023) Plant and microbial features governing an endophytic lifestyle. *Current Opinion in Plant Biology* **76**: 102483.

Wu T, Xu J, Liu J *et al.* (2019) Characterization and initial application of endophytic Bacillus safensis strain ZY16 for improving phytoremediation of oil-contaminated saline soils. *Frontiers in Microbiology* **10**: 991.

Wu Y, Liu Y, Kamyab H *et al.* (2023) Physico-chemical and biological remediation techniques for the elimination of endocrine-disrupting hazardous chemicals. *Environmental Research* **232**: 116363.

Xia Y, Liu J, Chen C *et al.* (2022) The multifunctions and future prospects of endophytes and their metabolites in plant disease management. *Microorganisms* **10(5)**: 1072.

Microbial Processes in Environmental Engineering: From Theory to Practice

Rajneesh Kumar, Ram Sharan Singh and Maulin P Shah
ISBN 978-1-83708-327-5
https://doi.org/10.1108/978-1-83708-326-820261011

Chapter 11
Harnessing microbial potential for climate change mitigation: innovations and applications

Komal Pandey and Shalini Singh Visen

Amity University Uttar Pradesh, Lucknow, India

Abstract

Microorganisms are at the forefront of innovative climate mitigation strategies, blending natural ecological processes with cutting-edge scientific advancements. Traditionally recognised for their central roles in biogeochemical cycles, such as carbon, nitrogen and sulfur cycling, microorganisms now serve as the backbone for advanced technological interventions aimed at combating climate change. Engineered microbial systems, including biofilms designed for carbon capture and storage, microbial electrosynthesis for renewable bioenergy production and metagenomic tools for optimising methane oxidation and nitrous oxide reduction, exemplify modern applications reshaping climate strategies. Marine and terrestrial microbiomes, particularly those in soils and oceans, function as essential regulators of greenhouse gas (GHG) fluxes, with capabilities to both sequester carbon dioxide and modulate emissions of methane and nitrous oxide. Recent developments in microbiome editing and the integration of biosensors are enhancing precision in environmental monitoring and intervention. Furthermore, synthetic biology advances are enabling the design of customised microbial consortia that optimise carbon sequestration and promote ecosystem resilience under climate stressors. Beyond mitigation, microorganisms are instrumental in adaptive climate strategies, including bioremediation of polluted environments and restoration of degraded ecosystems. By leveraging their unparalleled metabolic diversity and scalability, microorganisms offer transformative, nature-based solutions that align with global sustainability and net-zero carbon emission goals. This synthesis of traditional ecological functions with modern bioengineering underscores the immense potential of microorganisms as allies in building a resilient, climate-stable future.

Keywords: Biogeochemical cycles; carbon sequestration; methane oxidation; nitrous oxide reduction; microbial electrosynthesis.

11.1. Introduction

In the face of accelerating climate change, humanity stands at a critical crossroads. While technological solutions like renewable energy and carbon capture systems represent important pathways forward, nature offers potentially powerful allies in the form of microorganisms. These microscopic entities, which have been regulating Earth's biogeochemical cycles for billions of years, harbour remarkable capabilities that could transform our approach to climate mitigation.

Microorganisms – bacteria, archaea, fungi, algae and protists – are Earth's oldest and most diverse inhabitants, having evolved sophisticated metabolic strategies to thrive in virtually every environment on the planet. Their collective impact on global carbon and nitrogen cycles has shaped our atmosphere throughout geological history (Nielsen *et al.*, 2023).

Today, as we seek to manage climate risks and uncertainty, these microbial capabilities represent an untapped reservoir of potential solutions. The emerging field of microbial climate mitigation spans multiple domains. In agricultural systems, soil microorganisms enhance carbon sequestration while reducing emissions from fertilisers. In oceanic environments, certain phytoplankton communities serve as biological carbon pumps, transferring atmospheric carbon to deep-sea reservoirs. In industrial applications, engineered microorganisms transform waste carbon into valuable materials, offering pathways to circular bioeconomies that reduce fossil fuel dependence.

Recent advances in metagenomics, synthetic biology and microbial ecology have dramatically expanded our understanding of these communities and our ability to harness their functions.

Scientists can now identify, cultivate and, in some cases, enhance microbial populations with specific climate-beneficial properties. From methane-consuming bacteria that target greenhouse gases (GHGs) to specialised fungi that maximise soil carbon storage, microorganisms offer targeted approaches to address different aspects of climate mitigation (Favoino and Hogg, 2021).

This introduction explores the frontier of microbial climate solutions, examining their mechanisms, applications and potential integration into broader mitigation strategies. By understanding how these invisible allies function within ecosystems, we can develop approaches that work with natural processes rather than against them, potentially creating more resilient and sustainable pathways for climate action. As we will discover, the microscopic scale of these organisms belies their potential global impact on our shared climate future.

11.1.1 The climate crisis and its challenges

The Earth is subject to climate risks and uncertainties, characterised by fluctuating global temperatures, extreme weather events and profound ecological disruptions. Atmospheric carbon dioxide concentrations have surpassed 420 parts per million – levels unseen in over 800 000 years – driving cascading effects across Earth's systems (Srivastav *et al.*, 2019). This climate uncertainty presents multifaceted challenges: accelerating biodiversity loss, threatened food security, rising sea levels endangering coastal communities, and widening socioeconomic disparities as vulnerable populations bear disproportionate impacts.

Traditional mitigation approaches – renewable energy transition, improved energy efficiency and carbon capture technologies – while essential, are proving insufficient to meet ambitious targets established under international frameworks like the Paris Agreement. The scale and urgency of climate action requires innovation beyond conventional solutions, compelling scientists and policymakers to explore previously overlooked natural systems with potential to complement existing strategies (Salam and Obayori, 2023).

11.1.2 Microorganisms – nature's unsung climate heroes

In this critical search for solutions, microorganisms emerge as powerful yet underappreciated allies. These microscopic entities – bacteria, archaea, fungi and protists – have been shaping Earth's climate for billions of years through their fundamental roles in carbon, nitrogen and other biogeochemical cycles. Despite comprising only 0.5% of the planet's biomass, microorganisms collectively influence virtually every aspect of our biosphere (Berger *et al.*, 2023).

11.1.2.1 Evolutionary climate engineers

Microorganisms represent Earth's original climate engineers, having evolved sophisticated metabolic pathways for capturing, transforming and sequestering carbon and other greenhouse gases. From the cyanobacteria that first oxygenated our atmosphere to the myriad soil microorganisms that maintain carbon pools in terrestrial ecosystems, microbial communities have played decisive roles in establishing and maintaining planetary conditions conducive to complex life. This evolutionary history has equipped microorganisms with remarkable biochemical capabilities that now offer promising avenues for climate intervention (Timmis *et al.*, 2022).

11.1.2.2 Ubiquity and adaptability

The extraordinary ubiquity of microorganisms – thriving in environments from deep ocean trenches to Antarctic ice sheets – positions them as uniquely versatile climate mitigation agents applicable across diverse ecosystems. Their rapid reproduction rates and remarkable genetic adaptability enable microbial communities to respond to changing conditions at timescales relevant to human climate action. This responsivity, refined through billions of years of evolution, represents an invaluable resource for developing resilient, adaptable climate solutions (Liu *et al.*, 2022).

11.1.2.3 Technological potential

Recent advances in metagenomic sequencing, synthetic biology and microbiome engineering have dramatically expanded our capacity to harness microbial capabilities for targeted applications. These technologies enable identification, optimisation and deployment of microbial communities with enhanced capacities for carbon sequestration, greenhouse gas consumption and sustainable material production. From engineered soil amendments that maximise agricultural carbon storage to bioreactors that transform industrial emissions into valuable products, microbial technologies bridge natural processes with human innovation to address climate challenges across sectors (Deinove Biotechnology, 2023).

As we navigate the complexities of climate change mitigation, microorganisms offer complementary approaches that work with, rather than against, natural processes – potentially transforming our relationship with Earth's systems from exploitative to regenerative. This chapter explores the frontier of microbial climate solutions, examining both their remarkable potential and the scientific, technological and policy considerations essential for their responsible development and implementation.

11.2. Microbial functions in biogeochemical cycles

Biogeochemical cycles describe the movement and transformation of elements essential to life – such as carbon, nitrogen, sulfur and phosphorus – through the biological, geological and chemical processes of Earth's systems. Microorganisms play a central role in these cycles, catalysing key reactions that facilitate the transformation and mobility of these elements in the environment. This chapter highlights the crucial roles of microorganisms in maintaining the balance of the biosphere through their participation in the carbon, nitrogen, sulfur and phosphorus cycles.

11.2.1 Microbial roles in the carbon cycle

Microorganisms are integral to all major processes in the carbon cycle, which involves the fixation, transformation and release of carbon compounds in ecosystems.

11.2.1.1 Carbon fixation

Autotrophic microorganisms such as cyanobacteria, green sulfur bacteria and purple sulfur bacteria use light energy (photoautotrophs) or inorganic compounds (chemoautotrophs) to fix carbon dioxide into organic matter by way of

- Calvin-Benson cycle
- reverse TCA cycle
- 3-hydroxypropionate pathway.

These microbial pathways contribute significantly to primary productivity, especially in aquatic ecosystems and extreme environments where plants are absent.

11.2.1.2 Decomposition and mineralisation

Heterotrophic microorganisms such as fungi, actinomycetes and bacteria (e.g. *Bacillus*, *Pseudomonas*) decompose complex organic matter (e.g. cellulose, lignin, proteins) into simpler compounds, releasing carbon dioxide (CO_2) through respiration and fermentation processes. This mineralisation is vital for nutrient recycling.

11.2.1.3 Methanogenesis

In anaerobic environments such as wetlands, rice paddies and ruminant guts, methanogenic archaea (e.g. *Methanobacterium* and *Methanococcus*) produce methane (CH_4) by reducing CO_2 with hydrogen (H_2) or by fermenting acetate. Methane is a potent greenhouse gas and a substrate for methanotrophic microorganisms.

11.2.1.4 Methanotrophy

Methanotrophs such as *Methylococcus capsulatus* oxidise methane to CO_2 under aerobic conditions. This process reduces atmospheric methane and contributes to global carbon regulation.

11.2.2 Microbial roles in the nitrogen cycle

The nitrogen cycle encompasses transformations between various chemical forms of nitrogen, largely driven by microbial metabolism. These transformations ensure the availability of nitrogen in forms usable by living organisms.

11.2.2.1 Nitrogen fixation

Nitrogen-fixing bacteria and archaea convert atmospheric nitrogen (N_2) into ammonia (NH_3), which can be assimilated by plants. Examples include,

- free-living bacteria: *Azotobacter*, *Clostridium*
- symbiotic bacteria: *Rhizobium* (in legume nodules), *Frankia*
- cyanobacteria: *Anabaena*, *Nostoc*.

The enzyme nitrogenase catalyses this process, which is energy-intensive and oxygen-sensitive.

11.2.2.2 Nitrification

Nitrifying bacteria oxidise ammonia to nitrate in two steps:

- ammonia to nitrite: *Nitrosomonas*, *Nitrosococcus*
- nitrite to nitrate: *Nitrobacter*, *Nitrospira*.

This aerobic process is crucial in soil and aquatic ecosystems for converting ammonia, a potential toxin, into nitrate, a more mobile form.

11.2.2.3 Assimilation

Plants and microorganisms assimilate ammonium and nitrate into organic nitrogen compounds (amino acids and nucleotides), supporting growth and biomass production.

11.2.2.4 Ammonification

Decomposer microorganisms degrade organic nitrogen from dead biomass and waste into ammonia. Bacteria like *Proteus* and *Bacillus* contribute to this process.

11.2.2.5 Denitrification

Facultative anaerobes such as *Pseudomonas*, *Paracoccus* and *Thiobacillus* reduce nitrate back to gaseous nitrogen forms (N_2, N_2O) under anaerobic conditions. This step closes the nitrogen cycle and prevents nitrogen accumulation in ecosystems.

11.2.2.6 Anammox (anaerobic ammonium oxidation)

Anammox bacteria (e.g. *Brocadia* and *Kuenenia*) oxidise ammonium using nitrite as an electron acceptor, producing N_2 gas. This process is significant in nitrogen removal from marine and wastewater environments.

11.2.3 Microbial roles in the sulfur cycle

Sulfur exists in both organic and inorganic forms and undergoes complex transformations mediated by diverse microbial taxa.

11.2.3.1 Sulfate Reduction

Sulfate-reducing bacteria (SRB) such as *Desulfovibrio* and *Desulfotomaculum* use sulfate (SO_4^{2-}) as a terminal electron acceptor under anaerobic conditions, reducing it to hydrogen sulfide (H_2S).

11.2.3.2 Sulfide oxidation

Sulfur-oxidising bacteria like *Thiobacillus*, *Beggiatoa* and *Acidithiobacillus* oxidise H_2S, elemental sulfur (S^0) or thiosulfate to sulfate under aerobic or microaerophilic conditions. This process occurs in sulfur springs, marine sediments and hydrothermal vents.

11.2.3.3 Organic sulfur mineralisation

Decomposer microorganisms (e.g. *Bacillus* and *Aspergillus*) degrade organic sulfur compounds (e.g. cysteine and methionine) in plant and animal biomass, releasing sulfate into the environment.

11.2.4 Microbial roles in the phosphorus cycle

Phosphorus is a key element in adenosine triphosphate (ATP), nucleic acids and membranes, but is typically present in insoluble mineral forms or bound in organic matter.

11.2.4.1 Mineralisation of organic phosphorus

Microorganisms including *Bacillus*, *Pseudomonas* and *Penicillium* secrete phosphatases that hydrolyse organic phosphorus compounds, releasing inorganic phosphate (Pi) into the soil.

11.2.4.2 Solubilisation of inorganic phosphates

Phosphate-solubilising microorganisms (PSM) produce organic acids (e.g. gluconic acid and citric acid) that chelate cations bound to phosphate, thereby making phosphorus available to plants. Examples include

- bacteria: *Bacillus megaterium*, *Pseudomonas fluorescens*
- fungi: *Aspergillus niger*, *Penicillium* spp.

11.2.4.3 Immobilisation and storage

Some microorganisms assimilate and store phosphorus intracellularly as polyphosphate granules, contributing to temporary removal from the available pool and influencing ecosystem phosphorus dynamics.

11.3. Microbial applications in climate mitigation technologies

Microorganisms, due to their immense metabolic diversity and adaptability, are increasingly recognised as powerful tools in addressing the global challenge of climate change. Through innovative applications in bioengineering, renewable energy generation and bioremediation, microbial systems are being harnessed to reduce greenhouse gas emissions, sequester carbon and develop sustainable energy alternatives (Zhang *et al.*, 2022). This section explores current and emerging microbial technologies that contribute to climate change mitigation.

11.3.1 Bioengineering microbial systems

Bioengineering involves the deliberate modification of microbial genomes to enhance or introduce new functions useful for environmental sustainability and climate resilience (Johnson and Johnson, 2023).

11.3.1.1 Synthetic biology for carbon capture

Engineered microorganisms such as cyanobacteria, *E. coli* and yeast can be modified to enhance CO_2 fixation and convert it into value-added products such as biofuels, bioplastics and organic acids. Techniques include

- introduction of synthetic carbon fixation pathways (e.g. the Calvin-Benson cycle or the 3-hydroxypropionate bicycle)
- overexpression of carbonic anhydrases and Rubisco for improved CO_2 assimilation.

11.3.1.2 Engineered methanotrophs and methanogens

Genetically enhanced methanotrophs can be used to convert methane from agricultural, landfill or oilfield sources into biomaterials or biofuels, reducing methane emissions while generating economic value (Mendoza-Arroyo *et al.*, 2022).

11.3.1.3 Microbial production of bioplastics

Bacteria such as *Cupriavidus necator* are engineered to convert CO_2, waste carbon sources or methane into polyhydroxyalkanoates (PHAs), a family of biodegradable plastics that replace petroleum-based polymers (Bhattacharya and Shaw, 2023).

11.3.1.4 Carbon-negative fermentation processes

Industrial fermentative microorganisms (e.g. *Clostridium* and *Lactobacillus*) are being tailored to operate on non-food biomass and capture carbon during fermentation, leading to carbon-neutral or carbon-negative biomanufacturing systems (Wang and Song, 2023).

11.3.2 Microbial electrosynthesis for renewable energy

Microbial electrosynthesis (MES) is a cutting-edge approach that uses microorganisms to convert electrical energy (often from renewable sources) into chemical energy stored in biofuels or organic compounds (Pattanaik *et al.*, 2023).

11.3.2.1 Principles of microbial electrosynthesis

MES uses electroactive microorganisms that receive electrons from a cathode (or transfer them to an anode) in a bioelectrochemical system to reduce CO_2 into multi-carbon compounds. These microorganisms form biofilms on electrode surfaces and mediate electron transfer through cytochromes or nanowires (Hernández *et al.*, 2022).

11.3.2.2 Key microorganisms in MES

- Acetogenic bacteria (e.g. *Sporomusa ovata*, *Clostridium ljungdahlii*) reduce CO_2 to acetate and other volatile fatty acids.
- Methanogens (e.g. *Methanobacterium palustre*) convert CO_2 and electrons to methane, acting as a biological power-to-gas solution.
- Photosynthetic cyanobacteria and algae can also be engineered to produce hydrogen gas or biohydrocarbons when interfaced with light-driven electrode systems.

11.3.2.3 Applications and benefits

- Production of renewable biofuels like methane, ethanol, butanol and hydrogen.
- Valorisation of industrial CO_2 emissions.
- Integration with solar or wind energy systems for grid-scale carbon capture and storage.

11.3.3 Bioremediation of greenhouse gases

Bioremediation refers to the use of living organisms to neutralise or remove environmental pollutants, including GHGs such as carbon dioxide (CO_2), methane (CH_4) and nitrous oxide (N_2O) (Xue *et al.*, 2022).

11.3.3.1 Methane bioremediation

- Methanotrophic bacteria like *Methylosinus*, *Methylococcus* and *Methylomonas* oxidise methane into CO_2 and water.
- Applications include landfill covers, agricultural biocovers and natural gas leak management (Osman *et al.*, 2022).

11.3.3.2 Nitrous oxide reduction

- Certain denitrifying bacteria (e.g. *Paracoccus denitrificans*) can reduce N_2O to N_2, helping to mitigate this highly potent greenhouse gas.
- Enhanced wetland management and microbial inoculants in agriculture are strategies being explored to stimulate in situ N_2O reduction (Jiang *et al.*, 2023).

11.3.3.3 CO_2 sequestration by way of microbial carbonates

- Ureolytic bacteria like *Sporosarcina pasteurii* induce microbial-induced calcite precipitation (MICP), transforming CO_2 into stable carbonate minerals.
- These microorganisms are being trialled in carbon-storing concretes, mine tailings and ocean alkalinity enhancement (Sanchez-Andrea *et al.*, 2023).

11.3.3.4 Marine and soil microbiomes

- Marine cyanobacteria (*Prochlorococcus*, *Synechococcus*) and soil microorganisms contribute to natural carbon sinks by way of photosynthesis and biofilm formation.
- Restoring soil microbial health through regenerative agriculture can enhance carbon storage capacity in soils (Pathak and Kumar, 2022).

11.4. Marine and soil microbiomes as climate regulators

Microbial communities in marine and terrestrial environments are central to the Earth's climate regulation systems. They drive essential biogeochemical processes, mediate greenhouse gas fluxes and influence the carbon balance of ecosystems. Understanding their ecological roles and interactions is critical for developing climate-resilient environmental management practices (Lüdeke-Freund *et al.*, 2023).

11.4.1 Marine microbial communities

Marine microorganisms, including bacteria, archaea, viruses and microalgae, constitute the largest reservoir of biomass in the ocean and form the foundation of the oceanic food web. These organisms play crucial roles in carbon cycling, nutrient dynamics and climate modulation (Pang *et al.*, 2023).

11.4.1.1 Oceanic carbon sequestration

Marine microorganisms contribute significantly to the biological carbon pump, a process where atmospheric CO_2 is captured by way of photosynthesis by phytoplankton (e.g. *Prochlorococcus*, *Synechococcus* and *Thalassiosira*) and transferred to the deep ocean through

- sinking particulate organic carbon (POC)
- microbial loop processes involving zooplankton grazing and microbial degradation (Chen *et al.*, 2023).

This long-term carbon sequestration process helps mitigate global warming by reducing atmospheric CO_2 concentrations.

11.4.1.2 Microbial roles in oceanic nutrient cycles

The principal microbial roles in oceanic nutrient cycles are as follows.

- Nitrogen fixation by marine diazotrophs like *Trichodesmium* supports primary production in nutrient-poor areas.
- Nitrification and denitrification by ammonia-oxidising archaea and bacteria regulate the nitrogen balance and release or consume nitrous oxide (N_2O), a potent greenhouse gas.
- Sulfur cycling involves the microbial transformation of dimethylsulfoniopropionate (DMSP) into dimethyl sulfide (DMS), a compound that affects cloud formation and Earth's radiative balance (Yeung and Richardson, 2023).

11.4.1.3 Marine microbiomes and climate feedback

Marine microbial activity responds to ocean warming, acidification and deoxygenation. Climate change impacts include

- community composition (e.g. shift from diatoms to cyanobacteria)
- metabolic pathways (e.g. enhanced respiration increasing CO_2 release)
- feedback loops that may either amplify or buffer climate effects.

11.4.2 Soil microbial dynamics

Soils harbour some of the most diverse microbial communities on Earth, encompassing bacteria, fungi, archaea, protozoa and viruses. These microorganisms are key to regulating greenhouse gas fluxes and maintaining soil carbon and nutrient balances (Verma *et al.*, 2023).

11.4.2.1 Soil microorganisms and carbon sequestration

Soil microorganisms regulate the decomposition of organic matter and the stabilisation of carbon in soil aggregates. Important processes include

- lignocellulose degradation by fungi (e.g. *Trichoderma* and *Penicillium*) and actinobacteria
- formation of stable humic substances, leading to long-term carbon storage
- contribution to soil organic carbon pools through microbial necromass.

11.4.2.2 Nitrogen and methane cycling in soils

Some key aspects of nitrogen and methane cycling in soils are as follows.

- Nitrifying bacteria (e.g. *Nitrosomonas* and *Nitrobacter*) and denitrifiers (e.g. *Pseudomonas* and *Paracoccus*) control emissions of nitrous oxide (N_2O).
- Methanotrophs oxidise methane in upland soils, acting as a biological filter for atmospheric CH_4.
- In wetlands, methanogens (e.g. *Methanosaeta* and *Methanobacterium*) produce CH_4 under anaerobic conditions, contributing to natural methane emissions

Figure 11.1 Innovative strategies of microbial biotechnology to mitigate climate change

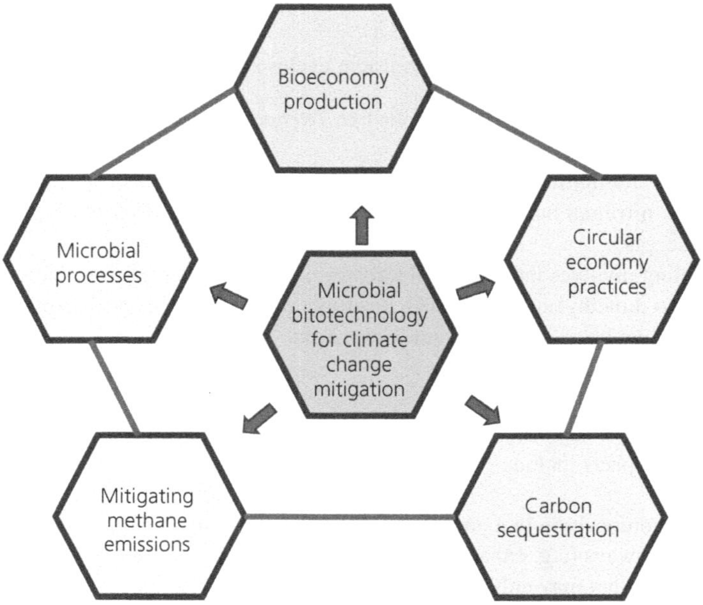

11.4.2.3 Soil health and climate resilience

Soil microbial biodiversity supports plant productivity, drought tolerance and disease suppression. Practices such as reduced tillage, cover cropping and organic amendments enhance microbial activity and carbon retention. Mycorrhizal fungi (e.g. *Glomus* and *Rhizophagus*) improve soil structure and carbon input through symbiosis with plant roots. Figure 11.1 summarises the opportunities for microbial biotechnology to mitigate climate change.

11.4.2.4 Microbial indicators of climate change

Shifts in microbial community composition and function serve as early warning indicators of climate change impacts (Table 11.1), indicated by

- increased thermophilic or acidophilic microorganisms with warming and acidification
- reduced decomposer diversity linked to soil degradation or drought.

Table 11.1 Summarises the innovations and applications of microbial potential for climate change mitigation.

11.5. Cutting-edge innovations and future directions

As the urgency to mitigate climate change accelerates, the integration of advanced technologies into microbial sciences opens new frontiers in environmental management and climate regulation. From precise microbiome editing to artificial intelligence (AI)-enhanced biosensing, these innovations hold transformative potential (Manefield *et al.*, 2023). However, their implementation raises challenges and ethical dilemmas that must be addressed in a responsible and inclusive framework.

Table 11.1 Summary of innovations and applications of microbial potential for climate change mitigation (continued on next page)

Microbial innovation/ application	Mechanism	Climate change impact	Example/case study
Soil carbon sequestration	Microbial decomposition and humification of organic residues into stable soil organic matter	Long-term storage of atmospheric CO_2 in soils	Fungal–bacterial consortia in regenerative agriculture systems
Methanotrophic bacteria	Aerobic methane oxidation by methanotrophs	Reduction of atmospheric CH_4 emissions	Methane-oxidising bacteria in forest soils and tree bark ecosystems
Nitrogen-fixing microorganisms	Biological nitrogen fixation replacing synthetic fertilisers	Reduced N_2O emissions and fossil fuel dependence	*Rhizobium* in legumes; *Azospirillum* in cereals
Microbial carbon capture and storage (MCCS)	Engineered microorganisms converting CO_2 into biomass or bioproducts	Direct reduction of atmospheric CO_2 levels	Algae-based photobioreactors for CO_2 fixation
Microbial fuel cells	Extracellular electron transfer generating electricity from organic substrates	Renewable bioenergy production reducing fossil fuel use	*Geobacter* and *Shewanella* spp. in bioelectrochemical systems
Microbial biochar enhancement	Inoculation of biochar with beneficial microorganisms	Increased soil carbon stability and fertility	Biochar + PGPR applications in agroforestry
Compost microbiology	Controlled microbial degradation of organic waste	Reduction of landfill methane emissions	Vermicomposting systems integrating microorganisms and earthworms
Biological soil crusts (BSCs)	Cyanobacteria and fungi stabilising arid soils	Carbon storage and desertification prevention	Dryland restoration using biocrust inoculation
Denitrifying microorganisms	Conversion of nitrates to N_2 gas in low-oxygen environments	Mitigation of nitrous oxide emissions	*Pseudomonas* spp. in rice paddies

Table 11.1 Continued

Microbial innovation/ application	Mechanism	Climate change impact	Example/case study
Microalgae for biofuels	Photosynthetic CO_2 fixation and lipid production	Replacement of fossil fuels; reduced carbon footprint	*Chlorella* and *Spirulina* in biodiesel production
Lignocellulolytic microorganisms	Enzymatic breakdown of agricultural residues	Reduced open-field burning and associated GHG emissions	*Trichoderma* and *Phanerochaete* in biomass conversion
Enteric methane mitigation	Modulation of rumen microbiota using probiotics or inhibitors	Decreased livestock methane emissions	Targeting methanogenic archaea in cattle
Phosphate-solubilising microorganisms (PSMS)	Increased phosphorus bioavailability in soil	Reduced phosphate mining and indirect CO_2 emissions	*Bacillus* and *Aspergillus* in sustainable farming
Marine microbial carbon pump	Microbial transformation of dissolved organic carbon into recalcitrant forms	Enhanced long-term ocean carbon sequestration	Deep-sea microbial carbon storage processes
Biocorrosion control	Management of microbially influenced corrosion (MIC)	Reduced infrastructure loss and carbon-intensive repairs	Sulfate-reducing bacteria control in pipelines

11.5.1 Microbiome editing and synthetic biology

11.5.1.1 Microbiome editing technologies

Recent advances in gene-editing tools, especially CRISPR-Cas systems, have enabled researchers to selectively manipulate microbial genomes or entire microbiomes to enhance desired traits (Manzoor *et al.*, 2023). Applications include

- improving carbon sequestration capabilities of soil or marine microorganisms
- engineering microorganisms to degrade greenhouse gases like methane and nitrous oxide
- enhancing plant-microorganism interactions for climate-resilient agriculture.

Microbiome editing also allows for the removal of pathogenic or harmful microbial strains and introduction of beneficial consortia, tailored to specific environmental goals (Torres-Tiji *et al.*, 2023).

11.5.1.2 Synthetic biology in microbial design

Synthetic biology focuses on constructing novel biological parts, systems and organisms (Dubey *et al.*, 2022). In climate-related applications, synthetic biology enables

- creation of artificial microbial consortia for biogeochemical optimisation
- development of CO_2-fixing autotrophs with enhanced efficiency
- construction of biosynthetic pathways for converting waste into biofuels or carbon-neutral products.

Examples include engineered cyanobacteria for direct fuel production and yeast strains capable of sequestering CO_2 while producing industrial precursors.

11.5.2 Integration of biosensors and AI

11.5.2.1 Microbial biosensors

Microbial biosensors are living or semi-synthetic systems that detect and report environmental changes, such as gas fluxes or pollutant levels (Thomas *et al.*, 2022). These are

- engineered with reporter genes like green fluorescent protein (GFP) or electrochemical signals
- used for real-time monitoring of microbial activity in soil, water and air
- essential in carbon accounting, nutrient cycling studies and early detection of climate-related anomalies.

11.5.2.2 Role of artificial intelligence and machine learning

AI is revolutionising the way we analyse complex microbial datasets (Ghosh *et al.*, 2023). With metagenomics, transcriptomics and metabolomics generating large volumes of data, AI and machine learning (ML) offer

- pattern recognition in microbial community shifts
- predictive modelling for microorganism–climate interactions
- optimisation of microbial strain design for synthetic biology.

AI-driven climate modelling that integrates microbial data could improve the accuracy of greenhouse gas emission forecasts, ecosystem resilience predictions and climate change mitigation strategies.

11.5.3 Challenges and ethical considerations

11.5.3.1 Technical and ecological risks

While microbiome editing and synthetic biology promise significant benefits, risks include

- horizontal gene transfer to native or non-target species
- unintended ecological imbalances or collapse of natural microbial networks
- development of resistant or invasive microbial strains (Huang *et al.*, 2023).

11.5.3.2 Societal and regulatory concerns

Such concerns of the introduction of microbial technology into the field include

- lack of international regulatory standards for synthetic microorganisms or edited microbiomes
- concerns about biosecurity and potential misuse in bioterrorism
- need for transparent, inclusive public engagement to build trust and acceptance (Maroušek *et al.*, 2023).

11.5.3.3 Ethical considerations

The key ethical considerations for use of microbial technology are as follows.

- Ownership and access to engineered microbial technologies raise questions of equity and justice, especially in resource-poor nations.
- The precautionary principle must be applied to prevent long-term ecological and evolutionary impacts.
- Ethical deployment requires respecting indigenous knowledge systems, ensuring environmental stewardship and promoting sustainability over profit (Rajaee *et al.*, 2023).

11.6. Practices integrating microorganisms to reduce agricultural emissions

Agriculture is a major contributor to GHG emissions, especially methane (CH_4) from livestock and rice paddies, nitrous oxide (N_2O) from fertilised soils and carbon dioxide (CO_2) from land-use change and soil degradation. Microbial interventions offer sustainable and biologically intelligent strategies to reduce these emissions while enhancing soil fertility, plant health and productivity (Figure 11.2) (Morelli *et al.*, 2023).

11.6.1 Biofertilisers and nitrogen-fixing microorganisms

11.6.1.1 Reduction in nitrous oxide emissions

Biofertilisers containing nitrogen-fixing bacteria (e.g. *Rhizobium*, *Azotobacter* and *Frankia*) reduce dependence on synthetic nitrogen fertilisers, which are a major source of N_2O emissions. By biologically fixing atmospheric nitrogen, these microorganisms

- provide plants with a sustainable nitrogen source
- minimise nitrogen leaching and denitrification losses
- reduce energy inputs associated with fertiliser production.

Figure 11.2 Revolutionary approaches in microbial biotechnology to enhance food security

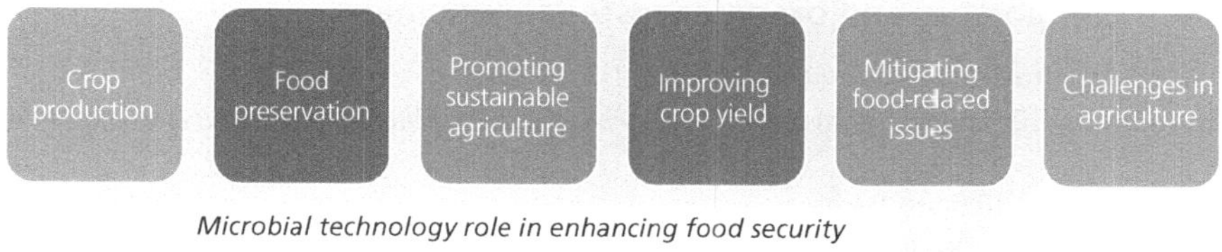

11.6.1.2 Slow-release and controlled mineralisation

Phosphate-solubilising and potassium-mobilising microorganisms facilitate nutrient availability in a more controlled manner, leading to efficient nutrient use and lower GHG output (Yadav *et al.*, 2022)

11.6.2 Methane-reducing microorganisms in paddy fields and livestock

11.6.2.1 Methanotrophs in rice cultivation

In flooded rice paddies, methanogens (anaerobic archaea) generate methane during organic matter degradation. Methanotrophs, such as *Methylococcus* and *Methylocystis*, consume this methane before it escapes into the atmosphere (Nielsen *et al.*, 2023). Practices include

- inoculating soils with methanotroph-enriched consortia
- managing water levels (alternate wetting and drying) to support aerobic zones favouring methanotroph activity.

11.6.2.2 Microbiome engineering in ruminants

Enteric fermentation in livestock produces substantial CH_4. Modifying the rumen microbiome can be achieved using

- probiotics (e.g. *Propionibacterium* and *Bacillus subtilis*)
- methanogen inhibitors (e.g. 3-nitrooxypropanol targeting *Methanobrevibacter*)
- dietary microbial additives (e.g. yeast and algae) that shift fermentation away from methanogenesis.

11.6.3 Microbial consortia for carbon sequestration

11.6.3.1 Enhancing soil organic carbon

- Mycorrhizal fungi (e.g. *Glomus* and *Rhizophagus*) enhance carbon input from plant roots and contribute to soil aggregation.
- Actinobacteria and saprophytic fungi decompose residues into stable humic substances.
- Biochar inoculated with microbial consortia creates a long-lasting carbon sink.

11.6.3.2 Plant growth-promoting rhizobacteria (PGPR)

PGPR such as *Pseudomonas*, *Bacillus* and *Azospirillum* improve root biomass, leading to greater carbon input into soil systems and indirectly reduce GHG emissions by enhancing nutrient efficiency (Favoino and Hogg, 2021).

11.6.4 Composting and organic waste management

11.6.4.1 Microbial inoculants in composting

Use of thermophilic microorganisms (e.g. *Thermus* and *Bacillus licheniformis*) in composting

- enhances decomposition rate
- reduces CH_4 and N_2O emissions from improperly managed manure and crop residues
- converts waste into high-quality organic fertiliser, reducing synthetic input needs (Salam and Obayori, 2023).

11.6.4.2 Anaerobic digesters

Microbial communities in biogas plants convert livestock waste into methane and digestate.

- Methane can be captured and used as renewable energy.
- Digestate is nutrient-rich and reduces the need for chemical fertilisers.

11.6.5 Biochar–microorganism interactions

Biochar provides habitat and attachment surfaces for microorganisms. When co-inoculated with beneficial microorganisms it

- enhances microbial stability and activity
- improves nutrient retention
- reduces N_2O emissions by promoting complete denitrification (conversion of nitrate to N_2, not N_2O) (Berger *et al.*, 2023).

11.6.6 Precision agriculture with microbial sensors

Some considerations in the use of microbial sensors include

- use of biosensors to monitor microbial activity and nutrient cycling in real time
- application of AI-based decision tools to apply microbial products when and where they are most effective
- reduction in over-application of inputs, minimising emissions and environmental contamination.

11.7. Conclusions

Microbial interventions represent a transformative shift in sustainable agriculture, offering eco-friendly alternatives to conventional practices that often contribute heavily to greenhouse gas emissions. Through nitrogen fixation, methane mitigation, carbon sequestration and organic waste recycling, microorganisms act as natural allies in reducing agriculture's environmental footprint.

By leveraging biofertilisers, methanotrophs, mycorrhizae and composting inoculants, farmers can enhance nutrient efficiency, improve soil health and reduce reliance on synthetic inputs – all while cutting emissions of potent greenhouse gases like methane and nitrous oxide.

Emerging strategies, including microbiome engineering and biochar–microorganism synergy, further enhance the potential of microbial solutions. The integration of these microbial practices, supported by modern technologies such as biosensors and AI-driven precision agriculture, signals a paradigm shift toward climate-smart farming. However, widespread adoption requires collaborative efforts in research, policy support, farmer training and public awareness. Ultimately, harnessing the power of microorganisms not only mitigates climate change but also fosters resilience and regeneration in agricultural ecosystems – paving the way for a more sustainable and productive future.

REFERENCES

Berger S, Jang I and Seo J (2023) Microbial carbon capture: from lab to industrial scale implementation. *Applied Energy* **331**: 120355.

Bhattacharya P and Shaw R (2023) Climate-resilient agriculture through microbial interventions. *Frontiers in Microbiology* **14**: 1209574.

Chen X, Wang H and Li P (2023) Engineered microorganisms for carbon dioxide capture and conversion: progress and challenges. *Biotechnology Advances* **62**: 108052.

Deinove Biotechnology (2023) Advances in microbial carbon capture and utilization. *Biotechnology for Biofuels* **16**: 84.

Dubey A, Malla MA, Khan F *et al.* (2022) Microbes as catalysts for environmental sustainability: biotransformation and bioremediation. *Frontiers in Microbiology* **13**: 963769.

Favoino E and Hogg D (2021) Composting and greenhouse gases: effects of materials, management and policy. *Waste Management and Research* **39(7)**: 889–905.

Ghosh P, Shah G, Sahota S *et al.* (2023) Advances in biochar-based microbial fuel cells for waste-to-energy production and carbon sequestration. *Renewable and Sustainable Energy Reviews* **173**: 113086.

Hernández M, Conrad R, Klose M *et al.* (2022) Methanotrophs as biocatalysts for conversion of methane to high-value products. *Trends in Biotechnology* **40(9)**: 1115–1130.

Huang J, Cao Y and Zhan D (2023) Novel approaches for enhancing microbial methanotrophy in landfill cover soils. *Waste Management* **162**: 24–37.

Jiang Z, Huang Y and Xu X (2023) Microbiomes of coastal marine ecosystems: key players in blue carbon sequestration and climate regulation. *Nature Ecology and Evolution* **7**: 1139–1153.

Johnson KA and Johnson DE (2023) Methane emissions from ruminant livestock: microbial interventions and dietary strategies. *Journal of Animal Science* **101(5)** 1–18.

Liu C, Wang K and Zheng X (2022) Biochar application and methane emission: a meta-analysis. *Global Change Biology* **28(4)**: 1470–1486.

Lüdeke-Freund F, Gold S and Bocken NMP (2023) Circular business models for microbial bioremediation: economic and ecological perspectives. *Journal of Cleaner Production* **402**: 136523.

Manefield M, Gut F and Jones MA (2023) Microbial approaches to methane emission reductions in rice paddies. *Science of the Total Environment* **863**: 160348.

Manzoor M, Ahmad I and Haider MS (2023) Microalgae-bacteria consortia: synergistic approach for enhanced carbon capture and biomass production. *Algal Research* **71**: 103010.

Maroušek J, Stehel V and Vochozka M (2023) Fermentation processes in biogas production: a techno-economic assessment of GHG reduction potential. *Journal of Cleaner Production* **398**: 136507.

Mendoza-Arroyo GE, González-Figueroa AC and Cuervo-López FM (2022) Advanced microbial processes for N2O emission reduction in wastewater treatment. *Bioresource Technology* **361**: 127576.

Morelli B, Cashman S and Arden S (2023) Life cycle assessment of microbial carbon capture technologies in industrial applications. *Environmental Science and Technology* **57(12)**: 4824–4836.

Nielsen J, Archer J, Essack M *et al.* (2023) Building a sustainable future with microbial biotechnology. *Nature Reviews Microbiology* **21**: 351–365.

Osman AI, Hefny M, Maksoud MIA *et al.* (2022) Recent advances in carbon capture storage and utilisation technologies: a review. *Environmental Chemistry Letters* **20**: 2551–2604.

Pang N, Wang T and Hu X (2023) Cyanobacteria-based technologies for sustainable CO2 fixation and biomaterial production. *Trends in Biotechnology* **41(8)**: 859–874.

Pathak VM and Kumar A (2022) Microbial degradation of plastics: a comprehensive review. *Environmental Pollution* **310**: 119841.

Pattanaik L, Pradhan D, Halder PK and Biswal B (2023) Value-added products from microalgae biomass for efficient carbon sequestration and sustainable development. *Bioresource Technology* **369**: 128339.

Rajaee M, Kermani M and Driscoll CT (2023) Microalgae-based biofertilizers: a sustainable alternative to reduce agricultural greenhouse gas emissions. *Journal of Cleaner Production* **409**: 137191.

Salam LB and Obayori OS (2023) Microbial remediation of anthropogenic pollutants: current status and future prospects. *Environmental Pollution* **319**: 120879.

Sanchez-Andrea I, Guedes IA and Hornung B (2023) The microbiome of renewable energy: optimization strategies for biogas production. *Renewable Energy* **204**: 355–371.

Srivastav A (2019) *The Science and Impact of Climate Change*. Springer, Singapore.

Thomas SC, Fricker GA and Sanchez-Azofeifa GA (2022) Biochar amendments for enhanced soil carbon sequestration: a meta-analysis of microbial activity and climate effects. *Biochar* **4**: 43.

Timmis K, de Vos WM, Ramos JL *et al.* (2022) The urgent need for microbiology solutions to global problems. *Environmental Microbiology* **24(7)**: 2992–3010.

Torres-Tiji Y, Fields FJ and Mayfield SP (2023) Industrial microalgae as a novel approach to atmospheric carbon mitigation. *Trends in Biotechnology* **41(1)**: 43–57.

Verma S, Bhattacharya A and Singh A (2023) Purple non-sulfur bacteria: A sustainable approach for carbon sequestration and waste valorization. *Renewable and Sustainable Energy Reviews* **178**: 113138.

Wang J and Song Y (2023) Microbial manipulation of soil organic carbon for climate change mitigation. *Soil Biology and Biochemistry* **176**: 108862.

Xue X, Yuan H and Kang J (2022) Advances in microbial electrochemical systems for carbon capture and conversion. *Renewable and Sustainable Energy Reviews* **167**: 112729.

Yadav AN, Singh J, Singh S *et al.* (2022) Recent advances in microbial technologies for sustainable agriculture and environment management. *Environmental and Sustainability Indicators* **16**: 100198.

Yeung ACY and Richardson JS (2023) Microbial responses to climate change in freshwater ecosystems: From metabolic adaptations to ecosystem services. *FEMS Microbiology Reviews* **47(2)**: fuad003.

Zhang Y, Kumar M and Guo M (2022) Microbially induced calcium carbonate precipitation for carbon sequestration: Mechanisms and applications. *Nature Communications Earth and Environment* **3**: 245.

Microbial Processes in Environmental Engineering: From Theory to Practice

emerald PUBLISHING ice Publishing

Rajneesh Kumar, Ram Sharan Singh and Maulin P Shah
ISBN 978-1-83708-327-5
https://doi.org/10.1108/978-1-83708-326-820261012

Chapter 12
Microbial plastic and microplastic bioremediation and innovations

Aubrey Dickon Chigwada and Memory Tekere

Department of Environmental Science, College of Agricultural and Environmental Sciences, University of South Africa, Roodepoort, Gauteng, South Africa

Abstract

Plastics have become an indispensable part of our daily lives and, despite their negative effects on the environment, they are unavoidable. A continual accumulation of plastic waste has led to plastic pollution and this environmental problem has increased pressure on scientists to develop alternatives for plastic biodegradation and bioremediation. This chapter will examine the methods of action of microorganisms and their enzymatic pathways in the biodegradation of plastic polymers, which will also highlight recent findings of multiple microbial species that may biodegrade plastic and the distinct enzymes they produce throughout the bioremediation process. Also highlighted are new methods in engineering, biotechnology and nanotechnology that may increase the pace of microbial plastic biodegradation by maximising microbial efficacies. This chapter will also discuss the limitations, obstacles and difficulties encountered in the implementation of plastic biodegradation, as well as the ecological effects of engineering solutions and the efficient promotion of natural plastic biodegradation. Finally, the significance of interdisciplinary research in promoting sustainable plastic biodegradation within waste management, addressing the constraints of plastic bioremediation and providing avenues for future research while enhancing the potential for large-scale industrial plastic biodegradation, is highlighted.

Keywords: Plastic and microplastic biodegradation; enzymatic remediation of plastic; hazards; environmental pollution; microbial elimination of plastic waste; waste management.

12.1. Introduction

Plastics are composed of polymers which are repeating units of monomer molecules that are chemically bound during polymerisation, which eventually form lengthy chains of distinct networks that give plastics their distinctive properties (Lear *et al.*, 2021). The chemical makeup of plastic is influenced by the degree of polymerisation as well as the stabilising, plasticising, flame-repelling, antioxidant and pigment additives that are frequently added to plastic to enhance its performance (Zhang *et al.*, 2021).

The most widely used plastics are polystyrene (PS), polypropylene (PP), polyvinyl chloride (PVC) and polyethylene (PE) (Sangale *et al.*, 2019). These polymers are usually made from crude oil or natural gas (petrochemicals). Polybutylene succinate (PBS), polyvinyl alcohol (PVA), chitosan, rayon, polyhydroxy-butyrate-co-valerate (PHBV), cellulose acetates (CAs) and polylactic acid (PLA) are examples of biobased polymers that are derived from renewable resources like corn and sugarcane (Colachis *et al.*, 2024; Kwon *et al.*, 2023; Qi *et al.*, 2017). The ability to be moulded into any shape or form, durability, insulation, cost-effectiveness and partial recyclability make plastics valuable in a variety of industries, including electronics, medical equipment, packaging and construction as well as in our daily lives.

However, after-use plastics present a huge environmental problem: they can persist in the environment for long periods of time, as most plastics are non-biodegradable, leading to their accumulation in landfills and natural ecosystems (Biki *et al.*, 2021; Hou and Majumder, 2021). Microplastics are formed when plastics partially biodegrade, contaminating water, soil and air before making their way into food supplies and endangering the health of both humans and animals (Gong *et al.*, 2018).

An estimated 7 million tonnes of plastic debris enter the oceans each year through rivers, streams and maritime dumping; this contributes to entanglement, ingestion by marine life and degradation of oceanic habitats (Davidov *et al.*, 2020; Gao and Sun, 2021). The production of plastics is dependent on the use of fossil fuels leading to the production of large amounts of carbon dioxide (CO_2) and other greenhouse gases (Amobonye *et al.*, 2021). Approximately 1.5% of worldwide emissions in 2019 came from the manufacture and burning of plastic, which produced 850 000 000 t of greenhouse gases (Hillman *et al.*, 2015). Even though plastics can be recycled, only around 9% of all plastics manufactured have been recycled, according to current estimates. This is because of poor waste management, which leads to a significant amount of plastics being burned or ending up in landfills (Hillman *et al.*, 2015). Additionally, the use of non-renewable petroleum in the production of plastic will contribute to the depletion of limited resources; however, some plastic additives, like flame retardants, bisphenol A (BPA) and phthalates, can be toxic and may leak into the environment or food sources, posing health risks (Fauvelle *et al.*, 2021).

Since plastic pollution is still a major environmental problem, it is essential to comprehend the biological processes that can break down plastics naturally in order to create efficient bioremediation techniques. The process of turning plastics into ecologically benign end products by using microorganisms like fungi, bacteria, archaea and so on is known as bioremediation (Lear *et al.*, 2021). This method uses microbial innate metabolic powers to convert synthetic polymers into less toxic, simpler substances like biomass, carbon dioxide and water (Sheridan *et al.*, 2022).

Bioremediation is becoming increasingly popular as an environmentally sustainable solution to the growing problem of microplastic and plastic pollution. In a study by Liu *et al.* (2024), 29% biodegradation of PLA plastic was observed by species of *Thermotogota* and *Bacteroidota* in 30 days of incubation. Consortia of bacteria and fungi (*Bacillus subtilis*, *Fusarium oxysporum* and *Alternaria alternata*, and *Bacillus subtilis* and *Pseudomonas alloputida* formulated by Salinas *et al.* (2025)) achieved a multi-plastic biodegradation of up to 18% in weight loss.

Ibrahim *et al.* (2024) demonstrated the biodegradation of tyre rubber, polyurethane (PU) and PE plastics by a fungal consortium composed of *Fusarium*, *Penicillium*, *Botryotinia cinerea* and *Tri-choderma* isolated from lakes in Germany. Around 414 000 000 t of plastic are produced worldwide (Tournier *et al.*, 2020), polluting waterways and endangering aquatic life. The scientific world has given these plastic constituents a lot of attention because of their high chemical stability and persistence, which cause bioaccumulation and make them extremely recalcitrant in nature. Furthermore, it is a concerning problem for the ecosystem when nano- and micro-sized plastic particles are found in soils, lakes and even in areas like the Antarctic, the Arctic and clouds where there is little or very little human activity (Bacha *et al.*, 2023).

This chapter explores the science of microbial biodegradation and its possible application as a promising solution to reduce plastic pollution. It also looks at biodegradation mechanisms, bioremediation techniques, engineering innovation, nanotechnology and biotechnology in plastic and microplastic bioremediation. This chapter also addresses the difficulties encountered in the implementation of microbial biodegradation of plastics and explores novel approaches, such as (bio)electrochemical systems, which hold promise for improving the bioremediation of plastic waste.

12.2. Microbial mode of action and enzyme pathways in plastic degradation

The initial step in the process of plastic biodegradation is the adhesion of microorganisms to plastic surfaces, which results in the formation of biofilms that in turn result in the release of degradative enzymes (Deepika and Madhuri, 2015). Specialised enzymes break down long-chain polymers into oligomers, dimers and finally monomers as a result of the extracellular enzymes produced, which causes surface depolymerisation (Chigwada and Tekere, 2023). Subsequently, the monomers are absorbed by the microbial cells and undergo intracellular metabolism through many pathways, culminating in total mineralisation (Edwards *et al.*, 2022; Roberts *et al.*, 2020).

The driving forces behind the biodegradation of plastic polymers are enzymes, which are distinguished by their mode of action and the bonds they target.

Hydrolytic enzymes employ water molecules to break ester bonds in polymers. These enzymes include lipases, esterases and cutinases (Puiggené *et al.*, 2022; Ronkvist *et al.*, 2009; Temporiti *et al.*, 2022). Oxidative enzymes include laccases and peroxidases, which alternately add oxygen to polymer structures, weakening them for breakdown. Depolymerase enzymes exhibit action against both manufactured and naturally occurring polymers (Nakamura *et al.*, 1999; Pinnell and Turner, 2019; Zhang *et al.*, 2022).

Blázquez-Sánchez *et al.* (2022) showed that, at room temperature of 25°C, species of *Oleispira antarctica* and *Moraxella* bacteria produced psychrophilic hydrolytic enzymes that hydrolysed polyethylene terephthalate (PET) and polyester polycaprolactone and initiated biodegradation and ultimately resulted in films losing weight. Their bioinformatics study also confirmed the presence of psychrophilic PET hydrolase genes in the genomes of these bacteria (Blázquez-Sánchez *et al.*, 2022).

Similarly, metagenomically generated polyester hydrolase PHL7 was found by Richter *et al.* (2023) to have PET biodegradation activity, resulting in PET film biodegradation.

Ronkvist *et al.* (2009) showed that *Pseudomonas mendocina* and *Fusarium solani* species could hydrolyse PET films with cutinase, resulting in a 5% weight loss at 50°C. They also demonstrated that at 70°C *Humilica insolens* biodegraded PET film with low crystallinity by up to 97% using cutinase enzymes.

Tan *et al.* (2021) investigated the use of extracellular lipase derived from *Amycolatopsis mediteranei* species in the bioremediation of aliphatic polymers that were adjusted with 1,6-diisocyanatohexane, including poly(1,4-butylene succinate) and poly(e-caprolactone). According to their observations, these aliphatic polymers were completely hydrolysed.

The study by Li *et al.* (2025) presented a comprehensive mineralisation model of PE film, revealing that white rot fungus colonised PE film, secreted extracellular oxidative enzymes and accessed partially dissociated sub-microplastic debris. This process led to the oxidation and cracking of the PE film, providing new insights into the causes of polyethylene film cracking during biodegradation.

Table 12.1 shows a summary of microorganisms that degrade plastics, the enzyme mechanisms they utilise and the type of plastics they have been reported to biodegrade.

By utilising microorganisms like bacteria and fungi (Table 12.1) to break down plastic into smaller, less hazardous components, bioremediation is a promising strategy to combat plastic pollution. This environmentally friendly technique can be used to lessen the permanence and impact of contaminated areas such as soil, water bodies and landfills. Although bio-based polymers derived from sustainable biomass sources can either be natively biodegradable in natural environments or made to disintegrate more easily, plastic compositions can be improved by producing plastic with greater biodegradable qualities by adding biodegradable additives.

12.3. Microbial plastic and microplastic bioremediation design strategies

12.3.1 Bioreactors in plastic bioremediation

Bioreactors work by using microorganisms (fungi, bacteria and archaea) to degrade microplastics (Tekere, 2019). They are designed to support microbial activity and enhance degradation by providing a regulated and optimised environment that includes temperature, pH, oxygen levels and nutrition availability (Zhou *et al.*, 2024). Microplastics serve as the only source of carbon in ex situ bioremediation using bioreactors, which entails transporting plastic waste to the bioreactor site for offsite treatment (Tekere, 2019). Microplastics are then broken down by microorganisms to produce carbon dioxide, biomass and water as a source of energy and carbon. Using plastic as a carbon source, a minimum salt medium and laboratory flasks under controlled circumstances, microbial consortia have demonstrated good efficiency in plastic biodegradation in laboratory-scale bioreactors (Krainara *et al.*, 2023).

Table 12.1 Summary of microorganisms shown to degrade plastics and the confirmed enzymatic mechanisms they utilise

Organism	Plastic type degraded	Mechanism	Reference
Bacteria			
Ideonella sakaiensis	PET	Produces PETase and MHETase enzymes	Han *et al.*, 2017; Joo *et al.*, 2018; Tournier *et al.*, 2020
Pseudomonas species	PE, PU, PS	Oxidative degradation	Jiang *et al.*, 2021; Kawai, 2002; Priyanka and Archana, 2011
Rhodococcus ruber	PE, LDPE	Biofilm and enzymatic action	Mohamed *et al.*, 2024; Nair *et al.*, 2016; Rong *et al.*, 2024
Bacillus subtilis	PU, PS, PE	Secretes hydrolytic enzymes	Priyanka and Archana, 2011; Salinas *et al.*, 2023
Streptomyces species	PP, PU, PE	Enzymatic breakdown	Abraham *et al.*, 2017; El-Shafei *et al.*, 1998; Wichatham *et al.*, 2024
Comamonas testosteroni	PU, PET	Hydrolysis by way of esterase enzymes	Kumar *et al.*, 2021; Pathak and Navneet, 2017
Fungi			
Aspergillus niger	PE, LDPE, PU, PS, PVC	Secretes oxidases, esterases	Ali *et al.*, 2014; Priyanka and Archana, 2011; Saira *et al.*, 2022
Aspergillus flavus	PE, PS, LDPE	Laccases, hydrolytic enzymes	El-Shafe *et al.*, 1998; Saira *et al.*, 2022; Taghavi *et al.*, 2021
Penicillium simplicissimum	PE, PU	Oxidative enzymes, biofilm formation	Yamada-Onodera *et al.*, 2001
Fusarium solani	PET, PU	Esterases, ureases	Hellesnes *et al.*, 2023; Zahra *et al.*, 2010
Phanerochaete chrysosporium	PE, PVC	Ligninolytic enzymes (e.g. MnP, LiP)	Ali *et al.*, 2014; Lee *et al.*, 1991; Stoleru *et al.*, 2017
Trichoderma harzianum	PP, PE	Hydrolytic and oxidative enzymes	Sowmya *et al.*, 2014
Mucor circinelloides	PU, PS, PE	Esterases, proteases	Chaudhary and Vijayakumar, 2020; Cui *et al.*, 2017; Raaman *et al.*, 2012
Archaea			
Methanosaeta concilii	PLA	Anaerobic degradation	Zhen *et al.*, 2023
Methanobacterium petrolearium	PCL, PLA, PHB, PBS	Anaerobic degradation	Yagi *et al.*, 2014
Halobacterium species	PE	Enzymatic	Simó-Cabrera *et al.*, 2021

The controlled environment that several types of bioreactors offer for plastic microbial biodegradation makes microbial bioremediation an environmentally friendly substitute for vaporising plastic waste (Kumar *et al.*, 2021; Zahra *et al.*, 2010). There are several types of bioreactors including suspension bioreactors (Tournier *et al.*, 2020), enzymatic and anaerobic membrane bioreactors (Khanal *et al.*, 2017) and deep eutectic solvent bioreactors (Krainara *et al.*, 2023).

Suspension bioreactors, one of the earliest and most traditional types of modern bioreactors, provide a uniform distribution of microorganisms and plastic particles in a submerged liquid fermentation. They were used in the study by Moreno *et al.* (2024) that demonstrated low-density polyethylene (LDPE) biodegradation by fungal species of *Aspergillus brasiliensis*. This showed that the suspension bioreactor's aerobic or anaerobic operating conditions speed up the rates of LDPE biodegradation. Fachrul *et al.* (2024) utilised a rectangular laboratory-scale bioreactor in the degradation of PE by a consortium of bacteria and fungi and achieved over 60% degradation. It is noteworthy that a synthetic bacterial community successfully broke down ε-caprolactam and alkanes from plastic pyrolysis oil waste, eliminating ε-caprolactam and naphthalene completely in 7 days and removing alkanes with removal efficiencies ranging from 71% to 100% (Lü *et al.*, 2024).

Anaerobic membrane bioreactors coupled with or integrated into electrochemical systems are an additional kind of bioreactor. A study by Zhou *et al.* (2024) showed that anaerobic membrane bioreactors enhanced microbial activity and resistance and that hydrolytic-acidogenic bacteria, including *Firmicutes* and *Bacteroidota*, were able to degrade microplastics efficiently.

Bioreactors can also be in the form of composting systems that are often combined with deep eutectic solvent pretreatment. A study by Krainara *et al.* (2023) utilised a compost However and saw an improvement in PET and PLA biodegradation by a consortium of bacteria (*Mycolicibacterium parafortuitum*, *Achromobacter xylosoxidans*, *Stenotrophomonas pavanii*, *Chitinophaga jiangningensis*, *Gordonia desulfuricans* and *Nocardioides zeae*).

The most widely used bioreactors are enrichment bioreactors, which use plastic as the only carbon source and work under selective pressure to enrich microbial consortia that break down plastic. These bioreactors are primarily used to isolate plastic-degrading microorganisms from a variety of environments, including wastewater treatment facilities (Tuan Tran *et al.*, 2022), landfill soil (Chigwada *et al.*, 2023), cow dung (Skariyachan *et al.*, 2017), seawater (Chigwada *et al.*, 2025) and rivers (Rout *et al.*, 2024).

Bioreactors not only shield microorganisms from harsh conditions but also provide controlled optimised conditions, resulting in consistent and efficient plastic biodegradation. However, bioreactors are expensive, require a lot of infrastructure and management and they are not yet practical for large-scale use. Therefore, new innovations must be developed to improve their cost-effectiveness.

12.3.2 Bioaugmentation in plastic and microplastic remediation

One potential method of addressing plastic waste using microorganisms is through bioaugmentation, which is the introduction of microorganisms to polluted environments to hasten the degradation potential of pollutants. Bioaugmentation contributes to the degradation of plastic waste in diverse settings, with method, microbial consortia and environmental elements affecting the process.

In landfills, bioaugmentation by the introduction of potent consortia and/or isolates has shown significant improvement in plastic biodegradation in various plastic types (Baskaran and Sathiavelu, 2022; Nair *et al.*, 2016; Salinas *et al.*, 2025). By introducing potent microorganisms to a landfill for 30 days, Mohan and Usha (2018) demonstrated how LDPE sheets enhanced biodegradation under bioaugmentation, increasing PE biodegradation in the environment. The study's ecological importance addressed plastic waste by introducing bioaugmented strains into a landfill contaminated with plastic debris.

According to Muthukumaran *et al.* (2024), biotechnological techniques, including genetic engineering and the development of augmenting microbial consortia, that are effective at plastic biodegradation are crucial for increasing the effectiveness of PE degradation in the environment. Through the optimisation of microbial interactions and metabolic pathways, the bioaugmentation approach improves the rate of decomposition of plastic waste in a range of habitats. A study by Li *et al.* (2024) that examined the function of PET hydrolase and microbial communities in high-temperature composting for PET degradation showed more sophisticated bioaugmentation. This study involved the introduction of not only microorganisms to the environment but also enzymes to improve environmental biodegradation.

Technologies for biodegradation and bioremediation present viable ways to cut down on plastic waste and lessen its negative effects on the environment. The intricacy of polymeric polymers, scaling up bioremediation procedures and guaranteeing the safety of modified microorganisms are some of the major obstacles that still need to be overcome. It will take interdisciplinary cooperation, strong regulatory frameworks and continuous research and development to overcome these obstacles.

12.4. Strategies utilising (bio)electrochemical methods in bioremediation of micro- and nanoplastics

Plastic remediation by way of (bio)electrochemical systems refers to engineering small power plants driven by microorganisms that generate power using plastics as a source of carbon similar to bioreactors. The (bio)electrochemical method uses the interaction between microorganisms and electrical electrodes to transform chemical energy into electrical energy following the biodegradation of polymers by microorganisms (Bhadra and Sevda, 2024). (Bio)electrochemical setup is composed of a lower potential cathode and a high electrode potential anode where oxidation and reduction processes occur, respectively. This technology offers a bright future for energy production, environmental biodegradation, bioproduction and waste management (Wang *et al.*, 2024).

According to a review by Wang and Zhou (2024b), (bio)electrochemical systems can offer a promising method for (bio)electroremediation of plastic pollution by speeding up the aging of microplastics in wastewater, increasing the metabolic activity of degrading microorganisms and enhancing microbial resistance to contaminants. The potential use of bioelectrochemical systems for the removal of microplastics studied by Wang *et al.* (2024) emphasised system efficiency and sustainability in comparison to conventional techniques. In addition to encouraging methane recovery, the Wang and Zhou (2024a) study showed that the (bio)electrochemical system efficiently treats microplastics made of polyethylene and polyvinyl chloride. They also showed that applying voltage improved metabolic pathways and microbiological activity, reducing microplastic contamination in wastewater.

Furthermore, effective (bio)electroremediation of plastic was demonstrated in the study by Kim *et al.* (2022) which addressed the use of non-recyclable PET microplastics as an electron feedstock in photo-electrocatalytic biosynthesis, combining environmental remediation and biocatalytic photosynthesis to transform plastic waste into useful chemicals.

The study by Segundo *et al.* (2024) showed that the fungus *Pleurotus ostreatus* may efficiently reduce plastic waste while producing bioelectricity in microbial fuel cells. Further, the study highlighted the potential for (bio)electroremediation by changing the chemical structure of plastic and reducing its thickness considerably. Future research should concentrate more on the underlying issues with this approach to develop a comprehensive, functionally optimal remediation method.

Microbial plastic and microplastic full mineralisation can be accomplished through (bio) electroremediation.

12.5. Novel approaches in engineering, biotechnology and nanotechnology in plastic remediation

Recent developments in engineering, biotechnology and nanotechnology are speeding up the biodegradation of plastic by supporting optimising microbial efficacies, improving enzyme performance and increasing plastic accessibility.

12.5.1 Biotechnology application in microbial plastic biodegradation

Enzyme engineering has been transformed by computational redesign, which has made it possible to produce extremely effective PET-degrading enzymes (Huang *et al.*, 2023) – for example, TurboPETase was developed using a computational approach that enabled nearly full PET depolymerisation at settings relevant to the industry, with almost 89% conversion in 7 to 8 h (Cui *et al.*, 2024). FAST_PETase, which can completely break down untreated post-consumer PET sourced from over 50 different commercial products (e.g. bottles, trays and packaging) in 7 days, was developed using a deep learning-assisted redesign approach (Lu *et al.*, 2022). The strategy employed by Lu *et al.* (2022) overcame the drawbacks of conventional engineering techniques by optimising enzyme structure and function through the use of computational tools.

Enhancing the thermostability and catalytic efficiency of plastic-degrading enzymes has been achieved largely through directed evolution. Using a substrate known as BHET-OH and a novel fluorescence-based high-throughput screening assay, a thermostable variant called DepoPETase was identified. This enzyme demonstrated complete depolymerisation of 100 g of untreated PET in a litre-scale bioreactor at 50°C within 4 days (Gao *et al.*, 2024). This approach highlights how directed evolution can rapidly identify high-performing enzyme variants.

Furthermore, the activity and stability of enzymes such as HotPETase (an engineered variant of *Ideonella sakaiensis* PETase) and LCC mutants (improved variants of leaf-branch compost cutinase) have been further enhanced through techniques including error-prone polymerase chain reaction (PCR) and saturation mutagenesis (Bell et al., 2022), resulting in significantly improved PET biodegradation performance.

For industrial applications, increasing the stability and activity of enzymes is essential – for example, disulfide bridges, which are present in the BbPETase variants N364C and D418C, considerably raised the melting temperature of the enzyme by 14.8°C and enhanced its long-term degrading capabilities (Huang *et al.*, 2023), which is important for industrial application. Similarly, DuraPETase was developed using the GRAPE computational method, demonstrating a 40% increase in degradation activity towards high-crystallinity PET films and a 31°C increase in reactive temperature (Cui *et al.*, 2019). These developments demonstrate how crucial structural alterations are to producing strong and effective enzymes for the bioremediation of plastic waste industrially.

The use of consortia and a cocktail of enzymes in the biodegradation of plastics has shown some noteworthy outcomes in the depolymerisation of polymers. A review by Sánchez (2024) showed that the use of a cocktail of enzymes and consortia, such as cutinase, esterases and surfactants, significantly improved the biodegradation of PET and polyurethanes. The enzymes MHETase and FAST-PETase showed twofold PET biodegradation and increased activity following PET immobilisation with magnetic nanoparticles, according to a study by Kotnis (2025). Numoto *et al.* (2023) focused on engineering the cutinase (Cut190) enzyme, enhancing its thermal stability and activity for PET degradation through rational design and structural insights, ultimately aiming to improve enzymatic recycling of plastic waste with low environmental impact.

These strategies demonstrate how microbial engineering can be used to provide sustainable and scalable solutions for the treatment of plastic waste. Table 12.2 summarises the enzyme variants.

12.5.2 Nanotechnology application in microbial plastic biodegradation

Nanotechnology speeds up microbial breakdown rates by altering microbial metabolic pathways and improving plastic bioremediation. It increases microbial growth and promotes photocatalytic degradation, which yields low molecular weight intermediates for chemical companies, increasing the overall effectiveness of plastic degradation processes (Bhatia *et al.*, 2013; Malik, 2023).

According to a study by Xiao *et al.* (2024), nanomaterials can improve photocatalytic performance through element doping and heterojunction building, which can greatly increase degradation efficiency and remove microplastics from the environment. When combined with microbial biodegradation, such an experiment could produce significant outcomes – for example nanotechnology could improve microbial microplastic biodegradation, according to a 2017 study by Pathak and Kumar. Through bacterial growth acceleration and LDPE chemical structure modification, the use of SiO_2 nanoparticles in the study dramatically improved the biodegradation efficiency of LDPE polymer-degrading bacteria in *Bacillus* and *Pseudomonas* species (Pathak and Kumar, 2017).

It is noteworthy that a review by Pandey *et al.* (2015) provided creative solutions for the efficient bioremediation of LDPE plastics by demonstrating how nanoparticles, such as superparamagnetic iron oxide and fullerene-60, improved microbial growth profiles, decreased lag

Table 12.2 Key enzyme variants and their features presenting scalable opportunities for plastic/microplastic degradation

Enzyme variant	Key features	Reference
TurboPETase	High flexibility in PET-binding groove, near-complete depolymerisation	Cui *et al.*, 2024
DepoPETase	Complete depolymerisation of 100 g of untreated PET in a litre-scale bioreactor at 50°C within 4 days	Gao *et al.*, 2024
DuraPETase	Elevated temperature by 31°C, enhanced activity to PET	Cui *et al.*, 2019
MHETase and FAST-PETase	Superior 2fold PET-hydrolytic activity, thermostability and pH tolerance	Gulati and Sun, 2024; Kotnis, 2025; Palm *et al.*, 2019
N364C/D418C mutational variant	Increased the melting temp by 14.8°C, improved long-term degradation performance	Huang *et al.*, 2023
Cutinase Cut190 mutational variant	Improved thermal stability	Numoto *et al.*, 2023

phases and sped up breakdown rates. Using nanotechnology, plastics and microplastics can be transformed into bioenergy and bioproducts. In addition to addressing the issue of plastic pollution, this method offers a sustainable source of materials and energy.

A nanotechnology-enabled bio-methanation anaerobic reactor transformed plastic waste into energy, fertilisers and biodiesel (Acharya and Mishra, 2021). However, new approaches to improve the efficiency of bioremediation in nanotechnology must be explored and research gaps still need to be addressed. Research into the latest nanotech microplastic removal techniques is required because there is a dearth of long-term environmental evaluation of the impact of microbial biodegradation.

The remediation of plastics and microplastics using fungal nanotechnology has to be studied because no research has been done in this field. Lastly, a balance between microbiology, nanotechnology and biotechnology is needed to provide reliable, sustained restoration.

12.6. Limitations, obstacles and difficulties encountered in the implementation of microbial plastic biodegradation projects

Despite being a promising solution to the worldwide plastic pollution dilemma, biodegradation of plastics by microorganisms still confronts many obstacles that make it difficult to implement successfully. These difficulties are multifaceted and challenging, encompassing technological, environmental, economic and regulatory aspects. First, plastics are regarded as recalcitrant in nature, which means that they are naturally resistant to biodegradation (Thew *et al.*, 2023). This is primarily because of their extremely stable chemical structure, which is made up of carbon-to-carbon (C–C) and carbon-to-hydrogen (C–H) bonds (Colachis *et al.*, 2024).

Plastics are highly recalcitrant in natural environments due to their hydrophobicity, high crystallinity, lack of functional groups and stable C–C backbones, often resulting in very slow or incomplete microbial degradation that can take years or decades (Giyahchi and Moghimi, 2023). While specialised enzymes such as PETase and MHETase provide degradation of PET (Meyer *et al.*, 2020; Palm *et al.*, 2019), other plastics such as PS and PVC lack specific specialised enzymes for their degradation. Microorganisms find it difficult to degrade polyolefins because they are very hydrophobic and lack functional groups. Polyolefins make up a significant portion of the plastics produced.

Numerous issues, including efficiency and specificity, arise during enzymatic plastic biodegradation because there are not so many broad-spectrum enzymes that can biodegrade different kinds of plastics (Hou and Majumder, 2021; Palm *et al.*, 2019). Furthermore, enzymatic scaling is a major problem since industrial production is hindered by high costs and technical challenges that prevent widespread scalability (Kawai *et al.*, 2024). Enzymes also lack stability in settings like high pH, temperature and oxygen level, and in harsh environments (Hyder *et al.*, 2021).

Enzymatic and microbiological activity can be inhibited by other environmental contaminants, which can impact the implementation of microbial biodegradation (Karthick and Mohanraju, 2018; Pérez-García *et al.*, 2021; Yadav *et al.*, 2019). Intermediate byproducts of microbial plastic biodegradation can be toxic to the microbiome, reducing their efficacy. Furthermore, other byproducts like microplastics, nanoplastics and hazardous byproducts can pose a threat to both human and animal health (Cai *et al.*, 2023; Chigwada and Tekere, 2023; Nguyen *et al.*, 2021).

There are still many unanswered questions regarding the fate of microplastics and nanoplastics as well as their effects on the environment. Additionally, plastics contain additives, such as flame retardants and plasticisers, that might leach into the environment (Cai *et al.*, 2023; Carmen, 2021). The absence of specialist waste management infrastructure, such as industrial bioreactors for plastic biodegradation, in real-world situations is exacerbated by regulatory, social and legal issues (Harrison *et al.*, 2018). This is further fuelled by the lack of adequate waste management systems and public education on proper disposal practices, further complicating the implementation of biodegradation technologies.

Microbial plastic biodegradation lacks a standardised protocol for validating plastic degradation as reviewed by a number of papers (Ali *et al.*, 2021; Montazer *et al.*, 2020; Thew *et al.*, 2023). This has led to inconsistencies in assessing their environmental impact as well as confirming biodegradation rates. Validation and knowledge of the mechanisms by which microorganisms and enzymes function are crucial for optimising plastic biodegradation (Pires *et al.*, 2022); yet, the mechanisms by which specific microorganisms and enzymes degrade plastics are still poorly understood.

Aquatic environments present unique challenges in plastic biodegradation as conditions might not favour plastic biodegradation in these environments (de Villalobos *et al.*, 2022). Consequently, delayed microbial biodegradation is seen in marine habitats, which raises serious questions about the long-term effects on ecosystems. Furthermore, due to a dearth of study, little is known about the fate and effects of end products and microplastics in marine environments.

12.7. Sustainable plastic bioremediation: future perspectives

Bioremediation and plastic biodegradation's future depends entirely on creative approaches to the challenges and pollution caused by plastics. Microbial bioremediation, interdisciplinary research on catalytic enzyme breakthroughs, biotechnology, (bio)electrochemical, nanotechnology and chemistry should all be included in these strategies to lessen the negative health effects of plastic pollution.

Advances in synthetic biology and biotechnology should prioritise and speed up the development of new enzymes, microbial consortium optimisation and bioremediation technologies for plastic remediation. A circular economy model must also take the role of traditional linear waste management, which entails recycling and bioremediation in lieu of landfill dumping and plastic cremation, in order to provide a sustainable solution to plastic pollution (Obrador-Viel *et al.*, 2024; Priya *et al.*, 2022). Technological developments that measure the concentration of microplastics in the environment are crucial for the implementation and monitoring of focused remediation procedures. This will allow the implementation of emerging technologies like (bio)electrochemical (Wang *et al.*, 2024), photocatalysis (Kim *et al.*, 2022) and pyrolysis (Cupertino *et al.*, 2024; Kiener, 2024) that may convert plastics into valuable byproducts.

Significant progress has been achieved in increasing the efficiency of plastic decomposition by enzyme engineering. The creation of extremely stable and effective enzymes has been made possible by structural alterations, controlled evolution and computational redesigning. The potential of biotechnological techniques for sustainable plastic recycling is further highlighted by the employment of microbial consortia, whole-cell biocatalysts and synergistic enzyme systems. These developments offer environmentally beneficial answers to the world's plastic pollution problem and open the door to a circular plastic economy.

12.8. Conclusions

One promising approach to future plastic pollution clean-up and its effects on the environment is microbial plastic and microplastic biodegradation. Bioremediation offers an opportunity to use microorganisms and their enzymes to break down plastic while utilising biotechnological and nanotechnological advancements to restore the ecosystem's health.

Bioaugmentation and genetic engineering have shown potential with strains and consortia such as *Pseudomonas* and *Bacillus* (Roberts *et al.*, 2020). Enzymes such as TurboPETase have been shown to increase remediation up to 100% (Cui *et al.*, 2024). The infrastructure, the cost of scaling up, the regulations and social hurdles, the resistant nature of plastics, the toxicity of the final products, the problems with standardisation, the maritime environment and the limitations of the enzymes are still major obstacles.

To overcome these challenges, interdisciplinary collaboration and research and development efforts will be required. Approaches that will encompass the use of (bio)electrochemical, bioaugmentation, engineering potent broad-spectrum enzymes and the use of nanotechnological advances will help overcome plastic biodegradation challenges. Finally, the future bioremediation of plastic will depend on the development of new, sophisticated computational enzymes and the enhancement of bioreactor biodegradation consortiums.

REFERENCES

Abraham J, Ghosh E, Mukherjee P and Gajendiran A (2017) Microbial degradation of low density polyethylene. *Environmental Progress and Sustainable Energy* **36(1)**: 147–154, 10.1002/ep.12467.

Acharya SK and Mishra DP (2021) *Current Advances in Mechanical Engineering: Select Proceedings of ICRAMERD 2020.*

Ali MI, Ahmed S, Robson G *et al.* (2014) Isolation and molecular characterization of polyvinyl chloride (PVC) plastic degrading fungal isolates. *Journal of Basic Microbiology* **54(1)**: 18–27, 10.1002/jobm.201200496.

Ali SS, Elsamahy T, Al-Tohamy R *et al.* (2021) Plastic wastes biodegradation: mechanisms, challenges and future prospects. *Science of the Total Environment* **780**: 146590, 10.1016/j.scitotenv.2021.146590.

Amobonye A, Bhagwat P, Singh S and Pillai S (2021) Plastic biodegradation: frontline microbes and their enzymes. *Science of the Total Environment* **759(7):** 143536, 10.1016/j.scitotenv.2020.143536.

Bacha AUR., Nabi I, Zaheer M, Jin W and Ya L (2023) Biodegradation of macro- and microplastics in environment: A review on mechanism, toxicity and future perspectives. *Science of the Total Environment* **858(3)**: 160108, 10.1016/j.scitotenv.2022.160108.

Baskaran S and Sathiavelu M (2022) Bioaugmentation and biostimulation of dumpsites for plastic degradation. In *Advances in Environmental Pollution Research* (Kathi S, Devipriya S and KBT-CET for SW and WT Thamaraiselvi (eds)). Elsevier, Amsterdam, the Netherlands, pp. 9–23, 10.1016/B978-0-12-822933-0.00015-2.

Bell EL, Smithson R, Kilbride S *et al.* (2022) Directed evolution of an efficient and thermostable PET depolymerase. *Nature Catalysis* **5(8)**: 673–681, 10.1038/s41929-022-00821-3.

Bhadra S and Sevda S (2024) Integrated bioelectroremediation: simultaneous treatment of industrial effluents and bioenergy generation. *Environmental Engineering Research* **29(4)**, 10.4491/eer.2023.657.

Bhatia M, Girdhar A, Chandrakar B and Tiwari A (2013) Implicating nanoparticles as potential biodegradation enhancers: A review. *Journal of Nanomedicine and Nanotechnology* **4(4)**, 10.4172/2157-7439.1000175.

Biki SP, Mahmud S, Akhter S *et al.* (2021) Polyethylene degradation by Ralstonia sp. strain SKM2 and Bacillus sp. strain SM1 isolated from land fill soil site. *Environmental Technology and Innovation* **22**: 101495, 10.1016/j.eti.2021.101495.

Blázquez-Sánchez P, Engelberger F, Cifuentes-Anticevic J *et al.* (2022) Antarctic polyester hydrolases degrade aliphatic and aromatic polyesters at moderate temperatures. *Applied and Environmental Microbiology* **88(1)**: 1–14.

Cai Z, Li M, Zhu Z *et al.* (2023) Biological degradation of plastics and microplastics: a recent perspective on associated mechanisms and influencing factors. *Microorganisms* **11(7)**: 1661, 10.3390/microorganisms11071661.

Carmen S (2021) Microbial capability for the degradation of chemical additives present in petroleum-based plastic products: a review on current status and perspectives. *Journal of Hazardous Materials* **402(July 2020)**, 10.1016/j.jhazmat.2020.123534.

Chaudhary AK and Vijayakumar RP (2020) Studies on biological degradation of polystyrene by pure fungal cultures. *Environment, Development and Sustainability* **22(5)**: 4495–4508, 10.1007/s10668-019-00394-5.

Chigwada AD and Tekere M (2023) The plastic and microplastic waste menace and bacterial biodegradation for sustainable environmental clean-up a review. *Environmental Research* **231**(P1): 116110, 10.1016/j.envres.2023.116110.

Chigwada AD, Ogola HJO and Tekere M (2025) A catalog of metagenomes and metagenome-assembled genomes from culture-enrichment microcosms containing polyethylene as carbon source. *Microbiology Resource Announcements* **14(1)**: e00689-24.

Chigwada AD, Ogola OJH and Tekere M (2023) Multivariate analysis of enriched landfill soil consortia provide insight on the community structural perturbation and functioning during low-density polyethylene degradation. *Microbiological Research* **274(May)**: 127425, 10.1016/j.micres.2023.127425.

Colachis M, Lilly JL, Trigg E and Kucharzyk KH (2024) Analytical tools to assess polymer biodegradation: A critical review and recommendations. *Science of The Total Environment* **955**: 176920, 10.1016/j.scitotenv.2024.176920.

Cui Y, Chen Y, Liu X *et al.* (2019) Supplementary materials for computational redesign of a PETase for plastic biodegradation by the GRAPE strategy. *ACS Catalysis* **11(3)**, 1340–1350, https://doi.org/10.1021/acscatal.0c05126.

Cui Y, Chen Y, Sun J *et al.* (2024) Computational redesign of a hydrolase for nearly complete PET depolymerization at industrially relevant high-solids loading. *Nature Communications* **15(1)**: 1–12, 10.1038/s41467-024-45662-9.

Cui Z, Zhang X, Yang H *et al.* (2017) Bioremediation of heavy metal pollution utilizing composite microbial agent of Mucor circinelloides, Actinomucor sp. and Mortierella sp.. *Journal of Environmental Chemical Engineering* **5(4)**: 3616–3621, 10.1016/j.jece.2017.07.021.

Cupertino GFM, Silva PFR, Oliveira JE *et al.* (2024) Co-pyrolysis of biomass and polyethylene terephthalate (PET) as an alternative for energy production from waste valorization. *Fuel* **362**: 130761, https://doi.org/10.1016/j.fuel.2023.130761

Davidov K, Iankelevich-Kounio E, Yakovenko I *et al.* (2020) Identification of plastic-associated species in the Mediterranean Sea using DNA metabarcoding with Nanopore MinION. *Scientific Reports* **10(1)**: 1–11, 10.1038/s41598-020-74180-z.

de Villalobos NF, Costa MC and Marín-Beltrán I (2022) A community of marine bacteria with potential to biodegrade petroleum-based and biobased microplastics. *Marine Pollution Bulletin* **185(Part A)**: 114251, 10.1016/j.marpolbul.2022.114251.

Deepika S and Madhuri RJ (2015) Biodegradation of low density polyethylene by micro-organisms from garbage soil. *Journal of Experimental Biology and Agricultural Sciences* **3(1)**: 15–21. Available at: http://www.jebas.org.

Edwards S, León-Zayas R, Ditter R *et al.* (2022) Microbial consortia and mixed plastic waste: pangenomic analysis reveals potential for degradation of multiple plastic types via previously identified PET degrading bacteria. *International Journal of Molecular Sciences* **23(10)**, 10.3390/ijms23105612.

El-Shafei HA *et al.* (1998) Biodegradation of disposable polyethylene by fungi and Streptomyces species. *Polymer Degradation and Stability* **62(2)**: 361–365, 10.1016/S0141-3910(98)00019-6.

Fachrul MF, Fulazzaky MA, Rinanti A *et al.* (2024) Mass transfer kinetics of polyethylene degradation by bacterial-fungal consortium. *Cleaner Waste Systems* **7:** 100136, 10.1016/j.clwas.2024.100136.

Fauvelle V, Garel M, Tamburini C *et al.* (2021) Organic additive release from plastic to seawater is lower under deep-sea conditions. *Nature Communications* **12(1)**, 10.1038/s41467-021-24738-w.

Gao R and Sun C (2021) A marine bacterial community capable of degrading poly(ethylene terephthalate) and polyethylene. *Journal of Hazardous Materials* **416**: 125928, 10.1016/j.jhazmat.2021.125928.

Gao S, Shi L, Wei H *et al.* (2024) β -sheet Engineering of IsPETase for PET Depolymerization. *Engineering* **47**: 180–193, https://doi.org/10.1016/j.eng.2024.10.015

Giyahchi M and Moghimi H (2023) Aerobic biodegradation of untreated polyester-polyether urethanes by newly isolated yeast strains Exophilia sp. NS-7 and Rhodotorula sp. NS-12. *Scientific Reports* **13(1)**: 5016, 10.1038/s41598-023-31639-z.

Gong J, Kong T, Li Y *et al.* (2018) Biodegradation of microplastic derived from poly(ethylene terephthalate) with bacterial whole-cell biocatalysts. *Polymers* **10(12)**, 10.3390/polym10121326.

Gulati S and Sun Q (2024) Complete enzymatic depolymerization of polyethylene terephthalate (PET) plastic using a Saccharomyces cerevisiae-based whole-cell biocatalyst. *bioRxiv*: 2024.07.20.604433, 10.1021/acs.estlett.5c00190.

Han X, Liu W, Huang J-W *et al.* (2017) Structural insight into catalytic mechanism of PET hydrolase. *Nature Communications* **8(1)**: 2106, 10.1038/s41467-017-02255-z.

Harrison JP, Boardman C, O'Callaghan K, Delort A-M and Song J (2018) Biodegradability standards for carrier bags and plastic films in aquatic environments: a critical review. *Royal Society Open Science* **5(5)**, 10.1098/rsos.171792.

Hellesnes KN, Tufto HM, Bævre AS *et al.* (2023) Biochemical characterization and NMR study of a PET-hydrolyzing cutinase from Fusarium solani pisi. *Biochemistry* **62(8)**: 1369–1375, 10.1021/acs.biochem.2c00619.

Hillman K, Damgaard A, Eriksson O, Jonsson D and Fluck L (2015) *Climate Benefits of Material Recycling: Inventory of Average Greenhouse Gas Emissions for Denmark, Norway and Sweden.* TemaNord, 2015:547. Nordic Council of Ministers. https://doi.org/10.6027/TN2015-547

Hou L and Majumder ELW (2021) Potential for and distribution of enzymatic biodegradation of polystyrene by environmental microorganisms. *Materials* **14(3)**: 1–20, 10.3390/ma14030503.

Huang D, Zhang L and Sun Y (2023) Rational design of disulfide bridges in BbPETaseCD for enhancing the enzymatic performance in PET degradation. *Molecules* **28(8)**: 1–12, 10.3390/molecules28083528.

Hussain S and Khan M (2022) Biodegradation of low-density polyethylene (LDPE) bags by fungi isolated from waste disposal soil. *Applied and Environmental Soil Science* **2022(1)**: 8286344, 10.1155/2022/8286344.

Hyder A, Khan M, Khan S *et al.* (2021) Biodegradation of low-density polyethylene plastics by fungi isolated from waste disposal site at district Peshawar, Pakistan. *Pakistan Journal of Biochemistry and Biotechnology* **2(2)**: 127–133.

Ibrahim SS, Ionescu D and Grossart HP (2024) Tapping into fungal potential: biodegradation of plastic and rubber by potent Fungi. *Science of the Total Environment* **934**: 173188, 10.1016/j.scitotenv.2024.173188.

Jiang S, Su T, Zhao J and Wang Z (2021) Isolation, identification and characterization of polystyrene-degrading bacteria from the gut of Galleria Mellonella (Lepidoptera: Pyralidae) larvae. *Frontiers in Bioengineering and Biotechnology* **9**: 1–9, 10.3389/fbioe.2021.736062.

Joo S, Cho IJ, Seo H *et al.* (2018) Structural insight into molecular mechanism of poly (ethylene terephthalate) degradation. *Nature Communications* **9(1)**: 382, 10.1038/s41467-018-02881-1.

Karthick P and Mohanraju R (2018) Antimicrobial potential of epiphytic bacteria associated with seaweeds of little Andaman, India. *Frontiers in Microbiology* **9**: 1–11, 10.3389/fmicb.2018.00611.

Kawai F (2002) Microbial degradation of polyethers. *Applied Microbiology and Biotechnology* **58(1)**: 30–38, 10.1007/s00253-001-0850-2.

Kawai F, Iizuka R and Kawabata T (2024) Engineered polyethylene terephthalate hydrolases: perspectives and limits. *Applied Microbiology and Biotechnology* **108(1)**, 10.1007/s00253-024-13222-2.

Khanal SK, Giri B, Nitayavardhana S and Gadhamshetty V (2017) anaerobic bioreactors/digesters: design and development. In *Current Developments in Biotechnology and Bioengineering: Biological Treatment of Industrial Effluents* (Lee D-J, Jegatheesan V, Ngo HH, Hallenbeck PC and Pandey A (eds.)). Elsevier, Amsterdam, the Netherlands, pp. 261–279. 10.1016/B978-0-444-63665-2.00010-2.

Kiener R (2024) *Plastic Pollution*, 10.4135/cqrglobal20100700.

Kim J, Jang J, Hilberath T *et al.* (2022) Photoelectrocatalytic biosynthesis fuelled by microplastics. *Nature Synthesis* **1**(10): 776–786, 10.1038/s44160-022-00153-x.

Kotnis S, Gulati S and Sun Q (2025) High-efficiency PET degradation with a duo-enzyme system immobilized on magnetic nanoparticles. *Biotechnology and Bioengineering* **122(6)**: 1397–1401, 10.1002/bit.28963.

Krainara S, Mistry AN, Malee C *et al.* (2023) Development of a plastic waste treatment process by combining deep eutectic solvent (DES) pretreatment and bioaugmentation with a plastic-degrading bacterial consortium. *Journal of Hazardous Materials* **460**: 132507, 10.1016/j.jhazmat.2023.132507.

Kumar R, Pandit P, Kumar D *et al.* (2021) Landfill microbiome harbours plastic degrading genes: a metagenomic study of solid waste dumping site of Gujarat, India. *Science of the Total Environment* **779(336)**: 146184, 10.1016/j.scitotenv.2021.146184.

Kumari A, Bano N, Chaudhary DR and Jha B (2021) Transcriptome-guided insights into plastic degradation by the marine bacterium Bacillus sp. AIIW2. *Frontiers in Microbiology*, **12**, 751571. https://doi.org/10.3389/fmicb.2021.751571

Kwon S, Kim Y, Kim J *et al.* (2023) Aerobic aquatic biodegradation of bio-based and biodegradable polymers: Kinetic modeling and key factors for biodegradability. *International Biodeterioration and Biodegradation* **185**: 105671, 10.1016/j.ibiod.2023.105671.

Lear G, Kingsbury JM, Franchini S and Aydin E (2021) Plastics and the microbiome: impacts and solutions. *Environmental Microbiomes* **16(1)**: 1–19, 10.1186/s40793-020-00371-w.

Lee B, Pometto AL, Fratzke A and Bailey TB (1991) Biodegradation of degradable plastic Polyethylene by Phanerochaete and Streptomyces species. *Applied and Environmental Microbiology* **57(3)**: 678–685. https://journals.asm.org/journal/aem.

Li L, Wang Y, Zhang X, Li J and Li Y (2025) Phanerochaete chrysosporium hyphae bio-crack, endocytose and metabolize plastic films. *Journal of Hazardous Materials* **487(193)**: 137154, 10.1016/j.jhazmat.2025.137154.

Li X, Zhang Y, Wang J, Li H and Chen Y (2024) Hydrolase and plastic-degrading microbiota explain degradation of polyethylene terephthalate microplastics during high-temperature composting. *Bioresource Technology* **393**: 130108, 10.1016/j.biortech.2023.130108.

Liu W, Zhang L, Li J, Wang H and Chen X (2024) Enzyme modified biodegradable plastic preparation and performance in anaerobic co-digestion with food waste. *Bioresource Technology* **401**: 130739, 10.1016/j.biortech.2024.130739.

Lu H, Diaz JS, Zhang J *et al.* (2022) Machine learning-aided engineering of hydrolases for PET depolymerization. *Nature* **604(7907)**: 662–667, 10.1038/s41586-022-04599-z.

Lu H, Wang Y, Zhang L and Li J (2024) Microbial consortium degrading of organic pollutants: Source, degradation efficiency, pathway, mechanism and application. *Journal of Cleaner Production*: 141913.

Malik JA (2023) *Modern Nanotechnology*. Springer, Cham, Switzerland, 10.1007/978-3-031-31111-6.

Meyer IE, Franke L and Widdel F (2020) Synergistic biodegradation of aromatic-aliphatic copolyester plastic by a marine microbial consortium. *Nature Communications* **11(1)**, 10.1038/s41467-020-19583-2.

Mohamed DFMS, Al-Hussaini AS and El-Sayed MA (2024) Assessment of biodegradation and toxicity of alternative plasticizer di(2-ethylhexyl) terephthalate: Impacts on microbial biofilms, metabolism and reactive oxygen species-mediated stress response. *Environmental Pollution* **355**: 124217, 10.1016/j.envpol.2024.124217.

Mohan N and Usha R (2018) Bio-augmentation - Effective method of treating plastic waste - a field study. *Journal of Pure and Applied Microbiology* **12(3)**: 1641–1646, 10.22207/JPAM.12.3.71.

Montazer Z, Najafi M and Montazer M (2020) Challenges with verifying microbial degradation of polyethylene. *Polymers* **12(1)**: 123, 10.3390/polym12010123.

Moreno DG, Galeano GAN and Morales-Fonseca D (2024) Development of a bioreactor-based model for low-density polyethylene (LDPE) biodegradation by Aspergillus brasiliensis. *Universitas Scientiarum* **29(2)**: 127–144, 10.11144/JAVERIANA.SC292.DOAB.

Muthukumaran M, Narayana AA, Prabhavathi J, Malathi G and Debnath A (2024) Microbial perspectives on polythene biodegradation: Exploring the role of microorganisms in addressing plastic pollution. *Microbiology Research Journal International* **34(5)**: 18–28, 10.9734/mrji/2024/v34i51443.

Nair NR, Sekhar VC and Nampoothiri KM (2016) augmentation of a microbial consortium for enhanced polylactide (PLA) degradation. *Indian Journal of Microbiology* **56(1)**: 59–63, 10.1007/s12088-015-0559-z.

Nakamura Y, Sungusia MG, Sawada T and Kuwahara M (1999) Lignin-degrading enzyme production by Bjerkandera adusta immobilized on polyurethane foam. *Journal of Bioscience and Bioengineering* **88(1)**: 41–47, 10.1016/S1389-1723(99)80173-X.

Nguyen BAT, Hsieh JL, Lo SL and Huang SL (2021) Biodegradation of dioxins by Burkholderia cenocepacia strain 869T2: role of 2-haloacid dehalogenase. *Journal of Hazardous Materials* **401(July 2020)**: 123347, 10.1016/j.jhazmat.2020.123347.

Numoto N, Kamiya N and Oda M (2023) Improvement of the thermostability and activity of the PET-degrading enzyme Cut190 toward a detailed understanding and application of the enzymatic reaction mechanism. *ACS Symposium Series* **1450**, pp. 89–100, 10.1021/bk-2023-1450.ch006.

Obrador-Viel T, Zadjelovic V, Nogales B, Bosch R and Christie-Oleza JA (2024) Assessing microbial plastic degradation requires robust methods. *Microbial Biotechnology* **17(4)**: e14457, 10.1111/1751-7915.14457.

Palm GJ, Reisky L, Böttcher D *et al.* (2019) Structure of the plastic-degrading Ideonella sakaiensis MHETase bound to a substrate. *Nature Communications* **10(1)**: 1–10, 10.1038/s41467-019-09326-3.

Pandey P, Swati, Harshita *et al.* (2015) Nanoparticles accelerated in-vitro biodegradation of LDPE: a review. *Pelagia Research Library Advances in Applied Science Research* **6(4)**: 17–22. Available at: www.pelagiaresearchlibrary.com.

Pathak VM and Navneet (2017) Review on the current status of polymer degradation: a microbial approach. *Bioresources and Bioprocessing* **4(1)**: 15 10.1186/s40643-017-0145-9.

Pathak VM and Navneet Kumar (2017) Implications of SiO_2 nanoparticles for in vitro biodegradation of low-density polyethylene with potential isolates of Bacillus, Pseudomonas and their synergistic effect on Vigna mungo growth. *Energy, Ecology and Environment* **2(6)**: 418–427, 10.1007/s40974-017-0068-5.

Pérez-García P, Danso D, Zhang H *et al.* (2021) Exploring the global metagenome for plastic-degrading enzymes. *Methods in Enzymology* **648**: 137–157, 10.1016/bs.mie.2020.12.022.

Pinnell LJ and Turner JW (2019) Shotgun metagenomics reveals the benthic microbial community response to plastic and bioplastic in a coastal marine environment. *Frontiers in Microbiology* **10(June)**, 10.3389/fmicb.2019.01252.

Pires JR *et al.* (2022) Methodologies to assess the biodegradability of bio-based polymers—current knowledge and existing gaps. *Polymers*, 10.3390/polym14071359.

Priya A, Dutta K and Daverey A (2022) A comprehensive biotechnological and molecular insight into plastic degradation by microbial community. *Journal of Chemical Technology and Biotechnology* **97(2)**: 381–390, 10.1002/jctb.6675.

Priyanka N and Archana T (2011) biodegradability of polythene and plastic by the help of microorganism: a way for brighter future. *Journal of Environmental and Analytical Toxicology* **01**(02): 1–4, 10.4172/2161-0525.1000111.

Puiggené Ò, Ruiz C and Granda M (2022) Extracellular degradation of a polyurethane oligomer involving outer membrane vesicles and further insights on the degradation of 2,4-diaminotoluene in Pseudomonas capeferrum TDA1. *Scientific Reports* **12(1)**: 6558, 10.1038/s41598-022-06558-0.

Qi X, Ren Y and Wang X (2017) New advances in the biodegradation of poly(lactic) acid. *International Biodeterioration and Biodegradation* **117**: 215–223, 10.1016/j.ibiod.2017.01.010.

Raaman N, Muralidharan S, Selvakumar S and Sivaraman S (2012) Biodegradation of plastic by Aspergillus spp. isolated from polythene polluted sites around Chennai. *Journal of Academia and Industrial Research* **1(6)**: 313–316. http://jairjp.com/NOVEMBER%20 2012/08%20RAAMAN.pdf.

Richter PK, Blázquez-Sánchez P, Zhao Z *et al.* (2023) Structure and function of the metagenomic plastic-degrading polyester hydrolase PHL7 bound to its product. *Nature Communications* **14(1)**: 1903. https://doi.org/10.1038/s41467-023-37415-x;

Roberts C, Edwards S, Vague M *et al.* (2020) Environmental consortium containing pseudomonas and bacillus species synergistically degrades polyethylene terephthalate plastic. *mSphere* **5(6)**: 1–20, 10.1128/msphere.01151-20.

Rong Z, Wang Y, Zhang L, Li J and Chen Y (2024) Degradation of low-density polyethylene by the bacterium Rhodococcus sp. C-2 isolated from seawater. *Science of the Total Environment* **907**: 167993, 10.1016/j.scitotenv.2023.167993.

Ronkvist ÅM, Xie W, Lu W and Gross RA (2009) Cutinase-catalyzed hydrolysis of poly(ethylene terephthalate). *Macromolecules* **42(14)**: 5128–5138, 10.1021/ma9005318.

Rout AK, Behera BK, Mohanty BK, Acharya L and Singh S (2024) Metagenomic landscape of sediments of river Ganga reveals microbial diversity, potential plastic and xenobiotic degradation enzymes. *Journal of Hazardous Materials* **471(December)**: 134377, 10.1016/j.jhazmat.2024.134377.

Salinas J, Dörr T and Rühl M (2023) Development of plastic-degrading microbial consortia by induced selection in microcosms. *Frontiers in Microbiology* **14**: 1143769, 10.3389/fmicb.2023.1143769.

Salinas J, Dörr T and Rühl M (2025) Construction of versatile plastic-degrading microbial consortia based on ligninolytic microorganisms associated with agricultural waste composting. *Environmental Pollution* **366**: 125333, 10.1016/j.envpol.2024.125333.

Sánchez C (2024) Fusarium as a promising fungal genus with potential application in bioremediation for pollutants mitigation: a review. *Biotechnology Advances* **77(October)**, 10.1016/j.biotechadv.2024.108476.

Sangale MK, Deshpande SS and More SV (2019) Potential of fungi isolated from the dumping sites mangrove rhizosphere soil to degrade polythene. *Scientific Reports* **9**(1): 5390, 10.1038/s41598-019-41448-y.

Segundo R-F, Magaly DLCN, Luis C-C (2024) Reducing plastic waste and generating bioelectricity simultaneously through fuel cells using the fungus Pleurotus ostreatus. *Sustainability* **16(18)**: 7909, 10.3390/su16187909.

Sheridan EA, Fonvielle JA and Cottingham S (2022) Plastic pollution fosters more microbial growth in lakes than natural organic matter. *Nature Communications* **13(1)**: 1–9, 10.1038/s41467-022-31691-9.

Simó-Cabrera L, García-Chumillas S, Hagagy N *et al.* (2021) Haloarchaea as cell factories to produce bioplastics. *Marine Drugs* **19(3)**: 159, 10.3390/MD19030159.

Skariyachan S, Setlur AS, Naik SY *et al.* (2017) Enhanced biodegradation of low and high-density polyethylene by novel bacterial consortia formulated from plastic-contaminated cow dung under thermophilic conditions. *Environmental Science and Pollution Research* **24(9)**: 8443–8457, 10.1007/s11356-017-8537-0.

Sowmya H, Ramalingappa V, Krishnappa M and Thippeswamy B (2014) Degradation of polyethylene by Trichoderma harzianum—SEM, FTIR and NMR analyses. *Environmental Monitoring and Assessment* **186**: 6577–6586.

Stoleru E, Zaharia C, Hitruc EG, Dumitriu C and Brebu M (2017) Biodegradation of poly(lactic acid)/chitosan stratified composites in presence of the Phanerochaete chrysosporium fungus. *Polymer Degradation and Stability* **143**: 118–129.

Taghavi N, Singhal N, Zhuang W-Q and Baroutian S (2021) Degradation of plastic waste using stimulated and naturally occurring microbial strains. *Chemosphere* **263**: 127975, 10.1016/j.chemosphere.2020.127975.

Tan Y, Henehan GT, Kinsella GK and Ryan BJ (2021) An extracellular lipase from Amycolatopsis mediterannei is a cutinase with plastic degrading activity. *Computational and Structural Biotechnology Journal* **19**: 869–879, 10.1016/j.csbj.2021.01.019.

Tekere M (2019) Microbial bioremediation and different bioreactors designs applied. In *Biotechnology and Bioengineering*. InTechOpen, London, UK, pp. 1–19. https://doi.org/10.5772/intechopen.83661.

Temporiti MEE, Nicola L, Nielsen E and Tosi S (2022) Fungal enzymes involved in plastics biodegradation. *Microorganisms* **10(6)**: 1180, 10.3390/microorganisms10061180.

Thew CXE, Lee ZS, Srinophakun P, Ooi CW and Show PL (2023) Recent advances and challenges in sustainable management of plastic waste using biodegradation approach. *Bioresource Technology* **374**: 128772, 10.1016/j.biortech.2023.128772.

Tournier V, Topham CM, Gilles A *et al.* (2020) An engineered PET depolymerase to break down and recycle plastic bottles. *Nature* **580(7802)**: 216–219, 10.1038/s41586-020-2149-4.

Tran HT, Lin C, Bui X-T *et al.* (2022) Phthalates in the environment: characteristics, fate and transport and advanced wastewater treatment technologies. *Bioresource Technology* **344(Part B)**: 126249, 10.1016/j.biortech.2021.126249.

Wang H and Zhou Q (2024a) Bioelectrochemical anaerobic digestion mitigates microplastic pollution and promotes methane recovery of wastewater treatment in biofilm system. *Journal of Hazardous Materials* **472**: 134488, 10.1016/j.jhazmat.2024.134488.

Wang H and Zhou Q (2024b) Bioelectrochemical systems – a potentially effective technology for mitigating microplastic contamination in wastewater. *Journal of Cleaner Production* **450**: 141931, 10.1016/j.jclepro.2024.141931.

Wang S, Hadji-Thomas A, Adekunle A and Raghavan V (2024) The exploitation of bio-electrochemical system and microplastics removal: Possibilities and perspectives. *Science of the Total Environment* **930**: 172737, 10.1016/j.scitotenv.2024.172737.

Wichatham K, Piyaviriyakul P, Boontanon N *et al.* (2024) Biodegradation of polypropylene plastics in vitro and natural condition by Streptomyces sp. isolated from plastic-contaminated sites. *Environmental Technology and Innovation* **35**: 103681, 10.1016/j.eti.2024.103681.

Xiao Y, Tian Y, Xu W and Zhu J (2024) Photodegradation of microplastics through nanomaterials: insights into photocatalysts modification and detailed mechanisms. *Materials* **17(11)**, 10.3390/ma17112755.

Yadav M, Bista S, Yadav P and Yadav A (2019) Secretory laccase from Pestalotiopsis species CDBT-F-G1 fungal strain isolated from high altitude: Optimization of its production and characterization. *Applied Sciences (Switzerland)* **9(2)**: 1–15, 10.3390/app9020340.

Yagi H, Ninomiya F, Funabashi M and Kunioka M (2014) Mesophilic anaerobic biodegradation test and analysis of eubacteria and archaea involved in anaerobic biodegradation of four specified biodegradable polyesters. *Polymer Degradation and Stability* **110**: 278–283, 10.1016/j.polymdegradstab.2014.08.031.

Yamada-Onodera K, Mukumoto H, Katsuyaya Y, Saiganji A and Tani Y (2001) Degradation of polyethylene by a fungus, Penicillium simplicissimum YK. *Polymer Degradation and Stability* **72(2)**: 323–327.

Zahra S, Ahmad S and Khan MA (2010) Biodegradation of low-density polyethylene (LDPE) by isolated fungi in solid waste medium. *Waste Management* **30(3)**: 396–401, 10.1016/j.wasman.2009.09.027.

Zhang K, Peng J, Wang W, Li J and Xu X (2021) Understanding plastic degradation and microplastic formation in the environment: a review. *Environmental Pollution* **274**, 10.1016/j.envpol.2021.116554.

Zhang Z, Chen Y, Wang L, Li J and Liu Y (2022) Polyvinyl chloride degradation by a bacterium isolated from the gut of insect larvae. *Nature Communications* **13(1)**: 5360, 10.1038/s41467-022-32903-y.

Zhen ZG, Li J, Wang H, Zhang X and Chen Y (2023) Different responses of mesophilic and thermophilic anaerobic digestion of waste activated sludge to PVC microplastics. *Environmental Science and Pollution Research International* **30(58)**: 121584–121598, 10.1007/s11356-023-30935-5.

Zhou K, Li J, Wang H, Zhang X and Chen Y (2024) Electrochemically coupled anaerobic membrane bioreactor facilitates remediation of microplastic-containing wastewater. *Water* **(22)**: 3236, 10.3390/w16223236.

Microbial Processes in Environmental Engineering: From Theory to Practice

Rajneesh Kumar, Ram Sharan Singh and Maulin P Shah
ISBN 978-1-83708-327-5
https://doi.org/10.1108/978-1-83708-326-820261013

Chapter 13
Microbial processes in soil remediation

Rym Salah Tazdaït[1] and Djaber Tazdaït[1,2]

[1]Bioengineering and Process Engineering Laboratory (BIOGEP), École Nationale Polytechnique, Algiers, Algeria
[2]Department of Nature and Life Sciences, Faculty of Sciences, University of Algiers, Benyoucef Benkhedda, Algeria

Abstract

Soil pollution caused by anthropogenic activities is one of the most threatening issues that societies worldwide face. Implementing innovative and sustainable methods to treat contaminated soils is a challenging task. The bioremediation approach is a good alternative to conventional physicochemical treatment methods thanks to its ecological friendliness, cost-effectiveness and efficiency. The bioremediation approach uses different microorganisms (bacteria, archaea, fungi and algae) or the enzymes they produce (carboxylesterase, laccase, peroxidase, phosphotriesterase, peroxidase, enolase and so on) to degrade organic contaminants in different media, including soil, through biotransformation causing slight chemical modifications to the contaminant, or through mineralisation generating harmless mineral molecules (CO_2, H_2O, SO_4^{2-}, NH_3 and so on). Thanks to the rapid advances in genetic engineering and molecular biology, there has been a significant increase in the efficiency of microbial bioremediation. Indeed, many genetically engineered microbial strains have been successfully used to break down toxic compounds, such as pesticides, pharmaceuticals, polycyclic aromatic hydrocarbons, dioxins, plastics and flame retardants in soil with significant uptake efficiencies. This chapter describes the different microbial processes, including bioaugmentation, biostimulation and bioremediation techniques performed in the field in combination with process engineering used to treat various xenobiotics-contaminated soils, providing cost-effective and environmentally friendly solutions.

Keywords: Bioaugmentation; bioremediation; biostimulation; co-metabolism; soil pollution; xenobiotics.

13.1. Introduction

Pollution generated by anthropogenic activities is threatening life on Earth as never before. The sources of pollution are diverse; they include wastewater from different industries (chemical, pharmaceutical, agri-food and so on) and domestic activities.

The polluted effluents generated from industrial activity are particularly concerning due to the quantity and diversity of chemical compounds they contain, which include, among others, microplastics, plasticisers, polycyclic aromatic hydrocarbons (PAHs), heavy metals and dioxins, whose deleterious effects on health and ecosystems are proven. Their effects on human health go from benign symptoms (nausea, headache, irritation and so on) to very severe damage (mutagenic, teratogenic, cancerogenic, endocrine-disruption, severe respiratory damage and so on) (Tazdaït and Salah Tazdaït, 2021, 2023a). If they are not correctly managed, these contaminants integrate the water streams and end up in all environmental media, especially agricultural soils, where they accumulate and contaminate the crops, posing a serious threat to human health.

Additionally, overusing pesticides worldwide significantly increases the risk of their accumulation in food and different ecosystems (Ramadan *et al.*, 2016). To have an idea of the proliferation of using pesticides in agriculture, it has been estimated that the use of pesticides worldwide has increased by nearly 75% in a period of 20 years (1999–2019) (FAOSTAT, 2022). In order to mitigate chemical pollution in soil, different methods with differential efficiencies have been applied. These methods include those based on physicochemical principles (adsorption, ultrafiltration, reverse osmosis, advanced oxidation processes, ozonation and so on), those using microorganisms (bacteria, algae and fungi) and those using higher organisms (plants) or the enzymes they produce.

This bio-based approach, called bioremediation, is a cost-effective, easy-to-implement, environmentally friendly and sustainable solution for treating various contaminants for which the conventional physicochemical methods are less efficient (Salah Tazdaït *et al.*, 2025). Biological resources are virtually limitless, given the tremendous physiological and biochemical diversity within microorganisms, which could virtually degrade every artificial chemical compound (Becker and Seagren, 2010).

In the case of organic pollutants, the biodegradation of a certain pollutant can only occur at a defined concentration known as the minimum substrate concentration. The microbial processes implicated in the biodegradation of a given contaminant include mineralisation and biotransformation, depending on the extent to which it is degraded.

Mineralisation is a process that completely degrades organic pollutants into simple, non-toxic mineral molecules, such as water, ammonia, carbon dioxide and sulfate ions, without producing harmful metabolites. In this process, the degraded contaminant, referred to as the primary substrate (growth substrate), acts as energy and/or phosphorus, nitrogen, carbon and sulfur sources, therefore sustaining biomass growth. This process can be exemplified by a study in which *Shewanella oneidensis* MR-1 could mineralise 94% of 30 mg/L malathion and use it as a growth substrate under anaerobic conditions (Pan *et al.*, 2024) and another study, which highlighted the degradation of triclocarban (14.5 mg/L) by a co-cultivation of two bacterial strains (*Pseudomonas* sp. LY-1 and *Rhodococcus* sp. BX2), reaching 85.9% degradation efficiency within 7 days of incubation (Hong *et al.*, 2024).

On the contrary, biotransformation is a process that typically induces few chemical modifications in the contaminant co-metabolically. In this process, the contaminant, called a co-substrate, secondary substrate or co-metabolic substrate, does not serve as a growth substrate (Tazdaït and Salah-Tazdaït, 2023b). Its degradation can only take place in the presence of a primary substrate, which sustains microbial growth. The primary substrate required for co-metabolism can be either simple (fructose, glucose, butyrate, citrate, acetate, glycerol and so on) or complex (molasses, syrup, date, whey and so on), which could increase the process cost (Li *et al.*, 2023).

Using co-pollutants as co-metabolic substrates is garnering significant interest because of their ability to efficiently and cost-effectively degrade multiple pollutants simultaneously (Zhu *et al.*, 2024a). In this approach, microorganisms degrade a contaminant as a primary substrate in conjunction with another contaminant, playing the role of a secondary substrate, which is usually degraded through biotransformation. It should be pointed out that this process can often produce byproducts that are more hazardous than the mother pollutant. This results in a reduction in the performance of the co-metabolic process due to the inhibition of microbial growth.

This chapter describes the applicability of microbial bioremediation methods for effectively eliminating diverse organic pollutant contamination.

13.2. Main types of soil contaminants

A contaminant is defined as any substance of natural origin (metallic trace elements, crude oil, nitrate and so on) or anthropogenic origin (pesticides, organochlorine solvents, diesel, polychlorinated biphenyls (PCBs) and so on) that accumulates in the soil as a result of human activity and is likely to harm human health or the quality of the environment. A contaminated site is, therefore, a site containing polluting elements whose content exceeds the threshold value permitted and is defined by environmental standards (Mal Meena *et al.*, 2024).

The prominent families of contaminants in soils are as follows.

13.2.1 Heavy metals

Heavy metals are metals with a density superior to 5–6 g/cm^3. They form insoluble sulfides. Cadmium (Cd), manganese (Mn), cobalt (Co), chromium (Cr), copper (Cu), lead (Pb), mercury (Hg), nickel (Ni) and zinc (Zn) are the heavy metals most commonly found in soils. Some elements (Cd, Hg, Ni, Co, Pb, Cr, Cu, Mn, V, Sn, Zn, vanadium (V), tin (Sn) and thallium (Tl)) can be present in different valence states, with intense variations associated with their geochemical properties and, therefore, their potential toxicity. Soil metal content can be of natural or anthropogenic origin.

13.2.1.1 Natural sources of contamination

Soils have a natural concentration of metals resulting from altering the parent rock. The crumbling of the latter and erosion lead to the formation of soils. These geochemical processes are associated with other phenomena, particularly biological ones, called soil differentiation. It is the primary source of natural contamination of soil by heavy metals.

13.2.1.2 Anthropogenic sources of contamination

Soil pollution due to human activity occurs on a much shorter time-scale – a decade or less.

Table 13.1 summarises the primary anthropogenic sources of soil pollution by the most frequent heavy metals. These sources are primarily linked to primary metal processing activities (mining and metallurgical industries) (Briffa *et al.*, 2020; Zhao *et al.*, 2022).

Table 13.1 The primary anthropogenic sources of soil pollution by heavy metals (Briffa *et al.*, 2020; Rashid *et al.*, 2023; Zhao *et al.*, 2022)

Element		Industry	Combustion of fossil fuels (coal, fuel oil)	Household waste incineration	Traffic flow	Sewage treatment plant sludge	Others
As	Arsenic	X	X	X			Insecticides, herbicides
Cd	Cadmium	X	X	X	X	X	Batteries, phosphate fertilisers, pigments
Co	Cobalt	X					
Cr	Chromium	X	X	X		X	
Cu	Copper	X	X		X	X	Compost
Fe	Iron	X					
Hg	Mercury		X	X		X	Chemistry, chlorine production
Ni	Nickel	X	X	X	X	X	
Pb	Lead	X	X	X	X	X	Ammunition, insecticides
V	Vanadium	X	X	X			
Zn	Zinc	X	X	X	X	X	

13.2.2 Hydrocarbons or mineral oils

Hydrocarbons or mineral oils include different petroleum products (crude oil, refined petroleum, kerosene, gasoline, fuel, lubricants and motor oils). These families are based on petroleum refining cuts. Their presence in contaminated soil is linked to the refining and processing industries and hydrocarbon transport, storage and distribution. There is also pollution from discharged by products (in particular, used engine and cutting oils, degreasing residues and accompanying solvents). Some hydrocarbons are used as industrial solvents and/or as bases for chemical syntheses (cyclohexane, hexane and so on) (Ambaye *et al.*, 2022; Imam *et al.*, 2021).

Their main physicochemical properties are solubility and density. Most hydrocarbons are poorly miscible with water and when they reach groundwater, they tend to form distinct layers. The density of hydrocarbons varies with the length of the carbon chains. Most hydrocarbons (C5 to C26) selectively accumulate on the surface of the water (density less than 1) and the heaviest hydrocarbons (heavy fuel) can migrate under certain conditions into the unsaturated zone (density greater than 1). Aliphatic hydrocarbons in $20 > C > 9$ are readily biodegradable by bacteria and fungi under aerobic conditions. Hydrocarbons in $C < 9$ have a biodegradability strongly inhibited by their toxic effects on microorganisms. The long carbon chains make high molecular weight hydrocarbons (C > 20) challenging to biodegrade (Antón-Herrero *et al.*, 2023; Paddar *et al.*, 2019).

PAHs are generally defined as compounds containing at least two benzene rings fused and containing only hydrogen and carbon atoms. The American Environmental Protection Agency has selected sixteen forms of PAHs as priority pollutants because of the risk they pose to the environment. Many studies focus on these 16 PAHs, but the theoretical number likely to be encountered is more than 1000.

PAHs are formed mainly during the incomplete combustion of organic matter, whether recent or fossil. PAHs resulting from combustion are said to be of 'pyrolytic' origin. Industrial activity, particularly the coal industry (coking plants, gas plants), vehicle emissions and residential heating contribute significantly to the production of these compounds (Rusănescu *et al.*, 2025).

PAHs can also have a petroleum origin. Petroleum is formed by catagenesis at relatively low temperatures (50–150°C). Thus, the burial of organic matter in sedimentary basins and its slow maturation leads to the formation of complex PAH mixtures where alkylated derivatives predominate.

The position of the rings in a linear (anthracene) or angular (phenanthrene) way influences the stability of the molecules, with angular PAHs being the most stable. Similarly, the presence of 5-carbon rings among benzene rings would increase the stability of the molecules against attacks by microorganisms (Rusănescu *et al.*, 2025; Wang *et al.*, 2022).

The hydrophobicity of PAHs is characterised by a high octanol–water partition coefficient (Kow). This dimensionless constant varies between approximately 2000 for naphthalene and 10 000 times more for indeno-pyrene. Thus, the aromatic ring structure of PAHs directly influences their fate in the environment since it is responsible for their low solubility and high hydrophobicity and, consequently, their high adsorption on the solid phases of the soil (Sahu and Pandit, 2003).

13.2.3 Pesticides

Pesticides are a very heterogeneous family of substances used against animal or plant parasites of crops. Pesticides in soils are brought by agricultural activities but also by non-agricultural activities, such as controlling weeds on road and railway networks or maintaining green spaces and gardens. The chemical industries producing these compounds and their storage can constitute point or concentrated sources.

There are two categories of pesticides.

- Inorganic pesticides are used in large quantities, such as sulfur or copper. These are very old pesticides whose use appeared well before synthetic organic chemistry.
- Organic pesticides are generally classified into organochlorines, organophosphates and carbamates and belong to various chemical families. The great chemical diversity of pesticides is due to their chemical elements and the various chemical functions present in the molecules (Buerge *et al.*, 2024; Tudi *et al.*, 2021).

13.2.4 Plastics

Plastic pollution is anthropogenic pollution that is mainly present throughout the biosphere. This omnipresence results from multifactorial sources, including the production and massive use of plastics and the overconsumption of products packaged in plastic materials. It also involves the inadequate management of plastic waste discharged into rivers, oceans and other ecosystems. Traditional plastics made from crude oil and designed to be resistant make them a particularly persistent pollution in the environment, taking decades or even centuries to decompose. In addition, small fractions of plastic sizes, resulting from direct production or the fragmentation of larger ones, make these materials omnipresent in the environment.

There are several types of plastics, the nature of which differs according to their chemical and physical properties. The standard classification of plastics, known as a 'resin code', includes the following.

- Class 1. Polyethylene terephthalate (PET) is primarily used for water bottles, soft drink containers and food packaging.
- Class 2. High-density polyethylene (HDPE) is widely used for various applications, including milk bottles, cleaning product containers and some types of plastic bags.
- Class 3. Polyvinyl chloride (PVC) is the constituent of pipes, flooring, plastic windows and toys, and is used for various construction applications.
- Class 4. Low-density polyethylene (LDPE) is widely used, primarily for plastic bags, packaging films and coatings.
- Class 5. Polypropylene (PP) is used for various applications, including food containers, caps and automotive parts.
- Class 6. Polystyrene (PS) is used in some cups, fast food packaging, CD/DVD cases and other plastic products.
- Class 7. This category is called 'other' for 'other plastics' and includes specific plastic polymers that are not as common or standardised as the previous six categories. It includes a variety of other plastics, such as the following.

- Acrylonitrile butadiene styrene (ABS) is used in manufacturing automotive parts, toys, electronics housings and other products.
- Polylactic acid (PLA) is a biodegradable plastic obtained from renewable raw materials. PLA is used for applications such as biodegradable food packaging and bags.
- Polyhydroxybenzoic acid polycarbonate (PHB) is a biodegradable plastic used in applications similar to PLA. Polycarbonate is also used for reusable water bottles and eyeglass lenses.
- Nylon (polyamide) is used in the manufacture of textile fibres, mechanical parts, ropes and other products (Bułkowska *et al.*, 2024).

13.2.5 Some other contaminants

PCBs and polychlorinated terphenyls (PCTs) are also soil contaminants. There are over 200 possible chlorobiphenyl structures, called congeners, in which 1 to 10 chlorine atoms are bonded to the biphenyl group. The most toxic are those unsubstituted in the ortho position. They are called 'planar'. Their chlorine content varies from 20 to 60%. In industry, PCBs and PCTs are used as transformer and capacitor dielectrics, heat transfer and hydraulic fluids, plasticisers and lubricants. They are used in dyes, varnishes, inks, carbonless copy papers, cutting oils and so on (Seo *et al.*, 2025; Wichmann *et al.*, 2019).

Dioxins and furans are secondary compounds generated by some industrial combustion processes, such as waste incineration, cement production and the paper industry. These compounds are stable at high temperatures and highly lipophilic, which leads to their bioaccumulation in the food chain. Seventeen congeners are recognised as toxic compounds, the most toxic being 2, 3, 7, 8-TCDD (2,3,7,8-tetrachlorodibenzo-p-dioxin), identified during the chemical accident in Seveso (a town located in northern Italy) on 10 July, 1976 (Kirkok *et al.*, 2020; Warner *et al.*, 2020).

13.3. Bioremediation

13.3.1 Definition and principles

Bioremediation is an approach based on the use of eucaryotic (plants and fungi) and procaryotic (bacteria) organisms to eliminate or neutralise organic (PAHs, pesticides, pharmaceuticals, PCBs, detergents, microplastics and so on) or inorganic (heavy metals and radionuclides) contaminants from polluted environments (water, soil and air).

This ecologically friendly and cost-effective technique relies on the catabolic capabilities of the remediating organisms, in which the enzymes play a key role in the process of decontamination. The decontaminating organisms act through diverse mechanisms, including biomineralisation, bioreduction, biodegradation, biosorption, biotransformation, bioaccumulation (biocomplexation), phytovolatilisation and phytostabilisation (Salah-Tazdaït and Tazdaït, 2022; Tazdaït and Salah-Tazdaït, 2021).

During microbial degradation, particular microorganisms (bacteria, microalgae and fungi), whether isolated from natural environments or obtained through genetic or biochemical engineering, decompose pollutants, transforming them into less harmful compounds like water, carbon dioxide, ammonia and so on. This process occurs aerobically or anaerobically, depending on the contaminant's microorganisms, environment and chemical structure.

In order to make the process more performant, the use of purified degrading enzymes (laccase, carboxylesterase, enolase, peroxidase, phosphotriesterase and so on) from microorganisms, mainly bacteria and fungi, has been successfully applied to degrade various organic contaminants including pesticides. In comparison to the entire cell configurations, this approach presents some relevant advantages, which include (*a*) its innocuity – that is, there is no production of toxic metabolites; (*b*) high specificity in targeting contaminants and high degradation performance; and (*c*) possibility of using it in real bioremediation scale (in situ and ex situ strategies), thus enhancing the sustainability and cost-effectiveness of the environmental pollution treatment (Pashirova *et al.*, 2024).

Despite these advantages, the free enzymes have the major inconvenience of being affected by reversible or irreversible inhibitors present in the environment, which can seriously hinder their catalytic activity.

13.3.2 Mechanisms of microbial bioremediation

13.3.2.1 Mechanisms of microbial remediation of inorganic contaminants

The pollution caused by anthropogenically generated metals is as hazardous to human health as organic contaminants. This is because, in contrast to naturally occurring metals which are typically found in relatively stable states, anthropogenic metals exhibit significantly higher reactivity, rendering them more threatening to biological systems.

In this context, bioremediation offers a sustainable and environmentally friendly way to remove, detoxify or neutralise heavy metals, including radionuclides, from contaminated environments.

Metal resistance is the process behind which the metal-decontaminating microorganisms, especially fungi and bacteria, resist heavy metal toxicity, and it is their inherent genetic characteristics which enable them to grow in high metal-contaminated environments (Conteville *et al.*, 2023; Margaryan *et al.*, 2021). Microbial metal resistance is categorised into acquired resistance and natural resistance.

Although microorganisms cannot degrade metals, they will modify their chemical properties by way of various interaction mechanisms, such as biosorption, bioleaching, microbial enzymatic systems and biomineralisation. Thus, microorganisms will affect metal reactivity and mobility. These microorganisms can decontaminate certain metals and prevent the spread of particular contaminations (Sable *et al.*, 2024).

The objectives of bioremediation strategies for metal compounds are based on using bacteria to modify their physical and chemical properties.

The transformation of the metal into a non-toxic form (by reducing the solubility of the pollutant by biotransformation, its bioavailability is reduced as well as the level of exposure to this compound) is ensured

- either by enzymatic means – for example, the reduction of mercury-by-mercury reductase
- or by the action of a product of the cell's energy metabolism (the reduction of selenium by sulfur products such as thiosulfate or sulfites).

The immobilisation of metals in the soil matrix should limit the risks of transfer to the water table. This biological treatment technique of bioreduction applies mainly to metal species that are easily mobile because they are very soluble in water and whose reduced form is not very soluble (Panda *et al.*, 2022).

The mechanisms involved in the microorganism – metal interactions are discussed below and summarised in Figure 13.1.

13.3.2.1.1 BIOMINERALISATION

The biomineralisation of metals is a mechanism in which diverse bacteria, including iron-oxidising and sulfate-reducing bacteria and others such as *Lactobacillus sphaericus*, *Bacillus cereus*, *Bacillus subtilis* and *Pseudomonas stutzeri*, act by precipitating contaminating metal

Figure 13.1 Mechanisms involved in the interactions between microorganisms and metals

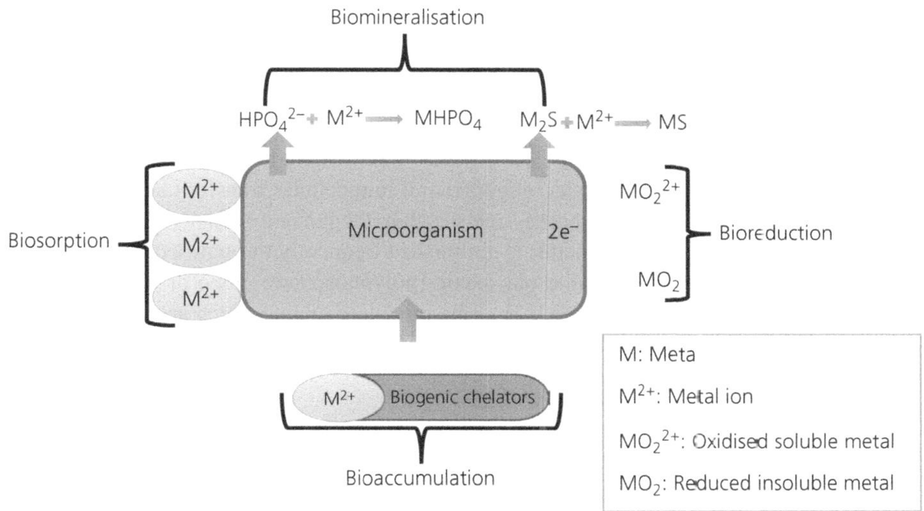

ions with the help of various terminal metabolites, including phosphate, carbonate, sulfate and hydroxide (Mallick *et al.*, 2025; Salah-Tazdaït and Tazdaït, 2022). It is worth noting that the nature of metal and environmental factors, such as nutrient availability, pH, oxygen concentration and temperature, determine the microorganisms engaged in this mechanism.

13.3.2.1.2 BIOSORPTION OR PASSIVE UPTAKE

The negative net charge of a cell's envelope enables some fungi and bacteria to adsorb heavy metal and radionuclide cations on their cell wall surface without metabolic involvement. This method effectively immobilises hazardous metals by obstructing their entry into microbial cells, hence averting their detrimental impact on intracellular enzymatic activities (Salah Tazdaït and Tazdaït, 2025). This phenomenon, called physisorption, is a non-metabolic-dependent mechanism, which initially entails the adsorption of the metal ions on the cell surface by way of weak van der Waals forces, succeeded by their stronger interactions (hydrogen and hydrophobic bonding) with the cell surface's biomolecules (such as proteins, peptidoglycan, lipopolysaccharide, teichoic acid and so on) through different chemical functional groups (such as sulfhydryl, carbonyl, amine, phosphoryl and so on) (Pandey and Keshavkant, 2021; Raji *et al.*, 2023). It should be pointed out that the differences in cell wall components between Gram-positive and Gram-negative bacteria appear to have minimal impact on metal biosorption performance (Ngwenya *et al.*, 2003).

However, biosorption is strongly influenced by abiotic conditions. This influence is particularly significant with the pH, which affects the net charge of the microbial wall surface through protonation and deprotonation reactions that affect its functional groups.

A biotechnological method based on cell surface modification by genetic engineering has recently emerged to improve the performance of the biosorption process. In this method, genetically engineered peptides with high affinity towards metal ions are attached to particular protein binding sites on the microbial cell surface, resulting in a significant enhancement in the biosorption process (Wang *et al.*, 2021).

13.3.2.1.3 BIOACCUMULATION (BIOCOMPLEXATION)

The process of metal microbial bioaccumulation relies on the ability of microorganisms to adsorb metals from the surrounding environment and sequester them intracellularly. Different mechanisms are involved in this process, which mainly include chelation using biogenic chelators (exopolysaccharides and siderophores) and intracellular transfer mediated by specialised ATP-dependent active transport systems (Salah-Tazdaït and Tazdaït, 2022). On entering microbial cells, the toxicity of metals is diminished frequently through sequestering them by intracellular polymers of inorganic phosphate (polyphosphate) alone or combined with polyhydroxybutyrate (Rogiers *et al.*, 2022), or by metallothioneins which are low molecular weight proteins that bind to metals (Salah-Tazdaït and Tazdaït, 2022).

13.3.2.1.4 BIOTRANSFORMATION (BIOREDUCTION)

This metabolically governed mechanism uses a specialised group of bacteria capable of reducing metals (metal-reducing bacteria) by changing their oxidation states through enzymatic systems that reduce different metal ions into their less soluble elemental forms, which can then precipitate. This mechanism is driven by the oxidative metabolism in which the electrons liberated from the oxidation of substrates (organic or inorganic) pass through the respiratory chain and serve to reduce the metal ions extracellularly, with simultaneous energy generation, in the form of adenosine triphosphate (ATP) by way of ATP synthase. This mechanism can be exemplified by a study that reported the ability of the bacterial strain *Cupriavidus metallidurans* MSR33, isolated from soil contaminated by toxic divalent mercury (Hg^{2+}), to reduce 82% Hg^{2+} (20 or 30 ppm) to less hazardous elemental mercury (Hg^0) after only 48 h of experiments (Bravo *et al.*, 2020).

13.3.2.2 Mechanisms of microbial remediation of organic contaminants

The microbial bioremediation process can be categorised into biotransformation and mineralisation by considering the biodegradation extent of a given organic pollutant.

Biotransformation is a process during which particular microorganisms act by catalysing minor chemical alterations (transformations) on the organic pollutants, generating complex metabolites that, in some cases, harbour the same or more significant toxicity than the parent contaminants, thus augmenting the risk for human health and the environment – for example, the abiotic and biotic degradation of malathion and parathion yield malaoxon and paraoxon, respectively, which are far more toxic than the parent insecticides (Angelini *et al.*, 2015). Therefore, the metabolites produced during the bioremediation process should be identified carefully to verify their innocuity prior to releasing them into the environment.

The degradation of the pollutants through biotransformation is possible only if growth substrates, such as lactate, glucose, molasses, whey and so on, are available. The biotransformed pollutants are thus co-metabolised and are called co-substrates or secondary substrates. The biotransformation reactions of pollutants in the environment are fundamentally influenced by their chemical structure and are catalysed by extra- or intra-cellular enzymes classified as transferases, dehalogenases, oxidoreductases, hydrolases and mono- or di-oxygenases (Doukani *et al.*, 2023).

The mineralisation process consists of degrading the organic contaminants into simple mineral compounds (H_2O, CO_2, PO_4^{3-} and so on) while serving as primary growth substrate (carbon, electron and energy sources). It is essential to emphasise that the biodegradation of organic contaminants occurs only at or above a specific threshold known as the minimum substrate concentration (S_{min}), at or over which the contaminants sustain microbial growth (Becker and Seagren, 2010).

To determine S_{min} and by considering that there is no biomass concentration variation ($dX/dt = 0$), it is possible to combine the Monod model, giving the variation of the pollutant degradation rate with respect to its initial concentration (Equation 13.1) and the general model describing the kinetics of microbial degradation of organic contaminants serving as a growth substrate (primary substrate) (Equation 13.2) (Becker and Seagren, 2010). This leads to Equation 13.3.

$$d = -d_{\max}\frac{S}{K_s + S} = \frac{r}{X} \tag{13.1}$$

where d is the specific pollutant consumption rate (h^{-1}), d_{max} is the maximum specific pollutant consumption rate (h^{-1}), S is the initial pollutant concentration (g/L), K_s is the half-saturation constant (g/L), which is a specific pollutant concentration for which d equals $d_{max}/2$, r is the pollutant consumption rate (g/L/h) and X is the pollutant-degrading biomass concentration (g/L).

$$q = \frac{dX}{dt} = Y(-r) - bX \tag{13.2}$$

where q, Y, dX/dt and b are the net microbial growth rate (g/l/h), the true yield, the variation of biomass concentration with time and the first-order decay coefficient (h^{-1}), respectively.

$$S_{\min} = bK_s / \left(Y\, d_{\max} - b\right) \tag{13.3}$$

As stated previously, S_{min} corresponds to the minimum concentration of the pollutant at which it is degraded and metabolised. This means that, on the condition that S is greater than S_{min}, the organic pollutant is metabolically degraded and serves as a primary substrate to support microbial growth. However, if S becomes less than S_{min}, the pollutant no longer supports the microbial growth; it acts, in lieu, as a co-substrate (secondary substrate) subjected to the biotransformation or even mineralisation processes through co-metabolism (Rittmann and McCarty, 2020) – for instance, trichloroethene (TCE) can be co-metabolised by different bacterial methanotrophs (e.g. *Methylococcus* sp., *Methyloversatilis* sp. and *Methylomonas* sp.) in the presence of methane, which serves as a growth substrate.

During this process, TCE is partially or totally degraded aerobically. First, it is subjected to an oxidative reaction catalysed by a particular methane monooxygenase (pMMO) or a soluble methane monooxygenase (sMMO) to form TCE-epoxide, which is then degraded either enzymatically or spontaneously to form different byproducts, including dichloroacetic acid, CO, formic acid and glyoxylic acid (Hwangbo *et al.*, 2023). Notably, phenol can be used as

Figure 13.2 Aerobic co-metabolic degradation of TCE in the presence of toluene

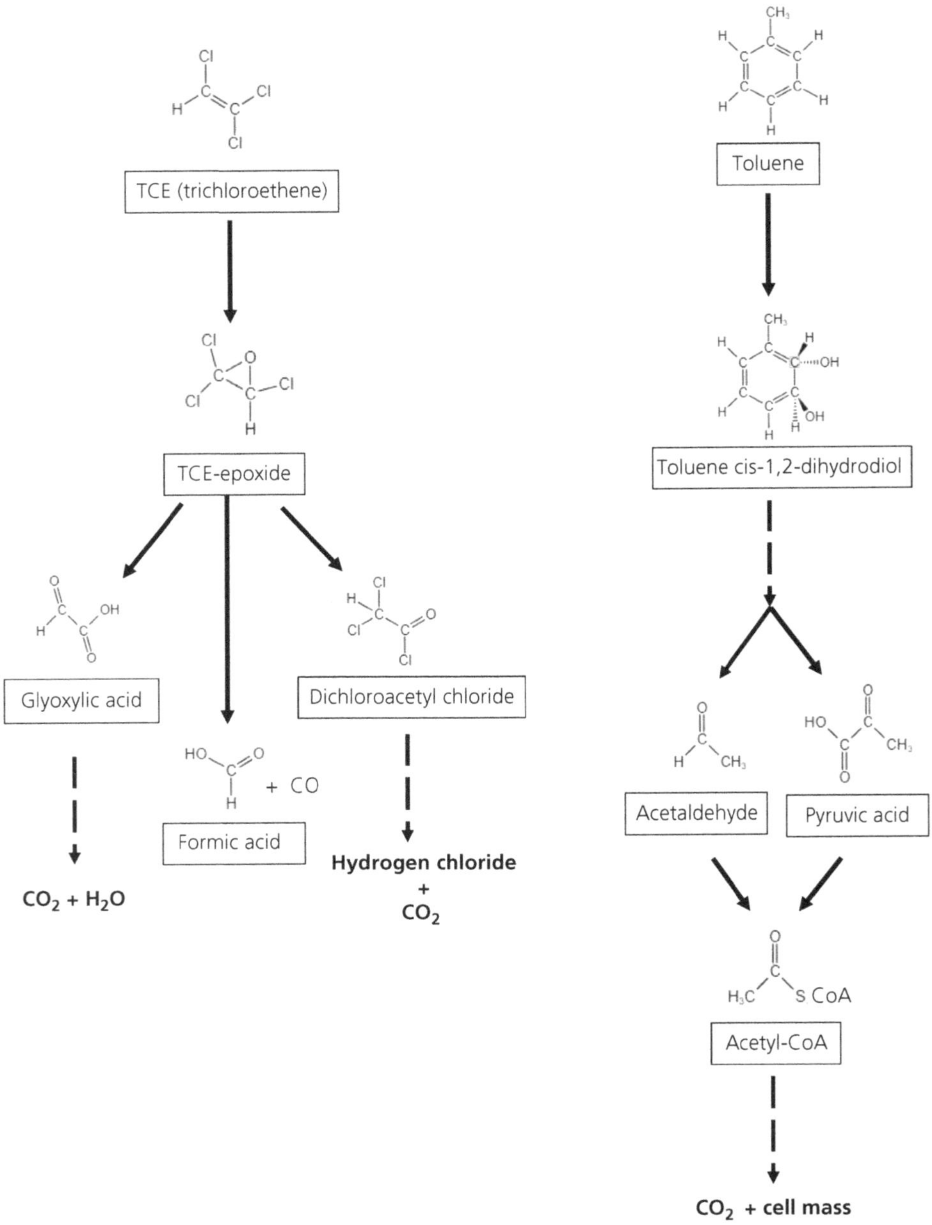

a co-substrate, or as a primary substrate during TCE and cis-1,2-dichloroethene (cDCE) degradation by the bacterial strain *Cupriavidus* sp. CY-1 and their conversion into biodegradable plastic (poly-β-hydroxybutyrate) (Chang *et al.*, 2021).

In another study, the co-metabolic degradation of the hazardous aromatic amine p-chloroaniline (PCA) in the presence of aniline, used as a co-substrate (primary substrate), was achieved using the bacterium *Pseudomonas stutzeri*. Aniline added at 20 mg/L effectively degraded PCA (50 mg/L), achieving 81.12% degradation efficiency after 7 days of experiments, under 170 rpm stirring, at 30°C and pH 8 (Zhu *et al.*, 2024a).

Figure 13.2 illustrates the co-metabolic degradation of TCE in the presence of toluene, another example of a co-contaminant used as a primary substrate.

Xenobiotic compounds, such as phenol and organophosphorous pesticides, can inhibit microbial growth at high concentrations. Among all existing mathematical models that describe microbial growth inhibition, the Andrew-Haldane model (Edwards, 1970) (Equation 13.4) is the most employed thanks to its simplicity and effectiveness in describing the inhibitory effect of diverse organic contaminants on the growth of different bacterial and fungal species used in bioremediation.

$$\mu = \mu_{\max} S / \left(K_s + S + \left(S^2 / K_i\right)\right) \tag{13.4}$$

where μ, μ_{max} and K_i are the specific growth rate (h^{-1}), the maximum specific growth rate (h^{-1}) and the inhibition constant (g/L), respectively, and K_s is a specific pollutant concentration for which μ equals $\mu_{max}/2$ when μ decreases with increasing pollutant concentration.

The Andrew-Haldane model can be rewritten by considering the variation of the pollutant degradation rate as a function of its initial concentration as follows (Equation 13.5).

$$d = -d_{\max} S / \left(K_s + S + \left(S^2 / K_i\right)\right) \tag{13.5}$$

13.4. Microbial processes

Microbiological processes stimulate the ability of microorganisms to use various types of organic and mineral compounds for their growth, energy needs and survival. These processes could allow the treatment of any contaminated site. However, the implementation of a bioremediation treatment based on the use of microorganisms requires the control not only of the availability of pollutants and the presence of purifying strains but also the continuous adjustment of a certain number of environmental conditions to maximise its effectiveness: quantity and type of nutrients, presence of different electron acceptors (oxygen, nitrate, sulfate and so on), pH, temperature and salinity (Novair *et al.*, 2024).

The main approaches used to achieve these objectives are as follows.

13.4.1 Natural attenuation

Natural attenuation is a process that uses the purifying capacities of microorganisms present in polluted ecosystems. Although this decontamination technique does not theoretically require human intervention, it is necessary to eliminate or neutralise the pollution source and continuously monitor the site until the end of the treatment. This type of bioremediation is very inexpensive because while it does not require many resources, it requires long treatment times. In this case, however, the degradation of xenobiotics cannot be attributed solely to microorganisms. Indeed, abiotic processes, such as evaporation, dissolution, dispersion, emulsification, adsorption or photooxidation can also come into play (Puigserver *et al.*, 2020).

13.4.2 Dynamic attenuation

Dynamic attenuation requires the addition of specific compounds to soils or groundwater to create favourable conditions for the activity of microorganisms responsible for the biodegradation of contaminants. To do this, microorganisms are maintained in optimal conditions (oxygen, pH, temperature and redox potential) and nutrients are added mainly in liquid form.

The different compounds are added in the unsaturated zone or the saturated zone. The nature and the concentration of the different compounds added depend on the environmental conditions, the contaminants and the biodegradation processes envisaged (aerobic oxidation, anaerobic reduction, co-metabolism and so on).

In some cases, non-endogenous bacteria are seeded. Nutrients are often injected in dissolved form by way of injection wells, drains or other means.

In the event of aerobic degradation, oxygen can be supplied in gaseous form (e.g. by pumping air or by adding pure oxygen), in liquid form (hydrogen peroxide (H_2O_2)) or solid form (magnesium peroxide (MgO_2)). This allows the oxygen to be released slowly and, therefore, has the advantage of creating more persistent conditions. In all cases, the treatment requires injection systems and sometimes recovery of groundwater and gases.

The in situ dynamised biodegradation system consists of

- a storage/production system for electron acceptors or donors of different configurations depending on whether they are in gaseous, liquid or solid form (suppressor, storage tank, water-mixing system, pump and so on)
- equipment relating to nutrients: storage tank, water-mixing system, distribution network
- a system of vertical (well injection) or horizontal (injection drains, trenches and so on) injection points
- a unit for recovering and treating gaseous effluents if necessary: extraction unit (volumetric, intermediate or centrifugal extractor), a condensate separator and a gas treatment line
- a unit for recovering and treating aqueous effluents: pumps, treatment line for wastewater retrieved on the surface and so on
- a system for storing solid and liquid waste resulting from the treatment
- a piezometric system for monitoring groundwater flows and quality (Barbero *et al.*, 2024; Zhu *et al.*, 2023).

The two most commonly used variants are biosparging and bioventing. These techniques combine two decontamination principles: biodegradation and ventilation. In the case of bioventing, there is forced aeration in the unsaturated soil (above the water table). This method combines venting and aerobic biodegradation treatments with reduced treatment duration. To further improve efficiency, bioventing is frequently coupled with air sparging (Anekwe and Isa, 2021).

13.4.3 Bioventing

Bioventing consists of stimulating the in situ biodegradation of pollutants in soils by providing the microflora in place with the oxygen necessary for bacterial metabolism. Oxygen, often the limiting element of microbial actions in aerobic conditions, is provided by injecting air into the contaminated area. Ventilation can be ensured by injecting air and/or extracting the gas phase from the unsaturated zone of the soil.

The air injection wells are located in highly contaminated areas, with the dual purpose of supplying oxygen where it is needed most and limiting the transfer of pollutants by volatilisation into the atmosphere. The extraction wells are located in uncontaminated areas on the site's periphery.

Atmospheric air flows continuously through passive or active aeration wells towards a suction (extraction). The oxygen thus transported is continually renewed and used for the respiration of organisms primarily consuming organic matter distributed in the soil.

The flow of air drawn into the system is kept as low as possible to obtain uniform laminar circulation throughout the contaminated area and avoid unnecessarily wasting energy. Therefore, to obtain a ventilation field that covers the entire contaminated area, the direction and intensity of air circulation in the soil can be adjusted at any time: passive or active infiltration and suction, horizontal, diagonal or circular currents, reversal of the direction of circulation and so on. In addition, it is possible to suck air from the soil through strainers placed at different depths or to infiltrate atmospheric air and other gases.

During venting, the airflow sets the volatile compounds in the soil in motion while biodegradation is carried out. In addition, air injection promotes the volatilisation of the liquid phase as the gaseous compounds are degraded.

Ventilation can treat all aerobically biodegradable constituents. It has proven particularly effective in remediating petroleum product discharges, including gasoline, kerosene and diesel fuel (Anekwe and Isa, 2021; Bhowmick *et al.*, 2024).

13.4.4 Biosparging

The principle of soil treatment by biosparging is the same as that of bioventing, except that it concerns pollution in the saturated zone.

Biosparging stimulates the in situ biodegradation of pollutants in soils by providing the microflora with the oxygen necessary for bacterial metabolism. Oxygen, which is often the limiting element for microbial actions in aerobic conditions, is provided by compressors injecting pressurised air into the contaminated zone in a saturated zone.

Biosparging is suitable for treating sources of pollution consisting of volatile organic compounds and petroleum hydrocarbons in a saturated zone.

In the case of volatile pollutants, extraction wells are located in uncontaminated areas on the site's periphery. This arrangement of the wells, as well as reasonable control of airflow rates, aims to optimise biodegradation and control the volatilisation of contaminants.

In biosparging, the injected air mobilises by vaporising the contaminants dissolved in the water or trapped by capillarity in the pores. The rising air bubbles carry this vaporised phase upwards, which will be biologically degraded in the unsaturated soil.

For this reason, several devices of bioventing systems are also used in biosparging.

The airflow injected into the subsoil will be recovered by suction. However, the injection/suction flow rate will be kept low enough to allow the soil microbial flora to degrade the polluting compounds volatilised and carried by the air current. This technique thus eliminates (or reduces) the need to treat the gaseous effluent extracted from the soil before its release into the atmosphere.

Technical design involves using only a small ventilation/suction device instead of the imposing air blowers and extractors usually used in venting or air vacuum techniques. The primary difficulty lies in punctually monitoring and controlling the processes that are only accessed indirectly. In addition, the injection of pressurised air into the subsoil can cause the formation of preferential circulation channels. If these are wide and few in number, or if they are not uniformly distributed in the volume to be treated, the efficiency of the process will be seriously impared. Poor soil structuring and its lack of homogeneity are, therefore, major limiting factors for using bioventing and biosparging techniques.

Currently, this technology is increasingly applied, particularly to decontaminate soils polluted by petroleum hydrocarbons. Excellent results are obtained, with more than 90% degradation rates in a few weeks (sometimes even without adding nutrients).

In the case of biosparging, bioavailability limits access to pollutants and, therefore, their biodegradation. However, this bioavailability can be increased by using surfactants that increase the solubility of pollutants.

Oxygen injection can cause the clogging of part of the soil pores by precipitation of metallic compounds, calcium carbonate or development of clogging microflora. However, techniques can be used to combat this clogging, including injection of hydrogen peroxide for bacterial clogging, alternating injection rhythm or use of nitrogen (Sanjana *et al.*, 2024; Sravya and Sangeetha, 2022).

13.4.5 Biopile

This technique is commonly used on sites with biodegradable volatile and semi-volatile hydrocarbons. The biopile is very widely used for on site decontamination. The treatment sites are called biocentres (biological treatment centres). The biopile applies to soils polluted by petroleum products such as diesel, fuel and kerosene. Under certain conditions, volatile organic halogenated compounds, pesticides and certain heavy petroleum fractions (PAH, organic oils and so on) can also be treated but with lower purification yields.

The biopile consists of putting polluted soils in piles for biological treatment. To do this, the polluted soils are generally amended and the conditions in the biopile are controlled (aeration, the addition of nutrients and so on). This requires prior excavation. Polluted soils are mixed with an amendment (structuring agent). They are then directed to a treatment area containing at least a leachate collection system and aeration units (extraction or air insufflation) to optimise oxygen transfer and stimulation of biodegradation. Biodegradation is controlled (temperature, humidity, mineral nutrients, microbial population, oxygen and pH).

An impermeable geomembrane most often covers biopiles to limit rainwater infiltration and volatilisation of pollutants, and to maintain or increase temperature. Leachates are partly recycled and partly treated on site before being discharged. Atmospheric discharges are treated if necessary (presence of volatile organic compound). Biological degradation is most often carried out by biostimulation.

The biopile system consists of

- a pre-treatment platform (homogenisation, screening, amendment of organic or structuring matter if necessary)
- equipment for adding nutrients: storage tank, leachate recovery and recirculation network
- equipment for adjusting humidity
- equipment required for aeration
- if necessary, a gas treatment system (biological treatment or biofilters most often)
- equipment for storing solid and liquid waste resulting from the treatment.

The largest biopiles have volumes that can reach several hundred cubic metres (length: several tens of metres, width: several metres (<10 m), height less than 4–5 m). Their height is limited to avoid problems of soil compaction that can occur over time, creating malfunctions in the distribution of gas flows. Aeration inside the biopile is ensured by continuous air suction through perforated pipes installed at intermediate heights and connected to a fan.

To avoid any pollution of the land on which it is installed, the biopile rests on a cement base or an impermeable membrane, allowing the collection and recycling of leachates.

To improve the performance of this type of biological treatment in which the porous solid material to be treated is static, it is crucial that the entire pile has uniform particle size distribution and compaction to guarantee that the circulation of air and nutrient solutions will affect the entire mass of material to be treated. Several preliminary operations can be considered, alone or in combination: screening or sieving to remove rubble and stones, crumbling and grinding or crushing to fragment the aggregates to increase the accessibility of pollutants to microorganisms and homogenisation by a mixer (during which it is possible to add nutrients and/or microorganisms and to modify, if necessary, the physicochemical context and, in particular, the pH).

The addition of compost benefits the microorganisms present, providing organic matter that is more easily assimilated than the pollutants to be degraded, thus promoting microbial growth and activity. Compost (or other granular materials such as wood chips) can also be an essential structuring element that ensures better circulation of gas and liquid flows. The efficiency of this process varies immensely depending on the environmental conditions. Sometimes, it can reach 90% to 99% if the treatment time is long enough.

This technique is widely used for heterogeneous soils containing readily biodegradable pollutants. However, it requires soil excavation and prior sorting; particle sizes greater than 60 mm are often excluded (Bhowmick *et al.*, 2024; Zhu *et al.*, 2024b).

13.4.6 Biostimulation

The intrinsic capacities of an environment to achieve natural attenuation are correlated with several parameters. The first is the bioavailability of the pollutant, which can be affected by bonds to the mineral and organic fractions of the soil. The quantity of organic matter, pH, mineralogy, temperature and the type of pollutant can influence this sorption. The second parameter is the redox environment of the system. Indeed, the availability of electron acceptors or donors can influence microbial activities. Finally, the bioavailability of nutrients (nitrogen and phosphorus) and microbial associations in the ecosystem can also control natural attenuation. Biostimulation will improve biodegradation by modifying the environmental conditions by adding surfactants, nutrients, water, electron donors or acceptors (Karishma *et al.*, 2024).

13.4.7 Bioaugmentation

Sometimes, the endogenous microbial populations of a polluted ecosystem do not have all the metabolic tools to achieve complete degradation of pollutants. Bioaugmentation can remedy this deficiency. It consists of incorporating microorganisms with adapted metabolisms into contaminated ecosystems. This contribution can be made by adding a single strain or a group of strains called a consortium. However, many factors, especially biotic (predation or competition), can jeopardise the effectiveness of such techniques by leading to a decline in the introduced populations.

To overcome these ecological constraints, physical protection systems for seeded microorganisms, such as gellan gum encapsulation, can be implemented. Genetic engineering can also solve these problems of stability of exogenous strains. In this context, endogenous microorganisms can be modified by providing them with the genes necessary to improve their purifying capacities (Wang, 2024).

13.5 Conclusions and perspectives

Bioremediation is currently recognised as a valuable, effective and ecologically friendly technique for restoring various contaminated media, including soil, air and water. This approach utilises taxonomically and physiologically diverse organisms, ranging from procaryotic (bacteria and archaea) and eucaryotic (fungi) microorganisms to eucaryotic superior organisms (plants), to remediate nearly all synthetic organic compounds. Enzymes extracted from these organisms can be used to break down contaminants, making them an efficient tool for bioremediation.

Despite all the advantages associated with the bioremediation method (cost-effectiveness, ecological friendliness, non-toxicity generation and easy implementation), there are still some limitations that need to be addressed, such as the time-scale of the biodegradation process for some contaminants, whose decomposition depends on their chemical structure and the action of different microorganisms acting complementarily. The process is also sensitive to different abiotic parameters, such as oxygen level, pH, temperature, nutrient availability, organic or inorganic inhibitors, which can hinder its effectiveness in harsh environments.

Bioremediation can be improved through nutrient enrichment (biostimulation) or bioaugmentation. The biostimulation process includes adding particular nutrients, such as phosphorus and nitrogen, to foster the growth of the pollutant-degrading microorganisms. Bioaugmentation is a technique that accelerates biodegradation by introducing highly performant degrading microorganisms into contaminated environments.

Another way to enhance bioremediation performance is to use co-metabolism degradation (non-growth-linked process), which has improved the degradation effectiveness of various synthetic organic contaminants (pesticides, detergents, hydrocarbons and so on) and even those known for their persistence in the environment. This process could also lower the overall cost of the bioremediation method by using other contaminants as co-substrates (primary substrates). However, further research efforts should be made to improve the performance of the microorganisms implicated in co-metabolism systems by applying biochemical and genetic engineering strategies.

On the other hand, screening and characterising new microbial strains with intrinsically improved degradative abilities for persistent polluting compounds could further enhance the bioremediation performance.

Finally, the innovative technology known as nanobioremediation, utilising organic or inorganic nanoparticles, could be a cost-effective and efficient approach to improve the co-metabolic decomposition of persistent contaminants and metal remediation in complex and harsh environments, by promoting microbial growth and enhancing the interactions between microbial cells and the contaminants. This approach could help achieve improved environmental stewardship and sustainability.

REFERENCES

Ambaye TG Formicola F, Sbaffoni S *et al.* (2022) Remediation of soil polluted with petroleum hydrocarbons and its reuse for agriculture: recent progress, challenges and perspectives. *Chemosphere* **293**, 10.1016/j.chemosphere.2022.133572.

Anekwe IMS and Isa YM (2021) Comparative evaluation of wastewater and bioventing system for the treatment of acid mine drainage contaminated soils. *Water-Energy Nexus* **4**: 134–140, 10.1016/j.wen.2021.08.001.

Angelini DJ Moyer RA, Cole S *et al.* (2015) The pesticide metabolites paraoxon and malaoxon induce cellular death by different mechanisms in cultured human pulmonary cells. *International Journal of Toxicology* **34(5)**: 433–441, 10 1177/1091581815593933.

Antón-Herrero R, Chicca I, García-Delgado C *et al.* (2023) Main factors determining the scale-up effectiveness of mycoremediation for the decontamination of aliphatic hydrocarbons in soil. *Journal of Fungi* **9**(12), 10.3390/jof9121205.

Barbero L, Martínez-Hernández V, Huidobro-López B, Meffe R and de Bustamante I (2024) Soil amendments and water management to improve attenuation and recovery of wastewater originated nutrients through a vegetation filter. *Journal of Environmental Management* **370**, 10.1016/j.jenvman.2024.122566.

Becker JG and Seagren EA (2010) Bioremediation of hazardous organics. In *Environmental Microbiology* (Mitchell R and Gu J-D (eds)). Wiley, Hoboken, NJ, USA, pp. 177–212, 10.1002/9780470495117.ch8.

Bhowmick K, Roy D, Rana D *et al.* (2024) Potential microbes in bioremediation: a review. *Materials Today Sustainability* **28**, 10.1016/j.mtsust.2024.101032.

Bravo G, Vega-Celedón P, Gentina JC and Seeger M (2020) Bioremediation by Cupriavidus metallidurans strain MSR33 of mercury-polluted agricultural soil in a rotary drum bioreactor and its effects on nitrogen cycle microorganisms. *Microorganisms* **8(12)**: 1952, 10.3390/microorganisms8121952.

Briffa J, Sinagra E and Blundell R (2020) Heavy metal pollution in the environment and their toxicological effects on humans. *Heliyon* **6(9)**, 10.1016/j.heliyon.2020.e04691.

Buerge IJ, Kasteel R and Poiger T (2024) Leaching of herbicides and their metabolites in lysimeters filled with soils from railway tracks. *Science of the Total Environment* **909**, 10.1016/j.scitotenv.2023.168396.

Bułkowska K, Zielinska M and Bułkowski M (2024) Blockchain-based management of recyclable plastic waste. *Energies* **17**, 10.3390/en17122937.

Chang Y-C, Venkateswar Reddy M and Choi D (2021) Cometabolic degradation of toxic trichloroethene or cis -1,2-dichloroethene with phenol and production of poly-β-hydroxybutyrate (PHB). *Green Chemistry* **23(7)**: 2729–2737, 10.1039/D1GC00265A.

Conteville LC, Oliveira-Ferreira J and Vicente ACP (2023) Heavy metal resistance in the Yanomami and Tunapuco microbiome. *Memórias do Instituto Oswaldo Cruz* **118**: e230086, 10.1590/0074-02760230086.

Doukani K Boukirat D, Boumezrag A, Bouhenni H and Bounouira Y (2023) Fundamentals of biodegradation process. In *Handbook of Biodegradable Materials* (Ali GAM and Makhlouf ASH (eds)). Springer, Cham, Swtizerland, pp. 57–83, 10.1007/978-3-031-09710-2_73.

Edwards VH (1970) The influence of high substrate concentrations on microbial kinetics. *Biotechnology and Bioengineering* **12(5)**: 679–712, 10.1002/bit.260120504.

FAOSTAT (2022) https://www.fao.org/faostat/en/#data/RP (accessed 02/01/2025).

Hong Y Sun G, Sun S *et al.* (2024) Enhancement of triclocarban biodegradation: metabolic division of labor in co-culture of Rhodococcus sp. BX2 and Pseudomonas sp. LY-1. *Environmental Pollution* **356**: 124346, 10.1016/j.envpol.2024.124346.

Hwangbo M Shao Y, Hatzinger PB and Chu K-H (2023) Acidophilic methanotrophs: occurrence, diversity and possible bioremediation applications. *Environmental Microbiology Reports* **15(4)**: 265–281, 10.1111/1758-2229.13156.

Imam A, Kanaujia PK, Ray A and Suman SK (2021) Removal of petroleum contaminants through bioremediation with integrated concepts of resource recovery: a review. *Indian Journal of Microbiology* **61**(3): 250–261, 10.1007/s12088-021-00928-4.

Karishma S, Saravanan A and Deivayanai VC (2024) Emerging strategies for enhancing microbial degradation of petroleum hydrocarbons: prospects and challenges. *Bioresource Technology Reports* **26**, 10.1016/j.biteb.2024.101866.

Kirkok SK, Kibet JK, Kinyanjui TK and Okanga FI (2020) A review of persistent organic pollutants: dioxins, furans and their associated nitrogenated analogues. *SN Applied Sciences* **2**, 10.1007/s42452-020-03551-y.

Li X, Tian T, Cui T *et al.* (2023) Alkaline-thermal hydrolysate of waste activated sludge as a co-metabolic substrate enhances biodegradation of refractory dye reactive black 5. *Waste Management* **170**: 40–49, 10.1016/j.wasman.2023.07.034.

Mal Meena P, Aggarwal RK, Meena R and Rathore MS (2024) Soil contamination, risk assessment and remediation. In *The Future of Risk Management* (Ivascu L *et al.* (eds)). IntechOpen, London, UK, 10.5772/intechopen.1005045.

Mallick S, Pradhan T and Das S (2025) Bacterial biomineralization of heavy metals and its influencing factors for metal bioremediation. *Journal of Environmental Management* **373**: 123977, 10.1016/j.jenvman.2024.123977.

Margaryan A, Panosyan H and Birkeland N-K (2021) Heavy metal resistance in prokaryotes: mechanism and application. In *Microbial Communities and their Interactions in the Extreme Environment* (Egamberdieva D *et al.* (eds)). Springer, Singapore, pp. 273–313, 10.1007/978-981-16-3731-5_13.

Ngwenya BT, Sutherland IW and Kennedy L (2003) Comparison of the acid-base behaviour and metal adsorption characteristics of a gram-negative bacterium with other strains. *Applied Geochemistry* **18(4)**: 527–538, 10.1016/S0883-2927(02)00118-X.

Novair SB, Biglari Z, Lajayer BA, Shu W and Price GW (2024) The role of sulphate-reducing bacteria (SRB) in bioremediation of sulphate-rich wastewater: focus on the source of electron donors. *Process Safety and Environmental Protection* **184**: 190–207, 10.1016/j.psep.2024.01.103.

Paddar K, Sarkar D and Sarkar A (2019) Construction of potential bacterial consortia for efficient hydrocarbon degradation. *International Biodeterioration and Biodegradation* **144**, 10.1016/j.ibiod.2019.104770.

Pan Q, Li Y, Zhang J *et al.* (2024) Mechanisms of oxidative response during biodegradation of malathion by S. oneidensis MR-1. *Environmental Science and Pollution Research* **31(11)**: 16832–16845, 10.1007/s11356-024-32283-4.

Panda JR, Sen S and Sarkar P (2022) Bioremediation of heavy metal pollutants in contaminated environment: principle, advantages, limitations and future. In *Developments in Applied Microbiology and Biotechnology, Synergistic Approaches for Bioremediation of Environmental Pollutants: Recent Advances and Challenges* (Kapoor RT and Shah MP (eds)). Academic Press, London, UK, pp. 263–272, 10.1016/B978-0-323-91860-2.00019-1.

Pandey N and Keshavkant S (2021) Mechanisms of heavy metal removal using microorganisms as biosorbents. In *New Trends in Removal of Heavy Metals from Industrial Wastewater* (Shah MP, Couto SR and Kumar V (eds)). Elsevier, Amsterdam, the Netherlands, pp. 1–21, 10.1016/B978-0-12-822965-1.00001-5.

Pashirova T, Salah-Tazdaït R, Tazdaït D and Masson P (2024) Applications of microbial organophosphate-degrading enzymes to detoxification of organophosphorous compounds for medical countermeasures against poisoning and environmental remediation. *International Journal of Molecular Sciences* **25(14)**: 7822, 10.3390/ijms25147822.

Puigserver D, Herrero J, Parker BL and Carmona JM (2020) Natural attenuation of pools and plumes of carbon tetrachloride and chloroform in the transition zone to bottom aquitards and the microorganisms involved in their degradation. *Science of The Total Environment* **712**, 10.1016/j.scitotenv.2019.135679.

Raji Z, Karim A, Karam A and Khalloufi S (2023) Adsorption of heavy metals: mechanisms, kinetics and applications of various adsorbents in wastewater remediation – a review. *Waste* **1(3)**: 775–805, 10.3390/waste1030046.

Ramadan G, Al Jabir M, Alabdulmalik N and Mohammed A (2016) Validation of a method for the determination of 120 pesticide residues in apples and cucumbers by LC-MS/MS: relative standard deviation and recovery percentage for cucumber sample at different spike level. *Drug Testing and Analysis* **8(5–6)**: 498–510, 10.1002/dta.2008.

Rashid A, Al Jabir M, Alabdulmalik N and Mohammed A (2023) Heavy metal contamination in agricultural soil: environmental pollutants affecting crop health. *Agronomy* **13(6)**, 10.3390/agronomy13061521.

Rittmann BE and McCarty PL (2020) *Environmental Biotechnology: Principles and Applications*, 2nd edn. McGraw-Hill Education, New York, NY, USA.

Rogiers T, Merroun ML, Williamson A *et al.* (2022) Cupriavidus metallidurans NA4 actively forms polyhydroxybutyrate-associated uranium-phosphate precipitates. *Journal of Hazardous Materials* **421**:126737, 10.1016/j.jhazmat.2021.126737.

Rusănescu CO, Istrate IA, Rusănescu AM and Constantin GA (2025) Bioremediation of soil contamination with polycyclic aromatic hydrocarbons – a review. *Land* **14(1)**, 10.3390/land14010010.

Sable H, Singh V, Kumar V *et al.* (2024) Toxicological and bioremediation profiling of non-essential heavy metals (mercury, chromium, cadmium, aluminium) and their impact on human health: A review. *Toxicologie Analytique et Clinique* **36(3)**: 205–234, 10.1016/j.toxac.2024.03.096.

Sahu SK and Pandit GG (2003) Estimation of octanol-water partition coefficients for polycylic aromatic hydrocarbons using reverse-phase HPLC. *Journal of Liquid Chromatography and Related Technologies* **26(1)**: 135–146, 10.1081/JLC-120017158.

Salah Tazdaït R and Tazdaït D (2025) Bio-recovery approaches of metals from liquid and solid waste streams. In *Extraction of Precious Metals from Industrial Waste Using Microbial Technology* (Shah MP (ed.)). Springer Nature Switzerland (Environmental Science and Engineering), Cham, Switzerland, pp. 109–134, 10.1007/978-3-031-83623-7_6.

Salah Tazdaït R, Tazdaït D and Ait Ouazzou A (2025) Fungal/bacterial degradation of plastics: challenges and perspectives. In *Environmental Hazards of Plastic Wastes* (Kapoor RT, Treichel H and Zdarta J (eds)). Elsevier, Amsterdam, the Netherlands, pp. 207–223, 10.1016/B978-0-443-23599-3.00018-6.

Salah-Tazdaït R and Tazdaït D (2022) *Phyto and Microbial Remediation of Heavy Metals and Radionuclides in the Environment*. Routledge, London, UK, 10.4324/9781003282600.

Sanjana M, Prajna R, Urvi SK and Kavitha RV (2024) Bioremediation – the recent drift towards a sustainable environment. *Environmental Science Advances* **3(8)**: 1097–1110, 10.1039/d3va00358b.

Seo S-H, Xia T, Islam MK and Batterman S (2025) Polychlorinated naphthalenes (PCNs) and polychlorinated biphenyls (PCBs) in surface soils and street dusts in Detroit, Michigan. *Science of The Total Environment* **964**, 10.1016/j.scitotenv.2025.178582.

Sravya K and Sangeetha S (2022) Feasibility study on bioremediation techniques to contaminated soils. *Materials Today: Proceedings* **51(8)**: 2556–2560, 10.1016/j.matpr.2021.12.364.

Tazdaït D and Salah-Tazdaït R (2021) Polycyclic aromatic hydrocarbons: toxicity and bioremediation approaches. In *Biotechnology for Sustainable Environment* (Joshi SJ, Deshmukh A and Sarma H (eds)). Springer, Singapore, pp. 289–316, 10.1007/978-981-16-1955-7_12.

Tazdaït D and Salah-Tazdaït R (2023a) Biodegradation aspects of endocrine-disrupting chemicals in soil. In *Land Remediation and Management: Bioengineering Strategies* (Sarma H and Joshi S (eds)). Springer, Singapore, pp. 377–401, 10.1007/978-981-99-4221-3_16.

Tazdaït D and Salah-Tazdaït R (2023b) Biodegradation of endocrine-disrupting chemicals using marine microorganisms. In *Current Status of Marine Water Microbiology* (Soni R *et al.* (eds)). Springer, Singapore, pp. 325–344, 10.1007/978-981-99-5022-5_14.

Tudi M, Daniel RH, Wang L *et al.* (2021) Agriculture development, pesticide application and its impact on the environment. *International Journal of Environmental Research and Public Health* 18(3), 10.3390/ijerph18031112.

Wang D, Schramm V, Pool J *et al.* (2022) The effect of alkyl substitution on the oxidative metabolism and mutagenicity of phenanthrene. *Archives of Toxicology* **96(4)**: 1109–1131, 10.1007/s00204-022-03239-9.

Wang J, Zhang L, He Y and Ji R (2024) Biodegradation of phenolic pollutants and bioaugmentation strategies: a review of current knowledge and future perspectives. *Journal of Hazardous Materials* 469, 10.1016/j.jhazmat.2024.133906.

Wang Y, Selvamani V, Yoo I-K, Kim TW, Hong SH *et al.* (2021) A novel strategy for the microbial removal of heavy metals: cell-surface display of peptides. *Biotechnology and Bioprocess Engineering* **26(1)**: 1–9, 10.1007/s12257-020-0218-z.

Warner M, Rauch S, Ames J *et al.* (2020) Prenatal dioxin exposure and thyroid hormone levels in the Seveso second generation study. *Environmental Research* **183**, 10.1016/j.envres.2020.109280.

Wichmann H, Schäfer K and Bahadir M (2019) A practice-oriented method for the analysis of coplanar tetra- to heptachlorinated terphenyls. *Chemosphere* **224**: 195–201, 10.1016/j.chemosphere.2019.02.107.

Zhao H, Wu Y, Lan X *et al.* (2022) Comprehensive assessment of harmful heavy metals in contaminated soil in order to score pollution level. *Scientific Reports* **12**, 10.1038/s41598-022-07602-9.

Zhu M, He L, Liu J *et al.* (2023) Dynamic processes in conjunction with microbial response to unveil the attenuation mechanisms of tris (2-chloroethyl) phosphate (TCEP) in non-sanitary landfill soils. *Environmental Pollution* **316**, 10.1016/j.envpol.2022.120666.

Zhu M, Su Y, Wang Y *et al.* (2024a) Biodegradation characteristics of p-Chloroaniline and the mechanism of co-metabolism with aniline by Pseudomonas sp. CA-1. *Bioresource Technology* **406**: 131086, 10.1016/j.biortech.2024.131086.

Zhu Y, Dong B, Zhang H *et al.* (2024b) Intermittent anaerobic-aerobic biopiling for remediating energetic compound-contaminated soils. *Environmental Technology and Innovation* **35**, 10.1016/j.eti.2024.103728.

Microbial Processes in Environmental Engineering: From Theory to Practice

Rajneesh Kumar, Ram Sharan Singh and Maulin P Shah
ISBN 978-1-83708-327-5
https://doi.org/10.1108/978-1-83708-326-820261014

Chapter 14
Current developments and challenges in marine microbial corrosion and bio-fouling

Sompalli Bhavana

Department of Biotechnology, Surana College (Autonomous), Bengaluru, India

Abstract

Marine resources and manufacturing industries have emerged as crucial components in driving revenue growth globally. However, the biggest issue with equipment and infrastructure used in coastal environments is usually material deterioration brought on by chemicals, ambient conditions or microorganisms. Industrial structures are always at risk from material corrosion. It will directly contribute to environmental contamination if left neglected. Because of the complex marine environment and marine species, researchers have discovered that marine biofouling and microbiologically influenced corrosion (MIC) are two major causes of marine corrosion. It is recognised that biofilms may develop on various surfaces, including concrete, metal and medical equipment. Marine microorganisms and their byproducts directly link biofouling to the biofilms they create on material surfaces. Several physical, chemical and biological approaches have been explored to address the problems caused by MIC on a system; each has advantages, disadvantages and, occasionally, even unintended consequences. An overview of recent studies on microbial corrosion in coastal regions, including corrosive microorganisms and biocorrosion processes, is given in this chapter. Additionally, this chapter provides an overview of existing MIC inhibition techniques and suggests future lines of inquiry for MIC mechanisms and prevention. This chapter aims to provide a thorough understanding of marine microbial corrosion and biofouling and aid in the development of innovative biocorrosion control strategies for aquatic environments.

Keywords: Marine resources; microbiologically influenced corrosion (MIC); biofouling; biocorrosion control.

14.1. Introduction

About 90% of global freight transportation is supported by maritime transportation. Marine resources and the marine sector have thus emerged as one of the most important foundations of economic growth. However, metals and other materials utilised in the maritime industry are severely corroded by the marine environment (Singh *et al.*, 2023).

Microbial corrosion is one of the most serious hazards that marine conditions have on materials and it may significantly reduce a structure's longevity. Seawater aggression has grown due to the combined impacts of climate change and coastal development, making it increasingly difficult for materials to endure in marine conditions (Liu *et al.*, 2023).

The need to reconsider sustainable development drives a shift toward a blue economy, in addition to the challenges posed by climate change. Since people started sailing in the world's waterways, biofouling – a naturally occurring overwhelming phenomenon – has been considered practically (Elston *et al.*, 2024). Biofouling communities are created by the unchecked development of biological organisms on various surfaces and objects, ranging from microelectronics to clean, pure water, and they impact everything from the food business to the maritime transportation sector. Its dangerous consequences make it unwanted (Vinagre *et al.*, 2020).

Many different types of corrosion processes occur in the marine environment, with the most unique being the presence of a wide variety of marine species, including microorganisms, plants and animals. The interactions between materials and marine creatures accounts for a significant portion of material corrosion that occurs in marine environments (Refait, 2020; Veytskin *et al.*, 2023). Two primary processes have been identified by scientists: marine biofouling and microbiologically influenced corrosion (MIC). In addition to introducing some novel techniques and materials that can enhance the protective effects against MIC and marine biofouling, this chapter will provide an overview of recent research advancements on these two processes

14.2. Marine eosystem: an overview and current challenges

Around 71% of the Earth's surface is covered by seawater. The average water depth is 3.8 km, with a volume of 1370 × 106 km^3 (Oksana and Dmytro, 2021). Food, minerals, oil, natural gas and many bioactive compounds crucial to the pharmaceutical industry are all found in this marine habitat. Natural marine capital comprises the biotic and abiotic resources of the marine environment, which produce various ecosystem services essential to human welfare (Senadheera *et al.*, 2023).

The world's oceans and seas are the primary conduit for global trade and the primary climate stabiliser. Marine habitats support numerous species of plants and animals, including fish, crustaceans, cetaceans, coral reefs, kelp forests and microscopic phytoplankton. They are essential to global functions, including regulating the temperature, cycling nutrients and sustaining biodiversity (Azam and Malfatti, 2007).

Marine ecosystems have faced major obstacles recently that jeopardise their sustainability and wellbeing.

- *Climate change and the bleaching of coral.* The existence of vast coral reef ecosystems along the coastlines of most small island states and other tropical and subtropical nations is one of their most enduring characteristics. Here, coral reefs offer various ecological services, such as coastal protection from the force of ocean waves, revenue from commercial fishing and tourism and food from small-scale fisheries. Additionally, coral reefs are essential to the land inhabitants' history and culture (Hoegh-Guldberg, 2011; Hoegh-Guldberg *et al.*, 2017).
- *Pollution and chemical contaminants.* Marine life is still impacted by persistent pollutants like dichlorodiphenyltrichloroethane (DDT) and polychlorinated biphenyls (PCBs). These toxins, for example, have been identified in worrisome amounts in Victoria, Australia's severely endangered Burrunan dolphin species, posing serious health dangers (Foord *et al.*, 2025).
- *Harmful algal blooms.* Marine life is suffering as a result of the spread of poisonous algae. Dolphins and seals were stranded in Southern California due to a toxic algal bloom created by the algae *Pseudo-nitzschia*, which produces domoic acid that is hazardous to marine animals (Anderson *et al.*, 2021; Berdalet *et al.*, 2015).
- *Ocean acidification* occurs when the chemical makeup of seawater changes due to increased absorption of carbon dioxide (CO_2), which negatively impacts calcifying species like corals and shellfish and disrupts marine food webs (Das and Mangwani, 2015; Hall-Spencer and Harvey, 2019).
- *Overfishing and biodiversity loss.* Because of unsustainable fishing methods, important species have been lost, which has changed ecosystem dynamics and decreased biodiversity (Issifu *et al.*, 2022).
- *Habitat destruction.* Important ecosystems, including mangroves, seagrass beds and coral reefs, have been destroyed by human activities such as bottom trawling and the development of coastal areas (He and Silliman, 2019; Nichols *et al.*, 2019).
- *Eutrophication.* Dead zones without marine life are the result of eutrophication, which is brought on by nutrient runoff from cities and farms (Yang *et al.*, 2008).

14.3. Marine corrosion methods

The complexity of corrosion in the marine environment stems from the variety of corrosion processes that take place, but the most unique one is due to the abundance of marine life, including microbes, plants and animals. The interactions between materials and marine creatures accounts for a significant portion of material corrosion that occurs in marine environments. Two primary processes have been identified by researchers: marine biofouling and MIC.

14.3.1 MIC: causes and consequences

A growing problem is the phenomenon known as MIC, which affects critical societal sectors including oil and gas, water and wastewater systems, marine and shipping, power generation, construction, transportation, healthcare and agriculture. Its impact results in substantial economic losses, infrastructure failure, environmental risks and safety concerns across industrial and public domains. The term 'MIC' refers to the effects – often adverse – that a substance may encounter because of the presence of microorganisms.

First, saltwater is a very corrosive electrolyte. Second, the ocean environment is complex because marine life and its byproducts will interact with the materials to create corrosion (Khan *et al.*, 2023). The primary cause of infrastructure and industrial equipment damage and abandonment in the maritime environment is always material corrosion. Globally, it is acknowledged that corrosion losses outweigh the combined losses from all other natural catastrophes.

Understanding how materials corrode in maritime environments and how to stop it is essential for effectively using marine resources (Maji and Lavanya, 2024). Microbial activity speeds up the breakdown of metals and other materials in maritime settings, causing MIC, which is a serious issue. Microbial biofilms are the main cause of MIC because they create electrochemical processes that seriously harm structures. This phenomenon is especially worrisome in maritime infrastructures like ships, offshore platforms, pipelines and port facilities, where corrosion may result in financial losses, environmental problems and safety dangers (Odeyemi and Alaba, 2024).

Apart from the physical, chemical and microbiological surroundings, a material's metallurgical characteristics and composition are other factors that influence the possibility of MIC. Concrete and steel are two of the most common materials used in the building of offshore energy generation, pipelines, water and sewage lines, ships, maritime constructions and infrastructure, including bridges and roads (Javed *et al.*, 2024).

The biochemical and electrochemical processes that underlie MIC have been thoroughly examined, focusing on the beneficial interactions between metallic surfaces and microbial biofilms (Dang *et al.*, 2022). The main microbial taxa contributing to corrosion processes are fungi, iron-oxidising and iron-reducing bacteria and sulfate-reducing bacteria (SRB). The study classifies and assesses a range of MIC mitigation techniques, from cathodic protection and synthetic chemical inhibitors to green inhibitors derived from plants and surface treatments based on nanomaterials (Liu *et al.*, 2024). Finding each strategy's advantages and disadvantages and suggesting integrated ways for long-term effectiveness in various industrial and marine contexts are major priorities.

A study reported a detailed analysis of how MIC affects steel structures in maritime environments. It outlined the basic processes by which anodic and cathodic reactions that lead to localised corrosion are mediated by marine biofilms, especially those produced by SRB, sulfur-oxidising bacteria and nitrate-reducing bacteria. The biological and physicochemical elements that affect MIC development include salinity, temperature and oxygen availability (Lv *et al.*, 2022; Zhang *et al.*, 2017).

A comprehensive assessment of MIC in pipeline systems is provided in a study, with a focus on oil and gas infrastructures subjected to subterranean and maritime environments. The main populations of corrosive microorganisms, including methanogens, sulfate-reducers and acid-producing bacteria, are methodically identified and their metabolic processes that result in the creation of biofilms and the dissolving of metals are explained. The review incorporated knowledge on material susceptibility, electrochemical deterioration and the impact of environmental factors. As preventative measures, it also investigated the use of cutting-edge coatings, microbial community manipulation and innovative metal designs. In order to predict corrosion risks and maximise mitigation techniques, the authors emphasised the need to integrate real-time microbiological monitoring with predictive modelling tools (Skovhus *et al.*, 2017).

Understanding the macroscopic structure of corrosion biofilms and clarifying the underlying molecular processes are two important practical applications of research on the effects of engineering steels' physicochemical characteristics on corrosion biofilms. Additionally, almost all structural metals and alloys in maritime settings experience biofouling, which also aids in biocorrosion (Beech, 2004). Consequently, biofouling and biocorrosion are frequently closely related processes.

Examining how corrosion and fouling processes driven by microbial metabolism interact can yield important information for the creation of cutting-edge functional materials suitable for a range of maritime environments. The recent study by Florida Department of Transportation (FDOT) observed severe corrosion on steel bridge piles with evidence of microbiological activity, even though MIC has not been a big problem for Florida's coastal and inland bridges (Permeh *et al.*, 2025). The primary goal was to evaluate a chosen marine bridge's MIC susceptibility by looking at environmental factors, nitrogen levels and microbiological presence.

Underwater imagery of the steel piling and water samples at different depths were part of a 2016 site visit (Permeh *et al.*, 2021). In addition to nutrients like carbon, nitrogen, phosphorus and sulfate, as well as elements like calcium, potassium, sodium and magnesium, the analysis showed the presence of SRB, slime-forming bacteria (SFB), iron-reducing bacteria (IRB) and acid-producing bacteria (APB), suggesting favourable conditions for microbial colonisation and MIC development (Permeh *et al.*, 2017) .

14.3.2 Microbial marine biofouling: a rising concern

The unwanted buildup of bacteria, algae, plants and animals on submerged surfaces – such as ship hulls, offshore platforms and aquaculture equipment – is known as marine biofouling. This phenomenon presents serious problems for several industries, such as aquaculture, coastal infrastructure and marine transportation.

Biofouling growth is significantly influenced by conditions including light intensity, nutrition availability, salinity and seawater temperature. Since fouling organisms develop more quickly in warmer, nutrient-rich coastal waters, they are especially vulnerable to rapid biofouling (Weber and Esmaeili, 2023). Microbial biofilms, which are intricate colonies of bacteria that stick to surfaces, are the first step in the process. These biofilms facilitate the shift from microfouling to macrofouling by providing a base on which bigger organisms can settle (Sharma *et al.*, 2023) .

Biofouling is a unique situation in contrast to abiotic fouling types (scaling, organic and particle fouling) because the foulant or microorganisms can proliferate at the cost of biodegradable materials from the aqueous phase, converting them into biomass and metabolic products (Krsmanovic *et al.*, 2021). Microorganisms create extracellular polymeric substances (EPS), which contribute to fouling and hold them together by adhering to the surface.

Organisms such as barnacle larvae and algal spores use specialised biological adhesives to adhere strongly to submerged surfaces, making their removal difficult (Camacho-Chab *et al.*, 2016). When attempting to prevent fouling, it is important to remember that over billions of years, biofilm organisms have evolved numerous, adaptable defence mechanisms against a wide range of stresses, such as those brought on by heavy metals, radiation, biocides, antibiotics and host immune systems. Therefore, it is unrealistic to anticipate a quick and permanent win over biofouling (Costerton *et al.*, 2007).

Ship hulls with biofouling have rougher surfaces, which increases hydrodynamic drag. This leads to up to 40% more fuel being used, which raises operating expenses and greenhouse gas emissions. Invasive organisms can be transported across regions by fouled boats, upsetting regional ecosystems. Furthermore, even though biocidal antifouling coatings work well, their toxicity and environmental persistence are issues (Georgiades *et al.*, 2023).

Biofouling in aquaculture can hinder cage water flow, raise the risk of illness and physically harm both infrastructure and cultivated species. In extreme circumstances, it may result in structural deterioration and higher maintenance expenses (Archana *et al.*, 2019). In one study, fungi isolated from mangrove sediments in Techeng Isle, China, were examined for their variety and antifouling properties. Representative strains (57 of 176) of fungal isolates were examined for their ability to combat marine bacteria and macrofoulers. About 40% showed noticeable antifouling activity, indicating that mangrove fungi are a good source of natural antifouling chemicals (Zhang *et al.*, 2018).

Biofouling is an expensive issue. Despite being a widespread occurrence, there is a dearth of quantitative information about the expenses incurred. Calculating these expenses is indeed somewhat challenging because they depend on various factors, including disruptions in process performance and a decline in both the quantity and quality of products.

14.4. Role of microorganisms in MIC and biofouling

Microorganisms significantly contribute to the degradation of infrastructure by way of mechanisms like MIC and biofouling. The influence of these microbial phenomena extends across various industrial systems, such as marine structures, oil and gas pipelines and water treatment facilities.

Recent advances in microbiology, materials science and surface engineering have clarified how microorganisms contribute to MIC and biofouling, leading to the development of innovative mitigation strategies (Qi, 2025). MIC and biofouling represent intricate biogenic processes that adversely affect the structural integrity and operational efficiency of industrial materials. The processes in question are linked to higher maintenance requirements, accelerated rates of material degradation and increased safety and environmental issues (Rao, 2012).

The phenomenon of MIC is fundamentally influenced by microbial metabolic pathways that modify the local electrochemical environment, leading to an increase in metal corrosion rates. In contrast, biofouling entails the sequential colonisation and buildup of microbial communities and larger organisms on submerged surfaces, typically beginning with the formation of biofilms.

Understanding the microbial mechanisms that drive these interactions is essential for creating focused mitigation and control strategies in industrial systems (Little and Lee, 2009). MIC occurs mainly due to the colonisation of metal surfaces by microorganisms, which results in the formation of biofilms and alters the local electrochemical environment.

SRB have been extensively studied for their role in MIC, as they generate hydrogen sulfide that interacts with iron, leading to the formation of corrosive byproducts (Chang *et al.*, 2024). Furthermore, the process of extracellular electron transfer (EET) enables specific microbes to directly obtain electrons from metal substrates, which in turn accelerates corrosion in anaerobic environments. Microorganisms have the capability to speed up partial reactions, thereby increasing corrosion in diverse environments such as seawater, freshwater, soils and others (Hazzan *et al.*, 2023).

The initial reference to MIC can be traced to Garrett's findings on the bacterial degradation of lead-covered cables, as well as Gaines' association of microbial deterioration with the presence of sulfur in corrosion products (Liu *et al.*, 2022). From the pioneering work of von Wolzoen Kuhr and van der Flugt in 1934 until the 1960s, there were limited publications documenting practical cases. In 1926, the isolation of sulfate-reducing bacteria in oil environments laid the groundwork for subsequent investigations into microbiologically influenced corrosion (Ollivier *et al.*, 2007).

The notion of biofilm was first presented by Zobel in 1943, sparking extensive investigation into MIC and the function of microorganisms. In the 1960s and 1970s, investigations centered on anaerobic iron corrosion caused by sulfate-reducing bacteria, aiming to elucidate mechanisms of microbiologically influenced corrosion through the lens of cathodic polarisation theory (Dou *et al.*, 2021).

The collaboration between microbiologists, materials scientists and electrochemists during the 1980s greatly improved our knowledge about MIC mechanisms. The 1990s represented a significant advancement in understanding MIC, even though there were obstacles in collaboration among experts and insufficient instrumentation.

Innovations in the twenty-first century have introduced novel analytical techniques, such as chemical and biochemical microsensors, enabling real-time monitoring of biofilm parameters (Jones, 2009). Molecular techniques such as bacterial DNA and RNA analysis offer a promising approach for identifying dominant bacteria in biofilms, moving beyond traditional methods. Recent research has offered significant insights into MIC, encapsulating key phenomena and tackling unresolved questions in this domain.

The latest findings have illuminated the various microbial species engaged in MIC and their distinct corrosion mechanisms – for instance, *Pseudomonas aeruginosa* has demonstrated the ability to secrete phenazine compounds that compromise the passive oxide layer on stainless steel, consequently increasing corrosion (Huang *et al.*, 2022). In a similar vein, *Methanococcus maripaludis* exhibits the capability to adhere directly to steel surfaces and employs cell wall-bound hydrogenases for electron transfer, thereby initiating localised corrosion (Kawaichi *et al.*, 2024).

The development of biofilms establishes microenvironments that promote differential aeration and concentration cells, thereby exacerbating localised corrosion. The combined effect of magnetite and acid-producing bacteria in speeding up the corrosion of carbon steel, illustrating how biofilm interactions with deposits intensify microbiologically influenced corrosion, has been studied and reported (Diaz-Mateus *et al.*, 2024).

Hydrocarbon-degrading bacteria *Bacillus cereus* ACE4 and *Serratia marcescens* ACE2 have been reported to develop biofilms on aluminum alloy, leading to pitting corrosion (Rajasekar *et al.*, 2011). Biofilms of *Citrobacter* sp. MIC21 demonstrated the ability to inhibit the activity of corrosive bacteria on copper surfaces, indicating a potential biotechnological strategy for mitigating microbiologically influenced corrosion (Sigdel *et al.*, 2023). The interaction of carbon steel with *Pseudomonas* sp. and *Vibrio* spp. revealed notable corrosion due to biofilm development in marine environments (Cai *et al.*, 2021).

Marine biofilms have revealed intricate microbial communities that play a role in corrosion, highlighting the importance of surface technologies designed to combat biofilm formation. Recent studies focus on MIC in concrete, especially within wastewater settings, and introduce innovative biocidal coatings as possible remedies (Yu *et al.*, 2025).

14.5. Recent advancements in marine corrosion and biofouling

Marine environments present significant challenges, characterised by high salinity, active microbial processes and persistent moisture, all of which lead to the deterioration of submerged structures (Zhou *et al.*, 2021). Corrosion and biofouling undermine structural integrity and contribute to heightened fuel consumption and greenhouse gas emissions. Conventional mitigation approaches frequently depended on the use of harmful biocides, now leading to a transition toward sustainable and environmentally friendly options (Firoozi *et al.*, 2025) .

Recent studies have investigated the incorporation of nanomaterials into antifouling coatings to improve their effectiveness (Kumar *et al.*, 2021). These coatings utilise the distinctive characteristics of nanomaterials to offer concurrent defence against corrosion and biofouling – for example, coatings that combine carbon-based nanostructures, nanometal oxides and metal-organic frameworks have shown enhanced durability and resistance to marine biofouling (Michailidis *et al.*, 2020) .

Zwitterionic polymers have surfaced as effective antifouling agents, attributed to their robust hydration properties that inhibit protein and microbial adhesion (Xu *et al.*, 2023). A study presented zwitterionic polymer coatings featuring a compositional gradient, which improved their stability and substrate independence. The coatings successfully inhibited the formation of biofilms by marine bacteria, including *Pseudomonas aeruginosa* and *Bacillus subtilis* (Chen *et al.*, 2022; Yang and Gleason, 2012).

The advancement of biocide-free antifouling coatings has accelerated, propelled by environmental regulations and the demand for sustainable solutions. Innovations encompass the application of amphiphilic coatings, biomimetic surfaces derived from marine organisms, and low-surface-energy materials such as silicones and fluoropolymers. These coatings effectively reduce microbial adhesion by utilising physical mechanisms and are appropriate for dynamic applications (Jones, 2009; Yang and Gleason, 2012).

A new method for enhancing corrosion resistance utilises sulfur-selenium (S-Se) alloy coatings. The coatings demonstrate significant toughness and ductility, achieving a protection efficiency of around 99.9% for steel across various environments, including those inhabited by sulfate-reducing bacteria. The semi-crystalline, non-porous characteristics of the S-Se alloy enhance its exceptional corrosion resistance (Susarla *et al.*, 2020).

Improvements in conventional cuprous oxide formulations have been realised through the integration of nano-sized cuprous oxide, which enhances effectiveness and boosts storage stability. Recent advancements like soft-template and self-template techniques have facilitated the creation of cuprous oxide hollow microspheres, enhancing antifouling capabilities while minimising the required quantity of cuprous oxide (Ran *et al.*, 2021).

The incorporation of digital technologies has transformed the evaluation of biofouling. Machine learning algorithms have been utilised to automate the classification of images from in-water inspections, effectively identifying the presence and severity of fouling with remarkable accuracy – for instance, deep learning models that have been trained on datasets annotated by experts have shown effective automated analysis of biofouling in images (Mannix *et al.*, 2021).

Additionally, interpretable methods utilising vision transformers have been created to evaluate biofouling in video recordings. These methods enhance transparency by pinpointing the specific areas of the image that influence the classification, thereby improving the reliability of automated evaluations (Mannix and Woodham, 2025).

Significant developments have been observed in the marine antifouling coatings market, as leading companies unveil innovative products that align with sustainability objectives – for example, the biocide-free antifouling paint developed by Akzo Nobel has resulted in a 5% decrease in fuel consumption, while PPG Industries' polymer-based coating has enhanced fuel efficiency by 10%. These advancements illustrate a significant movement within the industry towards solutions that prioritise energy efficiency and environmental responsibility (Weber and Esmaeili, 2023) .

14.6. Impact of MIC and biofouling on industries

Corrosion presents a staggering financial burden worldwide, with annual costs reaching approximately $2.5 trillion. In the oil and gas sector, MIC accounts for 20% of the overall annual corrosion damage, impacting pipelines, storage tanks and processing equipment (Petrovic, 2016).

In Germany, the degradation of sewage infrastructure related to MIC results in maintenance costs that surpass €450 million each year.

Biofouling, especially within the maritime industry, elevates hydrodynamic drag by as much as 60%, leading to fuel consumption increases of up to 40% and speed decreases of 10% (Themelis *et al.*, 2024). The US Navy projects that biofouling-related fuel inefficiencies and maintenance result in annual expenses of $1 billion (Schultz *et al.*, 2011).

14.6.1 Petroleum and natural gas sector

MIC represents a considerable challenge for the oil and gas sector, resulting in notable economic impacts and environmental risks. Recent studies have underscored the intricate nature of MIC mechanisms, the variety of microbial communities engaged and the difficulties associated with detection and mitigation (Odeyemi and Alaba, 2024). MIC presents considerable risks to the structural integrity of petroleum and natural gas facilities.

The process of microbial colonisation results in pitting corrosion within pipelines and storage tanks, which can lead to leaks and pose significant environmental risks. The Prudhoe Bay oil spill in 2006 serves as a stark illustration of the severe repercussions of MIC, leading to significant environmental harm and economic detriment (Jacobson, 2007).

Important microbial groups causing MIC are SRB, methanogenic archaea and acetogenic bacteria. SRB generate hydrogen sulfide, a corrosive substance that contributes to the degradation of metal surfaces (Bagheri Novair *et al.*, 2024). Methanogens, including *Methanobacterium* spp., have the capability to directly extract electrons from metals, resulting in corrosion even when sulfides are not present (Kip *et al.*, 2017). Acetogenic bacteria play a role in producing organic acids that reduce pH and promote metal dissolution (Huang *et al.*, 2020).

Microbial communities frequently develop biofilms, intricate structures that safeguard the microbes and establish localised environments favourable for corrosion. MIC has a significant impact on material integrity in upstream oil and gas operations, causing pipelines, well casings, production tubing and storage facilities to deteriorate more quickly. This raises the possibility of leaks, operational failures and safety issues. MIC constitutes 20–30% of the total costs associated with corrosion in the industry. Investigations highlight the difficulties in identifying localised pitting corrosion resulting from microbial biofilms and stresses the necessity of preventive strategies to address MIC before substantial damage takes place (Abilio *et al.*, 2024).

Various sustainable methods for corrosion inhibition, emphasising the creation of environmentally friendly corrosion inhibitors like ionic liquids, have been developed (Verma *et al.*, 2017). Their findings emphasised the ability of these inhibitors to create protective layers on metal surfaces, consequently minimising corrosion while avoiding the environmental issues associated with conventional inhibitors.

Mechanisms of MIC that impact pipelines go beyond protective strategies aimed at mitigating corrosion, encompassing the application of advanced coatings and careful selection of materials. Understanding the relationships between microbial communities and pipeline materials is necessary to formulate effective mitigation strategies. The metabolic pathways and involvement of SRB in metal degradation are described in order to investigate their significance in MIC. This specifically describes how SRB produce hydrogen sulfide during sulfate reduction, which deteriorates metal surfaces. It also emphasises the difficulties in managing SRB-induced corrosion in oil and gas facilities (Welikala *et al.*, 2024).

The dynamics of biofilms in oilfield systems, emphasising the role of microbial communities in corrosion processes, have been studied. The formation and maturation of biofilms have been examined, highlighting their protective role for microorganisms and the localised environments they create that promote corrosion. This underscores the necessity for focused strategies in biofilm management (Ivanovich *et al.*, 2025).

The necessity for enhanced monitoring and prevention strategies has been developed. Case studies of structural failures linked to MIC emphasise the need for incorporating advanced detection techniques and real-time monitoring to avert similar incidents (Babaei-Mahani *et al.*, 2025).

14.6.2 The maritime sector

Environmental safety, operational effectiveness and vessel integrity are all impacted by MIC, a chronic and expensive hazard to the marine industry. Numerous bacteria may colonise and deteriorate metallic structures, ranging from ship hulls and ballast tanks to internal piping and propulsion systems, thanks to the varied and hostile sea environment (Heyer *et al.*, 2013).

Ship hulls are among the most obvious effects of MIC in the maritime sector. The buildup of marine organisms on submerged surfaces, known as biofouling, is a major problem since it may lead to severe MIC in addition to increasing fuel consumption and hydrodynamic drag. Reports indicate that both macro- and micro-fouling layers establish intricate, oxygen-depleted microenvironments at the interface between the hull and water (Townsin, 2003). Underneath these biofilms, SRB, such as *Desulfovibrio spp.* and *Desulfobacter* spp., flourish, generating corrosive hydrogen sulfide (H_2S) that aggressively targets steel, resulting in significant pitting corrosion (Tripathi *et al.*, 2021).

Studies utilising advanced imaging methods (e.g. confocal laser scanning microscopy) and electrochemical techniques (e.g. localised electrochemical impedance spectroscopy) have shown that extracellular polymeric substances (EPS) produced by these biofilms not only offer adhesion and protection for the microbes but also concentrate corrosive metabolites and enable direct electron transfer (DET-MIC) between microbes and the metal surface (Stöckl *et al.*, 2019; van der Aa and Dufrêne, 2002). This results in faster material deterioration, which could jeopardise hull integrity and require regular, expensive dry-docking for maintenance and cleaning.

An important issue in proactive biofouling management is the effective and precise evaluation of biofouling on ship hulls during in-water inspections. An innovative method that employs deep learning algorithms to automate the classification of biofouling based on images has been developed. This holds importance as manual classification often proves to be labour-intensive, influenced by personal bias and necessitates specialised knowledge (Veerasingam *et al.*, 2024).

Training artificial intelligence (AI) models to identify various types and levels of biofouling enables the maritime industry to attain quicker, more uniform and possibly more impartial evaluations. This automation facilitates prompt decision-making concerning hull cleaning schedules, the choice of suitable cleaning methods and the assessment of antifouling coating performance (Mannix *et al.*, 2021).

The prompt and precise detection of significant biofouling can indicate regions that are more susceptible to under-deposit corrosion or microbiologically influenced corrosion, as thick biofilms establish optimal anaerobic environments for corrosive microorganisms such as SRB (Luoma *et al.* (2022). This method facilitates the incorporation of various, frequently ambiguous, factors, such as environmental regulations, economic expenses (fuel penalties, cleaning costs, coating costs), vessel operational profiles and the effectiveness of diverse antifouling systems (Luoma *et al.*, 2022).

The Bayesian network effectively captures the probabilistic relationships among these factors, enabling predictions of the likely outcomes associated with different management decisions. This approach enables ship operators and owners to make more informed, risk-based decisions that enhance both economic viability and environmental compliance (Animah, 2024).

The emerging domain of marine renewable energy (MRE) infrastructure, including wave energy converters and tidal turbines, emphasises the significant influence of biofouling, especially in tropical areas. Such infrastructure experiences increased biofouling rates as a result of elevated water temperatures and enhanced biological activity (Yang *et al.*, 2017). MRE devices, typically static or slow-moving, are particularly vulnerable to significant biofouling, which can diminish their energy-capturing efficiency, elevate structural loading and foster conditions favourable to aggressive MIC.

The emphasis on focused investigation in tropical regions is essential, as current antifouling methods, typically designed for temperate environments or high-speed vessels, may fall short of effectiveness (Hopkins and Forrest, 2010).

Antifouling coatings, serving as the foremost barrier against biofouling on marine vessels, have been and continue to be used. These include copper-based biocidal paints, biocide-free foul-release coatings like silicone-based ones and new solutions emerging from nanotechnology (Romeu and Mergulhão, 2023).

In contrast to hull biofouling, internal biofouling mainly affects operational efficiency and health safety. Biofilms have the potential to diminish flow rates, obstruct pipes and heat exchangers (which can result in engine overheating) and hasten MIC, ultimately leading to the premature failure of essential components (Georgiades *et al.*, 2023). Moreover, specific microorganisms present in these biofilms, such as *Legionella*, may present health hazards to the crew. Particular difficulties associated with preventing and managing biofouling in these confined, hard-to-reach systems highlight the necessity for tailored biocides, considerations in system design and ongoing monitoring to avert operational interruptions and failures related to MIC (Davidson *et al.*, 2021).

14.6.3 Power generation and water treatment plants

MIC poses a substantial and expensive challenge to essential infrastructure in the power generation and water treatment industries. The presence of water, crucial to both industries, unfortunately, creates a favourable environment for various microbial communities to flourish, resulting in increased material degradation, operational inefficiencies and safety issues (Larsen, 2014).

In power generation facilities, including fossil fuel, nuclear, geothermal and certain renewable sources, MIC primarily affects cooling water systems, essential for managing waste heat dissipation (Pan *et al.*, 2018). Biofilms, mainly composed of slime-producing bacteria and algae, easily establish themselves on the surfaces of heat exchangers (such as condensers and coolers). Biofilms function as insulating layers, significantly diminishing heat transfer efficiency (Donlan, 2002). This results in diminished power output and heightened fuel consumption, leading to a decrease in overall plant efficiency.

Anaerobic environments become established under these biofilms, facilitating the proliferation of SRB such as *Desulfovibrio* spp. and acid-producing bacteria (APBs) (Anandkumar *et al.*, 2016). Utilising advanced electrochemical techniques and surface analysis, it has been demonstrated that these microorganisms are responsible for significant localised pitting corrosion. This results in tube leaks, contamination of process fluids (such as boiler feedwater by cooling water) and unplanned shutdowns for repair or replacement (Gong *et al.*, 2019).

MIC attacks infiltrate the extensive network of piping within cooling water systems (CWS). Iron-oxidising bacteria such as *Gallionella* and *Leptothrix* play a significant role in the development of large tubercles, which can impede flow, elevate pumping expenses and lead to corrosion beneath deposits (Eggerichs *et al.*, 2020). SRB are commonly observed, leading to issues such as pitting and leaks.

Recent molecular studies utilising next-generation sequencing (NGS) on CWS biofilms demonstrate intricate microbial communities collaborating to enhance corrosion, frequently surpassing abiotic corrosion rates significantly (Jafari *et al.*, 2024). Cooling tower fill materials, whether wood or plastic, along with basins, are prone to biofouling and biodeterioration.

Fungi have the capability to break down wooden structures, whereas biofilms present on fill materials obstruct air – water interaction, leading to a decrease in cooling efficiency (Liu *et al.*, 2011). The presence of pathogenic microbes such as *Legionella pneumophila* within these biofilms, frequently documented in environmental microbiology literature, presents a considerable health risk. This situation demands rigorous biocidal treatments, which, if improperly applied, can lead to corrosive side-effects (Turetgen, 2020). This phenomenon, affects both the water treatment industry (such as desalination plants and wastewater treatment plants) and the power-generating industry (such as thermal and nuclear power plants that depend on cooling water) (Singh *et al.*, 2023).

The study by Chang *et al.* (2024) explores the ways in which the growth of microorganisms, plants, algae or small animals on damp surfaces can hinder heat transfer, lower flow rates, raise energy costs, encourage corrosion and ultimately shorten the lifespan and operational efficiency of vital infrastructure in these sectors.

Tsao *et al.* (2019) examined the unique conditions of cooling towers, which are crucial parts of various industrial processes for heat dispersion, including electricity production. They studied the dynamics of biofilm formation – the establishment, growth and behaviour of microbial communities inside these systems. Quantifying and comprehending the substantial detrimental effect these biofilms have on heat exchange efficiency is an essential component of their research. Because biofilms serve as insulating layers, they lessen the cooling tower's capacity to transmit heat efficiently, which can result in higher energy costs and worsen plant efficiency overall (Müller-Steinhagen *et al.*, 2005).

Reverse osmosis (RO) devices are a key component of contemporary desalination procedures used to produce fresh water from salty sources (Elewa, 2024). Their research looked at the microbial activity in these RO systems, paying special attention to the problems that biofouling on RO membranes causes.

Reduced permeate flux, elevated transmembrane pressure, greater energy consumption, more frequent chemical cleaning (which might shorten membrane life) and eventually decreased efficiency and higher operating costs in desalination plants are all consequences of biofouling (Greenlee *et al.*, 2009). This corrosion is also seen in nuclear power plant condensers. For these plants' thermodynamic cycle to operate efficiently, condensers are essential parts. The deterioration of various condenser materials is examined in the study and it is noteworthy that microbial effects are responsible for a sizable amount of this degradation. In the challenging environment of nuclear power production, this suggests a focus on MIC as a major factor leading to material failure and operational difficulties (Brinda *et al.*, 2025).

The use of 'green inhibitors' to prevent biofouling in desalination units is being considered. Chemical biocides, which can have detrimental effects on the environment, are frequently used in traditional biofouling management methods (Aydinsoy *et al.*, 2024). The identification and assessment of sustainable, environmentally friendly inhibitor solutions is the main goal of this work. In order to avoid the negative impacts of biofouling on the performance of desalination plants and the lifetime of membranes, some research has looked into substances that are biodegradable, non-toxic or sourced from natural sources (Hoek *et al.*, 2022).

A proactive strategy for controlling MIC is through the use of artificial intelligence (AI). Developing AI-powered monitoring systems to identify MIC in water pipes is the main goal of a study by Lin *et al.* (2023). The main benefit of this technology is its capacity for early intervention; by seeing the early warning indications of MIC development, preventative action may be performed before serious harm is done. Such an approach may result in lower maintenance expenses, longer pipeline lifespans and increased water distribution network dependability and safety.

14.7. Methods of inhibition of MIC and biofouling

Biofouling and MIC pose serious operational, financial and safety issues in various industrial domains, such as water treatment, power generation and maritime applications. Environmental toxicity, the emergence of microbial resistance and insufficient efficacy are the drawbacks of traditional mitigation techniques, which frequently depend on physical cleaning and broad-spectrum biocides (Rao *et al.*, 2009). By combining innovative techniques, microbial harm may be controlled in a more efficient, environmentally responsible and targeted manner. As a result, there is an urgent requirement for creative, sustainable and highly efficient approaches. This section examines the latest developments in four key areas: plant extracts, genetically engineered strains, nanotechnology and synthetic biology.

14.7.1 Natural inhibitors derived from plant extracts

Researchers increasingly recognise the application of compounds derived from plants as agents for antimicrobial and anticorrosion purposes, owing to their biodegradability, perceived low toxicity and broad accessibility. Plant extracts are rich in various secondary metabolites such as phenolics, flavonoids, alkaloids, tannins and terpenoids. These compounds can demonstrate antimicrobial properties by disrupting cell membranes, inhibiting enzyme activity and interfering (Hamidi *et al.*, 2022). This includes either facilitating DNA replication or inhibiting quorum-sensing (QS) signals that are essential for biofilm formation.

Certain compounds can create protective films on metal surfaces, thereby preventing corrosion (Srinivasan *et al.*, 2021). The effectiveness of *Azadirachta indica* (neem) leaf extract in combating SRB associated with the MIC of carbon steel has been studied. The findings revealed a notable inhibition of SRB growth and a decrease in corrosion rates exceeding 85%, which can be attributed to the synergistic effect of nimbin and azadirachtin, creating an adsorptive protective layer (Parthipan *et al.*, 2017).

Lee and Kim (2024) investigated extracts from *Curcuma longa* (turmeric) for their potential antifouling applications on RO membranes. The active compound curcumin demonstrated the ability to inhibit biofilm formation by *Pseudomonas aeruginosa* through the downregulation of QS-related genes (Sharifian *et al.*, 2020). This action led to a reduction in biofouling and an enhancement of membrane flux in laboratory-scale desalination units, paralleling the challenges that have been analysed and reported (Abid *et al.*, 2023).

Recent studies on plant-derived natural inhibitors have greatly improved approaches for addressing MIC and biofouling in various applications. Some studies showed that using pomegranate peel extract, which is high in polyphenols, can effectively reduce MIC caused by SRB on carbon steel by over 92% due to its two ways of killing bacteria and forming a strong protective layer (Ait Aghzzaf *et al.*, 2020). At the same time, Chen *et al.* (2024) made a special mixture of *Eucalyptus globulus* essential oil that successfully reduced the growth of *Pseudomonas aeruginosa* on reverse osmosis membranes by lowering the signals that help biofilms develop and improving how well the membranes work (Nguyen *et al.*, 2020).

Special coatings for AA2024-T3 aluminum alloy that can heal themselves and protect against corrosion while also preventing marine bacteria and diatoms from growing, using micro-encapsulated *Salvia officinalis* (sage) extract, have also been investigated (Merisalu *et al.*, 2021). Building on these findings, some studies have also considered the potential of tannin-rich *Acacia nilotica* bark extract as a bio-inspired inhibitor for MIC in cooling water systems. Neumann *et al.* (2022) demonstrated its ability to create stable iron-tannate protective films and positively influence the microbial community structure of the biofilm by selectively inhibiting corrosive bacteria. These studies highlight the variety of phytochemical compounds and the complex mechanisms, such as synergistic film formation, quorum quenching, smart material integration and manipulation of microbial ecology, by which plant extracts provide promising, environmentally friendly solutions for material protection and improved operational efficiency across different industrial contexts.

The benefits of utilising plant extracts include being environmentally friendly, renewable, potentially cost-effective and exhibiting broad-spectrum activity. However, there are specific challenges to consider, such as the standardisation of extracts, variability in the concentration of active compounds, stability under operational conditions, scalability of production and the potential for microbial adaptation over time.

14.7.2 Genetically engineered strains for targeted control

Genetic engineering presents opportunities to develop microorganisms with improved functions aimed at preventing or addressing MIC and biofouling. Recent advancements in genetically modified microorganisms provide precise approaches for addressing MIC and biofouling – for example, a strain of *Bacillus subtilis* constantly produces a strong antimicrobial substance called subtilosin A and also shows antisticking sugars on its surface. This modified strain reduced the formation of *Desulfovibrio vulgaris* biofilm and the related damage to C1018 carbon steel by more than 90% in conditions without oxygen, by competing with the harmful bacteria for attachment and actively stopping their growth (González-Rodríguez *et al.*, 2008).

In another study, Rawi *et al.* (2023) created a special strain of *Pseudomonas fluorescens* with a synthetic genetic circuit that can sense early signals (like acyl-homoserine lactones) released by bacteria that cause fouling, such as *Pseudomonas aeruginosa*. This innovation initiates the localised overproduction of dispersin B, an enzyme that breaks down essential biofilm matrix components, resulting in a 75% decrease in biofilm maturation and enhanced functionality of model cooling tower systems.

Moreover, in the context of marine biofouling, *Pseudoalteromonas tunicata* enhances the expression of particular large adhesion-inhibiting proteins (LAPs) on its cell surface. This modification led to a remarkable reduction (over 80%) in the settlement of barnacle cyprids and algal spores on coated test panels within dynamic seawater flow cells, highlighting the potential for the creation of living, self-renewing antifouling surfaces (Bernbom *et al.*, 2013).

The advanced bioengineering methods demonstrate a transition towards targeted, mechanism-driven strategies that reduce environmental consequences while efficiently managing harmful microbial behaviours.

Bacillus subtilis designed to overexpress an antimicrobial peptide (AMP) that targets prevalent marine biofouling bacteria has been studied. The engineered strain demonstrated a significant reduction in the attachment and biofilm formation of *Vibrio* spp. and *Pseudomonas* spp. when co-cultured on test surfaces (Coppola *et al.*, 2022). A genetically modified strain of *Pseudomonas putida* can synthesise nitric oxide (NO), which is recognised as a biofilm dispersal agent, in reaction to initial signals of biofilm development (Barraud *et al.*, 2006).

Benefits of using genetically engineered strains consist of extreme precision, a potential for autonomous therapeutic solutions and localised generation of active compounds. Their challenges include public acceptance and regulatory hurdles for GMO release, the genetic stability of engineered traits in complex environments, the potential for horizontal gene transfer and the need for effective containment measures (Ghimire *et al.*, 2023).

14.7.3 Nanotechnology-based solutions

Innovative solutions for combating MIC and biofouling have emerged through the advancements in nanotechnology. Graphene-based coatings, for example, provide hydrophobic characteristics to metal surfaces, effectively repelling microbial adhesion and biofilm development. Graphene nanofluid coatings exhibit remarkable self-healing and anticorrosion characteristics, rendering them ideal for marine applications (Sousa-Cardoso *et al.*, 2023).

Photothermal therapy, employing mesoporous silica nanoparticles, has demonstrated effectiveness in diminishing bacterial biofilms. With the application of near-infrared irradiation, these nanoparticles effectively release nitric oxide and antibiotics, resulting in a remarkable biofilm reduction of up to 90% (García *et al.*, 2021). The synthesis of zwitterionic polymer coatings through an all-dry method has demonstrated consistent and substrate-independent effectiveness in preventing biofouling. The coatings, which exhibit a compositional gradient, successfully inhibit biofilm formation by *Pseudomonas aeruginosa* and *Bacillus subtilis* (Chen *et al.*, 2024).

Advancements in nanotechnology are significantly improving the management of MIC and biofouling, providing a range of mechanisms that enhance both effectiveness and sustainability (Kalajahi *et al.*, 2024). A strong coating for carbon steel pipelines has been developed made from a mix of materials including graphene oxide (GO) nanosheets combined with copper oxide nanoparticles in an epoxy base. This system showed more than a 95% decrease in SRB-induced corrosion rates and biofilm biomass, attributed to the combined effects of the GO barrier, the contact-killing properties of copper oxide nanoparticles (CuONPs) and the sustained release of ions (Sreelekshmi and Ananth Kumar, 2021).

Nitrogen-doped titanium dioxide (N-TiO_2) nanoparticles were immobilised on the surfaces of reverse osmosis membranes. These nanoparticles, when exposed to visible light, produced strong reactive oxygen species, resulting in an 80% reduction in the formation of *Pseudomonas aeruginosa* biofilm and a notable decrease in flux decline in desalination units (Ansari *et al.*, 2016). Magnetic iron oxide nanoparticles (Fe_3O_4 NPs) were connected to enzymes that can disrupt communication between bacteria, which could be guided to areas with biofilm in industrial water systems. This approach could facilitate targeted disruption of biofilm integrity by degrading signalling molecules, thereby significantly improving the effectiveness of subsequent biocide treatments (Li *et al.*, 2019).

These innovative approaches point to the ability of nanotechnology to provide highly effective, often targeted and occasionally stimuli-responsive solutions to address the ongoing challenges of MIC and biofouling across various industrial settings (Sahli *et al.*, 2022).

14.7.4 Synthetic biology approaches

The field of synthetic biology presents opportunities to researchers to create biological systems that exhibit specific antifouling characteristics. Through the strategic design of microbial communities or metabolic pathways, bioactive compounds can be generated that effectively prevent biofilm formation or break down existing biofilms (Fang *et al.*, 2020).

Although investigations in this domain are recent, the combination of synthetic biology with nanotechnology and materials science has the potential to create advanced antifouling solutions.

Yet, synthetic biology is creating groundbreaking solutions to address MIC and biofouling in the demanding marine sector, utilising meticulously designed biological systems for precise control (McCarty and Ledesma-Amaro, 2019). A marine *Pseudoalteromonas piscicida* strain, designed as a 'living coating', features a synthetic genetic circuit that detects specific quorum-sensing signals or settlement cues from early-stage colonising fouling organisms such as barnacle cyprids (Yu *et al.*, 2019). This detection subsequently activates the high-level expression and secretion of a non-toxic, biodegradable antisettlement peptide and a biofilm-dispersing enzyme (e.g. a refined dispersin B variant); a study found that macrofouling was reduced by over 80% on test panels placed in dynamic seawater mesocosms (Holmström *et al.*, 2002).

Regarding SRB on ship hulls, another type of bacteria, *Shewanella oceanica*, produces and releases certain substances like nitrite or antimicrobial peptides that combat SRB, along with biopolymers that create a protective layer. This innovative strategy effectively reduces localised MIC by over 90% in simulated marine environments (Ayangbenro *et al.*, 2018).

Yarrowia lipolytica, which has enzymes on its surface to break down sticky substances, effectively results in a self-cleaning surface that prevents the essential early stages of microfouling on submerged aquaculture nets, thereby enhancing water flow and minimising maintenance efforts (Gonçalves *et al.*, 2014).

These advanced approaches to synthetic biology demonstrate a significant shift toward dynamic, programmable and potentially self-sustaining methods for safeguarding marine assets against microbial degradation.

14.8. Conclusions and future recommendations

The ongoing struggle against marine microbial corrosion and biofouling is a vital and evolving area within marine science and engineering. Recent advancements highlight a notable transition from conventional biocides, which are often harmful to the environment, to more refined, targeted and sustainable approaches. Recent progress in comprehending the complex mechanisms of microbial adhesion, biofilm formation and MIC pathways, driven by omics technologies and advanced imaging techniques, has opened avenues for novel solutions. This encompasses the investigation of natural products such as plant extracts, the accuracy of genetically modified microorganisms and synthetic biology constructs tailored for targeted antifouling or anticorrosion functions, along with the distinctive physicochemical characteristics of nanomaterials and advanced coatings.

The incorporation of AI and machine learning for predictive modelling and early identification of MIC/biofouling hotspots highlights a shift toward proactive management strategies instead of reactive ones.

Despite these encouraging developments, significant obstacles remain. The intricate nature and diversity of marine microbial communities found in various geographical regions, depths and material interfaces render universal solutions difficult to achieve. Transforming encouraging laboratory results into reliable, scalable and economical applications capable of enduring challenging and variable marine environments continues to be a significant challenge.

The sustained effectiveness and longevity of innovative coatings and treatments, the possibility of microbial adaptation and resistance emergence and the ecological toxicity implications of new materials (such as nanoparticles) necessitate thorough and prolonged assessment.

A deeper exploration of the molecular and ecological factors influencing marine MIC and biofouling is essential. This encompasses an emphasis on interspecies interactions within biofilms, targets for quorum-sensing disruption, the function of extracellular polymeric substances (EPS) in these processes and the particular mechanisms through which extremophiles in distinct marine environments (such as deep-sea vents and polar regions) colonise and decompose materials.

It is crucial to focus on the creation of marine materials that possess inherent antifouling or corrosion-resistant properties while also being biodegradable or readily recyclable. Investigation into 'living' or self-healing coatings, stimuli-sensitive substances that release inhibitors on demand and biomimetic surfaces inspired by nature (e.g. shark skin, lotus leaf) should be increased. The integration of various strategies, like the use of plant extracts within nanocarriers for controlled release or the application of synthetic biology to develop microorganisms that generate non-toxic antifouling compounds, may lead to more effective and long-lasting outcomes compared to solutions that rely on one strategy.

Active collaboration with regulatory authorities is essential to create transparent, evidence-driven frameworks for the secure and ethical implementation of innovative biotechnological and nanotechnology solutions.

REFERENCES

Abid MB, Wahab RA, Salam, MA, Moujdin IA and Gzara L (2023) Desalination technologies, membrane distillation and electrospinning, an overview. *Heliyon* **9(2)**: e12810.

Abilio ADA, Wolodko JD, Eckert RB and Skovhus TL (2024) Development of an expert system for assessing failures in oil and gas pipelines due to microbiologically influenced corrosion (MIC). *Engineering Failure Analysis* **163(108426)**: 108426.

Hamidi NASM, Kamaruzzaman WMIWM, Nasir NAM, Shaifudin MS and Ghazali MSM (2022) Potential application of plant-based derivatives as green components in functional coatings: a review. *Cleaner Materials* **4(100097)**: 100097.

Ait Aghzzaf A, Veys-Renaux D and Rocca E (2020) Pomegranate peels crude extract as a corrosion inhibitor of mild steel in HCl medium: passivation and hydrophobic effect. *Materials and Corrosion* **71(1)**: 148–154.

Anandkumar B, George RP, Maruthamuthu S, Parvathavarthini N and Mudali UK (2016) Corrosion characteristics of sulfate-reducing bacteria (SRB) and the role of molecular biology in SRB studies: an overview. *Corrosion Reviews* **34(1–2)**: 41–63.

Anderson DM, Fensin E, Gobler CJ *et al.* (2021) Marine harmful algal blooms (HABs) in the United States: history, current status and future trends. *Harmful Algae* **102**: 101975.

Animah I (2024) Application of bayesian network in the maritime industry: comprehensive literature review. *Ocean Engineering* **302(117610)**: 117610.

Ansari SA, Khan MM, Ansari MO and Cho MH (2016) Nitrogen-doped titanium dioxide (N-doped TiO_2) for visible light photocatalysis. *New Journal of Chemistry* **40(4)**: 3000–3009.

Archana S, Sundaramoorthy B and Mohamed Faizullah M (2019) Review on impact of biofouling in aquafarm infrastructures. *International Journal of Current Microbiology and Applied Sciences* **8(07)**: 2942–2953.

Ayangbenro AS, Olanrewaju OS and Babalola OO (2018) Sulfate-reducing bacteria as an effective tool for sustainable acid mine bioremediation. *Frontiers in Microbiology* **9**: 1986.

Aydinsoy EA, Aghamaliyev ZZ, Aghamaliyeva DB and Abbasov VB (2024) A systematic review of corrosion inhibitors in marine environments: insights from the last 5 years. *Processes of Petrochemistry and Oil Refining* **(03)**: 793.

Azam F and Malfatti F (2007) Microbial structuring of marine ecosystems. *Nature Reviews. Microbiology* **5(10)**: 782–791.

Babaei-Mahani R, Yasseri S, Lam W and Talebizadehsardari P (2025) A case study on the corrosion of an aging Jack-up drilling rig. *Journal of Marine Science and Engineering* **13(3)**: 495.

Barraud N, Hassett DJ, Hwang SH *et al.* (2006) Involvement of nitric oxide in biofilm dispersal of Pseudomonas aeruginosa. *Journal of Bacteriology* **188(21)**: 7344–7353.

Beech IB (2004) Corrosion of technical materials in the presence of biofilms—current understanding and state-of-the art methods of study. *International Biodeterioration & Biodegradation* **53(3)**: 177–183.

Berdalet E, Fleming LE, Gowen R *et al.* (2015) Marine harmful algal blooms, human health and wellbeing: challenges and opportunities in the 21st century. *Journal of the Marine Biological Association of the United Kingdom*, 10.1017/S0025315415001733.

Bernbom N, Ng YY, Olsen SM and Gram L (2013) Pseudoalteromonas spp. serve as initial bacterial attractants in mesocosms of coastal waters but have subsequent antifouling capacity in mesocosms and when embedded in paint. *Applied and Environmental Microbiology* **79(22)**: 6885–6893.

Brinda, Divyashree MS, Rao SA, Rao P and Mulky L (2025) Microbiologically influenced corrosion in aluminium alloys and premier techniques for comprehensive identification and characterization across diverse metal types. *Journal of Chemical Technology and Biotechnology* **100(6)**: 1143–1158.

Cai D, Wu, J and Chai K (2021) Microbiologically influenced corrosion behavior of carbon steel in the presence of Marine Bacteria sp. and sp. *ACS Omega* **6(5)**: 3780–3790.

Camacho-Chab JC, Lango-Reynoso F, Castañeda-Chávez MDR *et al.* (2016) Implications of extracellular polymeric substance matrices of microbial habitats associated with coastal aquaculture systems. *Water* **8(9)**: 369.

Chang W, Qian H, Li Z, Mol A and Zhang D (2024) Application and prospect of localized electrochemical techniques for microbiologically influenced corrosion: a review. *Corrosion Science* **236(112246)**: 112246.

Chang YS, Munro CJ, Fortunato L *et al.* (2024) Macrofouling remediation strategies for water intakes of desalination and other industrial plants – a review. *Desalination* **590(117987)**: 117987.

Chen P, Lang J, Zhou Y *et al.* (2022) An imidazolium-based zwitterionic polymer for antiviral and antibacterial dual functional coatings. *Science Advances* **8(2)**: eabl8812.

Chen P, Shu H, Tang W, Yu C and Yang R (2024) Zwitterionic polymer coatings with compositional gradient for stable and substrate-independent biofouling deterrence via all-dry synthesis. *arXiv [physics.chem-ph]*, 10.48550/ARXIV.2410.23552.

Coppola D, Buonocore C, Palisse M, Tedesco P and de Pascale D (2022) Exploring oceans for curative compounds: potential new antimicrobial and anti-virulence molecules against Pseudomonas aeruginosa. *Marine Drugs* **21(1)**, 10.3390/md21010009.

Costerton W, Veeh R, Shirtliff M *et al.* (2007) The application of biofilm science to the study and control of chronic bacterial infections. *Journal of Clinical Investigation* **117(1)**: 278–278.

Dang YT, Power A, Cozzolino D *et al.* (2022) Analytical characterisation of material corrosion by biofilms. *Journal of Bio- and Tribo-Corrosion* **8(2)**, 10.1007/s40735-022-00648-2.

Das S and Mangwani N (2015) Ocean acidification and marine microorganisms: responses and consequences. *Oceanologia* **57(4)**: 349–361.

Davidson I, Cahill P, Hinz A *et al.* (2021) A review of biofouling of ships internal seawater systems. *Frontiers in Marine Science* **8**, 10.3389/fmars.2021.761531.

Diaz-Mateus MA, Machuca LL, Farhat H and Salgar-Chaparro SJ (2024) Unveiling the effect of magnetite on the synergistic action of deposits and microorganisms on carbon steel corrosion. *Corrosion Science* **231(111940)**: 111940.

Donlan RM (2002) Biofilms: microbial life on surfaces. *Emerging Infectious Diseases* **8(9)**: 881–890.

Dou W, Xu, D and Gu T (2021) Biocorrosion caused by microbial biofilms is ubiquitous around us. *Microbial Biotechnology* **14(3)**: 803–805.

Eggerichs T, Wiegand M, Neumann K *et al.* (2020) Growth of iron-oxidizing Bacteria Gallionella ferruginea and Leptothrix cholodniiin oligotrophic environments: Ca, Mg and C as limiting factors and G. ferruginea Necromass as C-source. *Geomicrobiology Journal* **37(2)**: 190–199.

Elewa MM (2024) Emerging and conventional water desalination technologies powered by renewable energy and energy storage systems toward zero liquid discharge. *Separations* **11(10)**: 291.

Elston J, Pinto H and Nogueira C (2024) Tides of change for a sustainable Blue Economy: A systematic literature review of innovation in maritime activities. *Sustainability* **16(24)**: 11141.

Fang K, Park O-J and Hong SH (2020) Controlling biofilms using synthetic biology approaches. *Biotechnology Advances* **40**: 107518.

Firoozi AA, Oyejobi DO, Avudaiappan S and Flores ES (2025) Enhanced durability and environmental sustainability in marine infrastructure: innovations in anti-corrosive coating technologies. *Results in Engineering* **26(105144)**: 105144.

Foord CS, Robb K and Nugegoda D (2025) Legacy organic pollutants are still a threat for resident dolphins of south-east Australia. *Environmental Research* **270**: 121045.

García A, González B, Harvey C, Izquierdo-Barba I and Vallet-Regí M (2021) Effective reduction of biofilm through photothermal therapy by gold core@shell based mesoporous silica nanoparticles. *Microporous and Mesoporous Materials: the Official Journal of the International Zeolite Association* **328(111489)**: 111489.

Georgiades E, Scianni C and Tamburri MN (2023) Biofilms associated with ship submerged surfaces: implications for ship biofouling management and the environment. *Frontiers in Marine Science* **10**, 10.3389/fmars.2023.1197366.

Ghimire BK, Yu CY, Kim WR *et al.* (2023) Assessment of benefits and risk of genetically modified plants and products: current controversies and perspective. *Sustainability* **15(2)**: 1722.

Gonçalves FAG, Colen G and Takahashi JA (2014) Yarrowia lipolytica and its multiple applications in the biotechnological industry. *The Scientific World Journal* **2014**: 476207.

Gong Y, Ma FQ, Xue Y, Jiao CS and Yang ZG (2019) Failure analysis on leaked titanium tubes of seawater heat exchangers in recirculating cooling water system of coastal nuclear power plant. *Engineering Failure Analysis* **101**: 172–179.

González-Rodríguez CA, Rodríguez-Gómez FJ and Genescá-Llongueras J (2008) The influence of Desulfovibrio vulgaris on the efficiency of imidazoline as a corrosion inhibitor on low-carbon steel in seawater. *Electrochimica Acta* **54(1)**: 86–90.

Greenlee LF, Lawler DF, Freeman BD, Marrot B and Moulin P (2009) Reverse osmosis desalination: water sources, technology and todays challenges. *Water Research* **43(9)**: 2317–2348.

Hall-Spencer JM and Harvey BP (2019) Ocean acidification impacts on coastal ecosystem services due to habitat degradation. *Emerging Topics in Life Sciences* **3(2)**: 197–206.

Hazzan OO, Zhao B and Xiao Y (2023) Strategies for enhancing extracellular electron transfer in environmental biotechnology: a review. *Applied Sciences* **13(23)**: 12760.

He, Q and Silliman BR (2019) Climate change, human impacts and coastal ecosystems in the Anthropocene. *Current Biology: CB* **29(19)**: R1021–R1035.

Heyer A, D'Souza F, Morales CL *et al.* (2013) Ship ballast tanks a review from microbial corrosion and electrochemical point of view. *Ocean Engineering* **70**: 188–200.

Hoegh-Guldberg O (2011) The impact of climate change on coral reef ecosystems. In *Coral Reefs: An Ecosystem in Transition*. pp. 391–403.

Hoegh-Guldberg O, Poloczanska ES, Skirving W and Dove S (2017) Coral reef ecosystems under climate change and ocean acidification. *Frontiers in Marine Science* **4**, 10.3389/fmars.2017.00158.

Hoek EMV, Weigand TM and Edalat A (2022) Reverse osmosis membrane biofouling: causes, consequences and countermeasures. *npj Clean Water* **5(1)**, 10.1038/s41545-022-00183-0.

Holmström C, Egan S, Franks A, McCloy S and Kjelleberg S (2002) Antifouling activities expressed by marine surface associated Pseudoalteromonas species. *FEMS Microbiology Ecology* **41(1)**: 47–58.

Hopkins GA and Forrest BM (2010) A preliminary assessment of biofouling and non-indigenous marine species associated with commercial slow-moving vessels arriving in New Zealand. *Biofouling* **26(5)**: 613–621.

Huang C, Wang W, Sun X, Shen J and Wang L (2020) A novel acetogenic bacteria isolated from waste activated sludge and its potential application for enhancing anaerobic digestion performance. *Journal of Environmental Management* **255**: 109842.

Huang L, Chang W, Zhang D *et al.* (2022) Acceleration of corrosion of 304 stainless steel by outward extracellular electron transfer of Pseudomonas aeruginosa biofilm. *Corrosion Science* **199(110159)**: 110159.

Issifu I, Alava JJ, Lam VW and Sumaila UR (2022) Impact of ocean warming, overfishing and mercury on European fisheries: a risk assessment and policy solution framework. *Frontiers in Marine Science* **8**, 10.3389/fmars.2021.770805.

Ivanovich N, Marsili E, Shen X *et al.* (2025) Exploring the impact of flow dynamics on corrosive biofilms under simulated deep-sea high-pressure conditions using bio-electrochemostasis. *Frontiers in Microbiology* **16**, 1540664.

Jacobson GA (2007) Corrosion at Prudhoe Bay—a lesson on the line *Materials Performance* **46(8)**: 27–34.

Jafari M, Moghimi H, Tirandaz H and Ebrahim-Habibi MB (2024) Corrosion behavior of predominant Halodesulfovibrio in a marine SRB consortium and its mitigation using ZnO nanoparticles. *Scientific Reports* **14(1)**: 19545.

Javed MA, Ivanovich N, Messinese E *et al.* (2024) The role of metallurgical features in the microbially influenced corrosion of carbon steel: a critical review. *Microorganisms* **12(5)**, 10.3390/microorganisms12050892.

Jones G (2009) The battle against marine biofouling: a historical review. In *Advances in Marine Antifouling Coatings and Technologies*. Elsevier, Amsterdam, the Netherlands, pp. 19–45.

Kalajahi ST, Misra A and Koerdt A (2024) Nanotechnology to mitigate microbiologically influenced corrosion (MIC). *Frontiers in Nanotechnology* **6**, 10.3389/fnano.2024.1340352.

Kawaichi S, Kotoky R, Fiutowski J and Rotaru AE (2024) Adaptation of a methanogen to Fe corrosion via direct contact. *NPJ Biofilms and Microbiomes* **10(1)**: 100.

Khan MS, Yang K, Liu Z *et al.* (2023) Microorganisms involved in the biodegradation and microbiological corrosion of structural materials. *Coatings* **13(10)**: 1583.

Kip N, Jansen S, Leite MF *et al.* (2017) Methanogens predominate in natural corrosion protective layers on metal sheet piles. *Scientific Reports* **7(1)**: 11899.

Krsmanovic M, Biswas D, Ali H *et al.* (2021) Hydrodynamics and surface properties influence biofilm proliferation. *Advances in Colloid and Interface Science* **288**: 102336.

Kumar S, Ye F, Dobretsov S and Dutta J (2021) Nanocoating is a New Way for biofouling prevention. *Frontiers in Nanotechnology* **3**, 10.3389/fnano.2021.771098.

Larsen KR (2014) A closer look at microbiologically influenced corrosion. *Materials Performance* **53(1)**: 32–40.

Li J, Nickel R, Wu J *et al.* (2019) A new tool to attack biofilms: driving magnetic iron-oxide nanoparticles to disrupt the matrix. *Nanoscale* **11(14)**: 6905–6915.

Lin Z, Zhang W, Li J *et al.* (2023) Application of artificial intelligence (AI) in the area of corrosion protection. *Anti-corrosion Methods and Materials* **70(5)**: 243–251.

Little BJ and Lee JS (2009) Microbiologically influenced corrosion. In *Kirk-Othmer Encyclopedia of Chemical Technology*. Wiley, Hoboken, NJ, USA, 10.1002/0471238961.micrlitt.a01.

Liu L, Wu X, Wang Q *et al.* (2022) An overview of microbiologically influenced corrosion: mechanisms and its control by microbes. *Corrosion Reviews* **40(2)**: 103–117.

Liu N, Qiu L and Qiu L (2024) Carbon steel corrosion induced by sulfate-reducing bacteria: a review of electrochemical mechanisms and pathways in biofilms. *Coatings* **14(9)**: 1105.

Liu P, Zhang H, Fan Y and Xu D (2023) Microbially influenced corrosion of steel in marine environments: a review from mechanisms to prevention. *Microorganisms* **11(9)**, 10.3390/microorganisms11092299.

Liu Y, Zhang W, Sileika T *et al.* (2011) Disinfection of bacterial biofilms in pilot-scale cooling tower systems. *Biofouling* **27(4)**: 393–402.

Luoma E, Laurila-Pant M, Altarriba E *et al.* (2022) A multi-criteria decision analysis model for ship biofouling management in the Baltic Sea. *The Science of the Total Environment* **852**: 158316.

Lv M, Du M and Li Z (2022) Investigation of mixed species biofilm on corrosion of X65 steel in seawater environment. *Bioelectrochemistry*, **143**: 107951.

Maji K and Lavanya M (2024) Microbiologically influenced corrosion in stainless steel by Pseudomonas aeruginosa: an overview. *Journal of Bio- and Tribo-corrosion* **10(1)**, 10.1007/s40735-024-00820-w.

Mannix EJ and Woodham BA (2025) An interpretable approach to automating the assessment of biofouling in video footage. *arXiv [cs.CV]*, 10.48550/ARXIV.2503.12875.

Mannix EJ, Wei S, Woodham B, Wilkinson P and Robinson AP (2021) Automating the assessment of biofouling in images using expert agreement as a gold standard. *Scientific Reports* **11(1)**: 2739.

McCarty NS and Ledesma-Amaro R (2019) Synthetic biology tools to engineer microbial communities for biotechnology. *Trends in Biotechnology* **37(2)**: 181–197.

Merisalu M, Aarik L, Kozlova J *et al.* (2021) Effective corrosion protection of aluminum alloy AA2024-T3 with novel thin nanostructured oxide coating. *Surface & Coatings Technology* **411(126993)**: 126993.

Michailidis M, Gutner-Hoch E, Wengier R *et al.* (2020) Highly effective functionalized coatings with antibacterial and antifouling properties. *ACS Sustainable Chemistry & Engineering* **8(24)**: 8928–8937.

Müller-Steinhagen H, Malayeri MR and Watkinson AP (2005) Fouling of heat exchangers-new approaches to solve an old problem. *Heat Transfer Engineering* **26(1)**: 1–4.

Neumann N, Honke M, Povydysh M, Guenther S and Schulze C (2022) evaluating tannins and flavonoids from traditionally used medicinal plants with biofilm inhibitory effects against MRGN. *Molecules*, **27(7)**, 10.3390/molecules27072284.

Nguyen BV, Nagakubo T, Toyofuku M, Nomura N and Utada AS (2020) Synergy between sophorolipid biosurfactant and SDS increases the efficiency of biofilm disruption. *Langmuir: the ACS Journal of Surfaces and Colloids* **36(23)**: 6411–6420.

Nichols CR, Zinnert J and Young DR (2019) Degradation of coastal ecosystems: Causes, impacts and mitigation efforts. In *Tomorrow's Coasts: Complex and Impermanent*. Springer, Cham, Switzerland, pp. 119–136.

Novair SB, Biglari Z, Lajayer BA, Shu W and Price GW (2024) The role of sulphate-reducing bacteria (SRB) in bioremediation of sulphate-rich wastewater: focus on the source of electron donors. *Process Safety and Environmental Protection: Transactions of the Institution of Chemical Engineers, Part B* **184**: 190–207.

Odeyemi OO and Alaba A (2024) Microbiologically influenced corrosion in oil fields: mechanisms, detection and mitigation strategies. In *Corrosion Engineering - Recent Breakthroughs and Innovative Solutions*. IntechOpen, London, UK.

Oksana T and Dmytro G (2021) Earths water distribution. In *Encyclopedia of the UN Sustainable Development Goals*. Springer, Cham, Switzerland, pp. 1–14.

Ollivier B, Cayol J-L and Fauque G (2007) Sulphate-reducing bacteria from oil field environments and deep-sea hydrothermal vents. In *Sulphate-Reducing Bacteria* (Barton LL and Hamilton WA (eds)). Cambridge University Press, Cambridge, UK, pp. 305–328.

Olufemi Odeyemi O and Adeniyi Alaba: (2024) Microbiologically influenced corrosion in oil fields: mechanisms, detection and mitigation strategies. In *Corrosion Engineering - Recent Breakthroughs and Innovative Solutions*. IntechOpen.

Pan SY, Snyder SW, Packman AI, Lin YJ and Chiang PC (2018) Cooling water use in thermoelectric power generation and its associated challenges for addressing water-energy nexus. *Water-Energy Nexus* **1(1)**: 26–41.

Parthipan P, Narenkumar J, Elumalai P *et al.* (2017) Neem extract as a green inhibitor for microbiologically influenced corrosion of carbon steel API 5LX in a hypersaline environments. *Journal of Molecular Liquids* **240**: 121–127.

Permeh S, Reid C, Echeverría Boan M *et al.* (2017) Microbiological influenced corrosion (MIC) in Florida marine environment: a case study. In *Proceedings of Corrosion 2017,* New Orleans, LA, USA *2017*, NACE International, pp. 1–12.

Permeh S, Lau K, Boan ME and Duncan (2021) Influence of macro- and microfouling on corrosion of steel bridge piles submerged in natural waters. *Journal of Materials in Civil Engineering* **33(6)**: 04021105.

Permeh S, Reid C, Echeverría Boan M *et al.* (2025) Microbiological influenced corrosion (MIC) in Florida marine environment: a case study. In *Case Studies in Corrosion Failures*. AMPP, Association for Materials Protection and Performance, Houston, TX, USA, pp. 299–311.

Petrovic Z C (2016) Catastrophes caused by corrosion. *Vojnotehnicki Glasnik* **64(4)**: 1048–1064.

Qi P, Zeng Y, Zhang D, Sun Y and Wang P (2025) The biofilm-metal interface: a hotspot for microbiologically influenced corrosion. *Cell Reports. Physical Science* **6(3)**: 102500.

Rajasekar A, Balasubramanian R and VM Kuma J (2011) Role of hydrocarbon degrading BacteriaSerratia marcescensACE2 and Bacillus cereusACE4 on corrosion of carbon steel API 5LX. *Industrial & Engineering Chemistry Research* **50(17)**: 10041–10046.

Ran Y, Liu Y, Sun H *et al.* (2021) Surface modification of cuprous oxide nanoparticles with improved chemical stability and antibacterial activity. *Applied Surface Science* **565(150566)**: 150566.

Rao TS (2012) Microbial fouling and corrosion: fundamentals and mechanisms. In *Operational and Environmental Consequences of Large Industrial Cooling Water Systems*. Springer US, Boston, MA, USA, pp. 95–126.

Rao TS, Kora AJ, Chandramohan P, Panigrahi BS and Narasimhan SV (2009) Biofouling and microbial corrosion problem in the thermo-fluid heat exchanger and cooling water system of a nuclear test reactor. *Biofouling* **25(7)**: 581–591.

Rawi NN, Ramzi MM, Rahman NIA *et al.* (2023) Antifouling potential of ethyl acetate extract of marine bacteria Strain RLimb. *Life* **13(3)**, 10.3390/life13030802.

Refait P, Grolleau AM, Jeannin M, Rémazeilles C and Sabot R (2020) Corrosion of carbon steel in marine environments: role of the corrosion product layer. *Corrosion and Materials Degradation* **1(1)**: 198–218.

Romeu MJ and Mergulhão F (2023) Development of antifouling strategies for marine applications. *Microorganisms* **11(6)**, 10.3390/microorganisms11061568.

Sahli C, Moya SE, Lomas JS *et al.* (2022) Recent advances in nanotechnology for eradicating bacterial biofilm. *Theranostics* **12(5)**: 2383–2405.

Schultz MP, Bendick JA, Holm ER and Hertel WM (2011) Economic impact of biofouling on a naval surface ship. *Biofouling* **27(1)**: 87–98.

Senadheera TRL, Hossain A and Shahidi F (2023) Marine bioactives and their application in the food industry: a review. *Applied Sciences* **13(21)**: 12088.

Sharifian P, Yaslianifard S, Fallah P *et al.* (2020) Investigating the effect of nano-curcumin on the expression of biofilm regulatory genes of. *Infection and Drug Resistance* **13**: 2477–2484.

Sharma S, Mohler J, Mahajan SD *et al.* (2023) Microbial biofilm: a review on formation, infection, antibiotic resistance, control measures and innovative treatment. *Microorganisms* **11(6)**, 10.3390/microorganisms11061614.

Sigdel P, Kandadai A, Jawaharraj K *et al.* (2023) Microbial corrosion prevention by Citrobacter sp. Biofilms. *arXiv [cond-mat.mtrl-sci]*, 10.48550/ARXIV.2304.13862.

Singh BJ, Chakraborty A and Sehgal R (2023) A systematic review of industrial wastewater management: Evaluating challenges and enablers. *Journal of Environmental Management* **348**: 119230.

Singh S, Dwivedi A and Pratap S (2023) Sustainable maritime freight transportation: current status and future directions. *Sustainability* **15(8)**: 6996.

Skovhus TL, Eckert RB and Rodrigues E (2017) Management and control of microbiologically influenced corrosion (MIC) in the oil and gas industry-overview and a North Sea case study. *Journal of Biotechnology* **256**: 31–45.

Sousa-Cardoso F, Teixeira-Santos R, Campos AF *et al.* (2023) Graphene-based coating to mitigate biofilm development in marine environments. *Nanomaterials*, **13(3)**, 10.3390/nano13030381.

Sreelekshmi S and Ananth Kumar M (2021) Effect of reduced graphene oxide nanoparticles as anticorrosion material on mild steel substrate. *Journal of Physics. Conference Series* **2070(1)**: 012192.

Srinivasan R, Santhakumari S, Poonguzhali P *et al.* (2021) Bacterial biofilm inhibition: a focused review on recent therapeutic strategies for combating the biofilm mediated infections. *Frontiers in Microbiology* **12**: 676458.

Stöckl M, Teubner NC, Holtmann D, Mangold KM and Sand W (2019) Extracellular polymeric substances from Geobacter sulfurreducens biofilms in microbial fuel cells. *ACS Applied Materials & Interfaces* **11(9)**: 8961–8968.

Susarla S, Chilkoor G, Kalimuthu JR *et al.* (2020) Corrosion resistance of sulfur-selenium alloy coatings. *arXiv [cond-mat.mtrl-sci]*, 10.48550/ARXIV.2009.02451.

Themelis N, Nikolaidis G and Zagkas V (2024) Assessment of hull and propeller degradation due to biofouling using tree-based models. *Applied Sciences*, **14(20)**: 9363.

Townsin RL (2003) The ship hull fouling penalty. *Biofouling* **19** Suppl: 9–15.

Tripathi AK, Thakur P, Saxena P *et al.* (2021) Gene sets and mechanisms of sulfate-reducing bacteria biofilm formation and quorum sensing with impact on corrosion. *Frontiers in Microbiology* **12**: 754140.

Tsao HF, Scheikl U, Herbold C *et al.* (2019) The cooling tower water microbiota: seasonal dynamics and co-occurrence of bacterial and protist phylotypes. *Water Research* **159**: 464–479.

Turetgen I (2020) Reduction of biofilm formation on cooling tower heat exchangers using nano-silica coating. *Johnson Matthey Technology Review* **64(4)**: 419–424.

van der Aa BC and Dufrêne YF (2002) In situ characterization of bacterial extracellular polymeric substances by AFM. *Colloids and Surfaces. B, Biointerfaces* **23(2-3)**: 173–182.

Veerasingam S, Ranjani M, Asim FS *et al.* (2024) Development of an automated coastal biofouling detection system using artificial intelligence object detection. *Journal of Coastal Research* **113(sp1)**, 10.2112/jcr-si113-166.1.

Verma C, Ebenso EE and Quraishi MA (2017) Ionic liquids as green and sustainable corrosion inhibitors for metals and alloys: an overview. *Journal of Molecular Liquids* **233**: 403–414.

Veytskin YB, Kugler A, Brigmon RL, Burckhalter CE and Mickalonis J (2023) Evaluation of microbiologically influenced corrosion and biofouling in a nuclear storage basin. *Journal of Radiation Research and Applied Sciences* **16(3)**: 100641.

Vinagre PA, Simas T, Cruz E, Pinori E and Svenson J (2020) Marine biofouling: a European database for the marine renewable energy sector. *Journal of Marine Science and Engineering* **8(7)**: 495.

Weber F and Esmaeili N (2023) Marine biofouling and the role of biocidal coatings in balancing environmental impacts. *Biofouling* **39(6)**: 661–681.

Welikala S, Al-Saadi S, Gates WP, Panter C and Singh Raman RK (2024) Sulphate reducing bacteria (SRB) biofilm development and its role in microbial corrosion of carbon steel. *Frontiers in Materials* **11**, 10.3389/fmats.2024.1360869.

Xu K, Xie H, Sun C *et al.* (2023) Sustainable coating based on zwitterionic functionalized polyurushiol with antifouling and antibacterial properties. *Molecules*, **28(24)**, 10.3390/molecules28248040.

Yang R and Gleason KK (2012) Ultrathin antifouling coatings with stable surface zwitterionic functionality by initiated chemical vapor deposition (iCVD). *Langmuir: the ACS Journal of Surfaces and Colloids* **28(33)**: 12266–12274.

Yang SH, Ringsberg JW, Johnson E and Hu Z (2017) Biofouling on mooring lines and power cables used in wave energy converter systems—analysis of fatigue life and energy performance. *Applied Ocean Research* **65**: 166–177.

Yang XE, Wu X, Hao HL and He ZL (2008) Mechanisms and assessment of water eutrophication. *Journal of Zhejiang University. Science. B* **9(3)**: 197–209.

Yu S, Lu X and Lu H (2025) Marine microbial biofilms on diverse abiotic surfaces. *Frontiers in Marine Science* **12**, 10.3389/fmars.2025.1482946.

Yu ZL, Yu DL, Mao YF *et al.* (2019) Identification and characterization of a LuxI/R-type quorum sensing system in Pseudoalteromonas. *Research in Microbiology* **170(6-7)**: 243–255.

Zhang XY, Fu W, Chen X *et al.* (2018) Phylogenetic analysis and antifouling potentials of culturable fungi in mangrove sediments from Techeng Isle, China. *World Journal of Microbiology & Biotechnology* **34(7)**: 90.

Zhang Y, Wang X, Zhen Y *et al.* (2017) Microbial Diversity and Community Structure of Sulfate-Reducing and Sulfur-Oxidizing Bacteria in Sediment Cores from the East China Sea. *Frontiers in Microbiology* **8**: 2133.

Zhou J, Yin S, Fu Q *et al.* (2021) Microbial-induced concrete corrosion under high-salt conditions: Microbial community composition and environmental multivariate association analysis. *International Biodeterioration & Biodegradation* **164(105287)**: 105287.

Microbial Processes in Environmental Engineering: From Theory to Practice
Rajneesh Kumar, Ram Sharan Singh and Maulin P Shah
ISBN 978-1-83708-327-5
https://doi.org/10.1108/978-1-83708-326-820261015

Index

The manufacturer's authorised representative in the EU for product safety is Easy Access System Europe – Mustamäe tee 50, 10621 Tallinn, Estonia, easproject.com, +372 5696 8939, gpsr.requests@easproject.com

Printed and bound by CPI Group (UK) Ltd, Croydon, CR0 4YY

19/08/2026

02200684-0019